8th Asia Conference on Mechanical Engineering and Aerospace Engineering (MEAE 2022)

Journal of Physics: Conference Series Volume 2361

Wuhan, China
9-11 June 2022

ISBN: 978-1-7138-6469-1
ISSN: 1742-6596

Printed from e-media with permission by:

Curran Associates, Inc.
57 Morehouse Lane
Red Hook, NY 12571

CURRAN ASSOCIATES INC.
proceedings
.com

Some format issues inherent in the e-media version may also appear in this print version.

Institute of Physics
Dirac House, Temple Back
Bristol BS1 6BE UK

Phone: 44 1 17 929 7481
Fax: 44 1 17 920 0979

techtracking@iop.org

Additional copies of this publication are available from:

Curran Associates, Inc.
57 Morehouse Lane
Red Hook, NY 12571 USA
Phone: 845-758-0400
Fax: 845-758-2633
Email: curran@proceedings.com
Web: www.proceedings.com

TABLE OF CONTENTS

Preface

MEAE is an annual research conference aiming to provide a platform for engineers and scientists to present their ideas and experiences in the fields of Mechanical Engineering and Aerospace Engineering. 2022 8th Asia Conference on Mechanical Engineering and Aerospace Engineering (MEAE 2022) was held on June 9-11, 2022. It was organized by China University of Geosciences (Wuhan) and hosted by School of Mechanical Engineering and Electronic Information of CUG (Wuhan).

This year, MEAE organizing committee held the conference as remote meeting due to the instability of COVID-19. MEAE used ZOOM as the platform holding online conference. The conference was held virtually for 3 days. The test session on the first day was arranged to help authors learn some ZOOM basic functions. After ZOOM basic functions test, most of participants managed to use ZOOM smoothly. Keynote / Invited speeches and author parallel sessions were on the following two days. Presenters made their presentations on the topics of Materials and Mechanical Engineering, Industrial Manufacturing Systems and Control Technology, Mechanical and Electronic Technology and so on. Every presentation was about 15 minutes including 2-3 minutes for Q&A part. Though the authors and speakers couldn't communicate face to face, the passion for involvement wasn't affected.

The proceedings present a selection from papers submitted to the conference from universities, research institutes and industries. All the papers were double-blind reviewed. 28 papers were accepted from 63 submissions to be collected in the conference proceedings-Journal of Physics: Conference Series. Here we would like to thank all the technical program committee members who made great efforts on paper reviewing.

The appreciation also goes to other conference committee members, keynote/invited speakers, session chairs and all the authors, thanks for their understanding, support and efforts. Without their help, MEAE couldn't be held successfully.

Hope all the participants had wonderful time during conference and got fruitful inspirations from the presentations delivered by speakers and authors. Hope to see you next year onsite!

Committees

Advisory Committees
Prof. Maria Pia FANTI, Polytechnic of Bari, Italy (IEEE Fellow)

Conference Chair
Prof. Huafeng Ding, China University of Geosciences, China

Conference Co-chair
Prof. Fusheng Qiu, Shenyang Aerospace University, China
Prof. Chiharu Ishii, Hosei University, Japan

Program Chairs
Prof. Yang Yu, Beihang University, China
Prof. Guojun Wen, China University of Geosciences (Wuhan), China
Prof. Yanqiong Fei, Research Institute of Robotics, Shanghai Jiaotong University, China

Program Co-chairs
Prof. Jing Wang, Donghua Univeristy, China
Prof. Sumeet Aphale, University of Aberdeen, UK

Publication Chair
Prof. Ian McAndrew, Capitol Technology University, USA
Prof. Jie Luo, China University of Geosciences (Wuhan), China

Publicity Chairs
Prof. Bijay Kumar Rout, Birla Institute of Technology and Science Pilani, Pilani Campus, India
Prof. Jianjun Zhang, Qingdao University of Technology, China
Prof. Xianchang Wang, Jilin University, China

Local Organizing Committee
Assoc. Prof. Min Li, China University of Geosciences (Wuhan), China
Prof. Chaolong Song, China University of Geosciences (Wuhan), China
Prof. Tianye Huang, China University of Geosciences (Wuhan), China

Registration Chair
Changping Li, China University of Geosciences (Wuhan), China

Executive Secretary & Press Coordinator
Hong Xu, China University of Geosciences (Wuhan), China

Technical Committees
Prof. Chung-Hwei Su, National Kaohsiung University of Science and Technology, Taiwan, China
Prof. Lih-Wu Hourng, National Central University, Taiwan, China
Prof. Nader Asnafi, ÖREBRO UNIVERSITY, Sweden
Prof. Orlando Calogero, Kore University of Enna Cittadella Universitaria, Italy

Journal of Physics: Conference Series 2361 (2022) 011001 doi:10.1088/1742-6596/2361/1/011001

Statement of Peer Review

All papers published in this volume of Journal of Physics: Conference Series have been peer reviewed through processes administered by the proceedings Editors. Reviews were conducted by expert referees to the professional and scientific standards expected of a proceedings journal published by IOP Publishing.

Peer review declaration

All conference organisers/editors are required to declare details about their peer review. Therefore, please provide the following information:

- **Type of peer review: Double-blind**
- **Conference submission management system: Zmeeting submission system**
- **Number of submissions received: 63**
- **Number of submissions sent for review: 52**

  Number of submissions accepted: 28
- **Acceptance Rate (Number of Submissions Accepted / Number of Submissions Received X 100): 44.4%**
- **Average number of reviews per paper: 2**
- **Total number of reviewers involved: 29 reviewers**
- **Any additional info on review process:**

1. Preliminary Review
2. The papers passed the first review will be reviewed again from the following aspects: Originality, Innovation, Technical Soundness, Applicability, Presentation & English and Relevance

Contact person for queries: zengzhu@yaang.cn
Chengdu Young Academic Conference Co. Ltd., China

Please submit this form along with the rest of your files on the submission date written in your publishing agreement.

The information you provide will be published as part of your proceedings.

Diffusion-limited Nonequilibrium Phase Transformation of Nickel/Aluminum Dissimilar Materials during Laser Welding Part Ⅰ: Precipitation-strengthened Polycrystalline Nickel-based Superalloy

Zhiguo Gao

School of Mechanical Engineering, Anyang Institute of Technology, Huang He Da Dao (West Side), Anyang City, Henan Province, 455000, China

E-mail: gao_zhiguo@ayit.edu.cn

Abstract. Polycrystalline heterogeneities of grain growth, phase transformation and tensile properties are discussed by metallography, spectroscopy and fractography when choosing laser welding with butt-joint configuration to satisfactorily minimize effect of inappropriate weld pool shape on microstructure and mechanical properties of γ'' precipitation hardable nickel-based superalloy in aerospace industry. There is parabolic relationship between weld penetration and diffusion-limited nucleation, growth and coarseness of secondary dendrite arm spacing alongside fusion boundary through nonequilibrium solidification process, which is dendritically susceptible to grain growth variation and metallurgical discontinuities. The amount, size, morphology and distribution of nonequilibrium intermetallic phase near the dendrite boundaries are kinetically and thermodynamically rely on weld pool shape whose crystal structure is incoherent with γ solid solution austenite phase and increases preference for crack initiation and propagation of brittle intergranular and ductile dimple fracture failures to enormously contribute to losses of strength and ductility. Unsymmetrical keyhole weld is thermometallurgically inconvenient for amelioration of microstructure and mechanical properties on either side, and adversely mitigates resistance to hypereutectic-aided dendrite embrittlement. Beneficial grain refinement and suppression of detrimental nonequilibrium intermetallic phase at the same time are challenging. This problem is an integral part of inhomogeneous microstructure and mechanical properties. The finer grain size occurs, the larger grain boundary area is available for Laves/γ eutectic-type reaction and vice versa. Contributions of coarse grain kinetics and metallurgical reaction thermodynamics to weld disintegration and fracture failure mechanism of tensile properties are explained by microstructure characterization of multicomponent and multiphase weld. Finally, it is imperative to dendritically balance these important factors to minimize inevitable interface solidification products and anomalous substructure growth, and reasonably advance superior mechanical properties of reliable weld.

1. Introduction

Because of increasing thermal stability, oxidation- and corrosion-resistant Ni-based superalloys are promising materials for high temperature environment applications of electricity generation power plants and oil refineries, e.g. turbine components and heat exchangers, and they are prohibitively expensive. Correlation between grain growth, phase transformation and mechanical properties are metallurgically discussed in the below open literature. Shuai Guan *et al.* and Miao Wang *et al.* [1,2] analyzed interdependency of grain growth, precipitation behavior and tensile properties of high-strength Ni-Fe-Cr superalloy HT700 on alloying elements and rotation rate by friction stir processing (FSP). Tianfu Wang *et al.* and Linxu Li *et al.* [3,4] analyzed correlation between δ phase Ni_3Nb precipitation, grain growth, hardness and tensile properties of Ni-based superalloys GH4169 and MAR-M247 by tungsten inert gas (TIG) welding. L O Osoba *et al.* and Xinjie Di *et al.* [5,6] analyzed correlation between post-weld heat treatment(PWHT), carbide precipitation kinetics, grain growth and cracking susceptibility of Ni-based superalloy HY282 and high-chromium deposition metal during welding and Gleeble thermo-mechanical physical simulation. Jun Zhang *et al.* and Zijian Zhou *et al.* [7,8] analyzed interdependency of diffusion kinetics, grain growth, microsegregation, phase transformation and mechanical properties of dual-phase wrought Ni-Co superalloy on electric current pulses(ECP) and homogenization heat treatment. Anwen Zhang *et al.* and Dawei Han *et al.* [9,10] analyzed correlation between heat treatment, precipitation behavior, grain growth, microhardness and tensile properties of Ni-based superalloys IN718 and M718. Yang Liu *et al.* and K D Ramkumar *et al.* [11,12] analyzed interdependency of grain refinement, deleterious phases, residual stress and fatigue properties of IN718 superalloy on processing parameters by warm laser shock processing and multi-pass gas tungsten arc welding with filler wires. Yan Guo *et al.*, Xiangwei Li *et al.*, Tao Liu *et al.* and Enxiang Pu *et al.* [13-16] analyzed correlation between heat treatment, inter- and intra-granular precipitation behavior, grain growth as well as mechanical properties of Ni-based superalloys Inconel 617, K417G and Hastelloy C-276. Jinwen Zou *et al.* and Xu Zhang *et al.* [17,18] analyzed correlation between alloying elements, phase transformation, mechanical properties and ductility-dip cracking of powder metallurgy Ni-based superalloy FGH96. K S Prasad *et al.* and Shuang Gao *et al.* [19,20] analyzed interdependency of phase transformation, grain growth and ultimate tensile strength of Ni-based superalloys Inconel 625 and Rene 80 on process parameters of microplasma arc welding and solution heat treatment. Therefore, effect of nonequilibrium phase transformation and heterogeneous grain growth on inferior tensile properties during laser processing is macroscopically and microscopically elucidated in detail so as to ameliorate weldability.

2. Experiment procedure and material

Inconel 718 is subjected to fiber laser welding. Chemical composition is tabulated as follows. Length, width and thickness of rectangular wrought plates are 300mm, 100mm and 5mm, separately. These materials are welded together with butt-joint configuration. Optimal welding parameters are achieved in the following values. Laser beam diameter is 0.4 mm and focus point is located at substrate surface. Laser power is selected as 2.4kW, welding speed is 0.6m/min and Argon shielding gas flow rate is 20L/min. Weld samples are transversely sectioned and prepared by metallography and subsequent fractography procedures for multiscale microstructure characterization by optical microscope (OM) and scanning electron microscope (SEM) with energy-dispersive spectroscopy. For more accurate results, 3 sections of weld are indispensably measured. Polycrystalline microstructure and phase transformation of nonequilibrium solidification are more complicated than single-crystal superalloy, and valuable data and SEM images are obtainable. Additionally, strength and ductility of weld samples are evaluated by tensile test.

Table 1 Chemical composition of Ni-based superalloy Inconel 718 (in wt%)

Nickel	Iron	Chromium	Niobium+Tantalum	Molybdenum	Titanium	Aluminum	Silicon
Bal.	18.49	19.3	5.05	3.12	1.02	0.54	0.04

3. Results and discussion

Shape and size differences of weld is shown in figure 1. In (a), the weld is successfully free of internal hot cracking and porosity by using optimal welding parameters to improve weld integrity. Incomplete penetration is induced by periodic keyhole fluctuation and flow instability. The asymmetrical weld pool geometry is thermally and hydrodynamically attributed to different heat transfer and fluid flow in both sides of weld pool at macroscopic scale to severely deteriorate homogeneity of grain growth at microscopic scale because of inappropriate butt-joint configuration, which is closely related to sidewall gap tolerance and laser inclination angle, during interaction between laser and polycrystalline nickel-based superalloy. In (b), the transverse section of weld is essentially divided into three disparate parts with complex laser energy allocation. Wide upper portion by Marangoni convection and narrow bottom part by evaporation recoil pressure of keyhole weld pool are predominant factors for rapid solidification and grain refinement in the weld fusion zone. Middle neck transition region inevitably occurs during weld pool development.

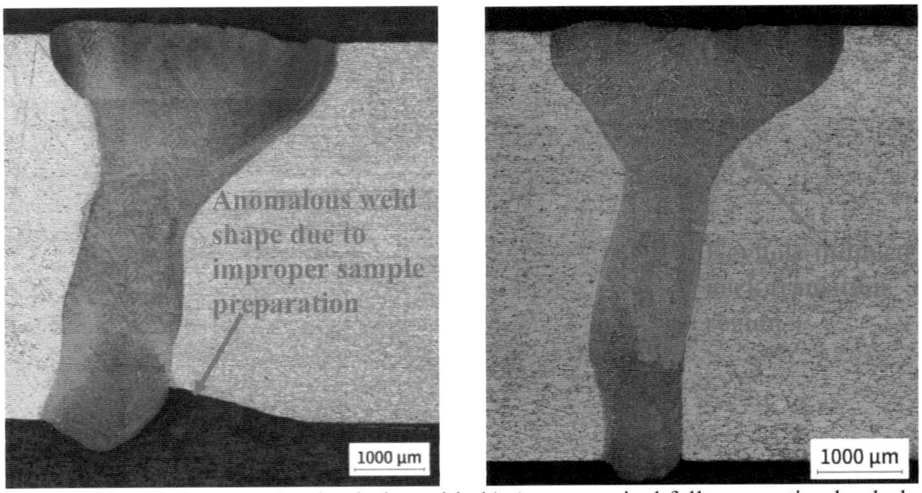

a) Infelicitous full penetration keyhole weld b) Asymmetrical full penetration keyhole weld

Figure 1. Comparison of weld geometry in the butt-joint configuration

Adequate penetration of defectless weld is prerequisite for reliable multiscale weldability evaluation. The irregular shape of weld is fairly vulnerable to stress concentration, segregation, brittle intermetallic phase and microstructure heterogeneity in the neck transition region, and increases heat-affected zone (HAZ) intergranular liquation cracking susceptibility. Even if weld geometry is improper, multiphysics mechanism of unsymmetrical weld pool phenomena provides another possibility for mitigation of unfavorable weld properties to maximize advantages of keyhole laser welding, simultaneously. Heat dissipation is prevalently accelerated on the both top and bottom surfaces of weld during weld cooling. However, significant middle neck transition region is subjected to slow stage of solidification as consequence of heat, solute and strain accumulations. It is metallurgically crucial to withstand harsh microstructure degradation by decrease of peak temperature and high temperature duration of welding thermal cycle. Considerable decline of heat input is advantageously capable of crack-insusceptible deep penetration to suppress keyhole collapse, stabilize weld pool dynamics, diminish fluid turbulence and flow disturbance, promote heterogeneous nucleation and grain growth, and hence open new opportunities for flawless high quality weld.

Dendrite growth at intermediate neck transition region is shown in figure 2. In (a), neck transition region, where last solidification liquid is located, augments diffusion-driven grain kinetics of nucleation and growth and is chemically and thermally prone to heterogeneity-susceptible grain growth, e.g. high-angle substructure intersection and interlock, to worsen severity of supersaturation, alloying partition, segregation and subsequent metallurgical reactions in the interdendrite region at central weld. In other words, grain growth is misorientated in weld center, but it is free of centerline cracking at final solidification shrinkage and thermal contraction stage of weld cooling. Solute is enriched in this location as a result of numerous alloying elements redistribution to contribute to chemistry inhomogeneity between grain boundaries and grain interior of polycrystal for coexistence of majority γ phase formation and minority intermetallic phase formation. Meanwhile, nuclei are heteriogeously dispersed to result in a faster nucleation process and larger grain density at neck transition region for abundant primary and secondary dendrite sidebranch development. Grain selection is insignificantly dependent on preferential crystallographic orientation with more competitive growth and conspicuous deviation angles, and involves excessive irregular protrusions.

No obvious relationship between weld pool shape, heat flow direction and crystal orientation is mainly attributed to polycrystalline morphology solidification with solid/liquid interface instability and is not prone to unidirectional grain growth any more.

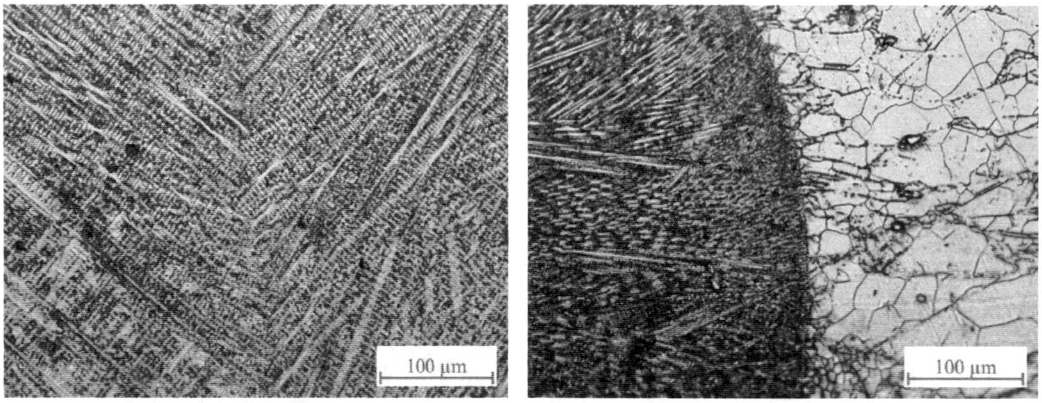

a) At central weld b) Near fusion boundary

Figure 2. Dendrite growth nearby intermediate neck transition region

In (b), although onset of substantial intergranular and intragranular liquation reactions for grain boundary degradation, which are initiated by constitutional liquation and grain boundary wetting as consequence of nonequilibrium segregation, no observable intergranular liquation cracking and grain boundary decohesion morphologically occur within the HAZ of dynamic recrystallization in the vicinity of fusion boundary, where grain boundaries are pinned by carbide, boride and δ phase. Pertinent low undercooling ameliorates morphology, size and orientation of γ phase solid solution with ability to accommodate shrinkage strain and tensile stress, thereby reduce thermal, metallurgical and mechanical driving forces for solidification cracking. It is of importance to noteworthy that loss of preferential <100> crystallographic orientation of grain growth is ascribable to material-related factors, such as polycrystalline properties instead of heat input, and this problem is further exacerbated when extensive insidious grain protrusion and deflection. Nonequilibrium partitionless solidification at solid/liquid interface satisfactorily increases microstructure stability for minimal weldability degradation by proper time-temperature adjustment of welding thermal cycle on-heating and on-cooling portions to mitigate peak temperature and shorten high temperature period over a range of cooling rates. Favorable thermal, chemical and mechanical conditions are available for useful control of grain growth and tensile properties to meet multiple conditions for improving weldability.

Substructure measurement around both sides of fusion boundary from top, middle to bottom of keyhole weld is shown in figure 3. Unsymmetrical weld pool shape is an important contributing factor to morphology, size and orientation of grain growth, which is interrelated to infelicitous heat transfer and fluid flow, untoward selective partition of alloying elements and inopportune dendrite tip undercooling through diffusion process on separate weld side. Flow convection in the weld pool results in dendrite tip fragmentation as nuclei for heterogeneous nucleation and growth of new grain. The curvature-induced weld pool shape significantly contributes to variation of γ phase grain growth. The severer variation of grain growth is produced, the more insidious propensity of microstructure degradation is developed to impair mechanical properties and vice versa. Neck part of nail-shaped weld is macroscopically and microscopically preferential sites for microstructure, chemistry, crystallography and metallurgy heterogeneity, regardless of high or low heat input welding conditions. Grain size within the eutectic-resistant bottom part is advantageously finer than middle neck region and top side of weld to keep epitaxial grain nucleation and growth below solid solubility rather than eutectic reaction of Laves phase formation under prevalent partitionless rapid cooling conditions. Variation of secondary dendrite arm spacing with location is intriguingly parabolic distribution. Unavoidable heterogeneity-related microstructure difference phenomenon is similar in both sides of fusion boundary. Overall, secondary dendrite arm spacing on the right side is coarser than left side, which is crystallographically close to fusion boundary. The range of secondary dendrite arm spacing is typical of 1-4 microns. Weld pool geometry on the right side is rather larger than left side and predominantly exacerbates grain growth kinetics across solid/liquid interface. The latter possesses attractive steep thermal gradient, slow solidification rate and concomitant fast cooling rate.

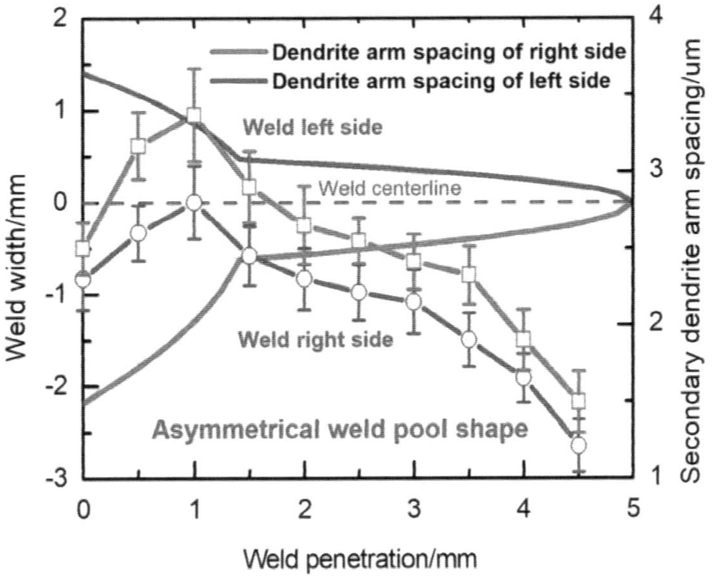

Figure 3. Substructure measurement around both sides of fusion boundary from top, middle to bottom of keyhole weld

Welding heat input is properly kept as low as possible to produce crackless weld, dendritically refine grain size, increase grain boundary surface area, improve hot ductility and hence augment resistance to hot cracking. The microstructure degradation, thermal and shrinkage stresses as well as inhomogeneous strain development are thermometallurgically and thermomechanically inadequate to incur centerline cracking at upper weld to totally limit degradation-susceptible region, notwithstanding weld fusion zone invariably experiences different peak temperatures and diverse durations with unsatisfactory weld pool shape. Moreover, Uncrackable liquid healing phenomenon is significantly prevalent within the fusion zone during rapid solidification.

Terminal solidification product of nonequilibrium phase transformation along dendrite interstice is shown in figure 4. Alloying elements, including Niobium, Molybdenum and Titanium, strongly enrich interdendrite liquid for preferential segregation and inhomogeneous distribution of intermetallic phase to increase phase instability in accordance with Ni-rich corner available data of binary equilibria phase diagrams. Niobium, Molybdenum and Titanium are thermodynamically tied up in the deleterious terminal solidification product of nonequilibrium phase transformation, and considerably deviate from critical equilibrium conditions. There is chemical partition between dendrite core and surrounding dendrite interstice to thermodynamically and kinetically maximize undercooling- and supersaturation-driven grain growth. The weld is primarily strengthened by precipitation of intermetallic gamma double prime γ''phase Ni_3Nb with body-centered tetragonal (BCT) crystal structure. Laves eutectic is hexagonal close-packed crystal structure (HCP). Primary γ phase is face-centered cubic crystal structure (FCC). Softening γ phase is compensated by strengthening γ'' precipitate. Chemistry heterogeneity and segregation increase primary γ phase instability. Laves phase is discontinuously dispersed at dendrite boundaries.

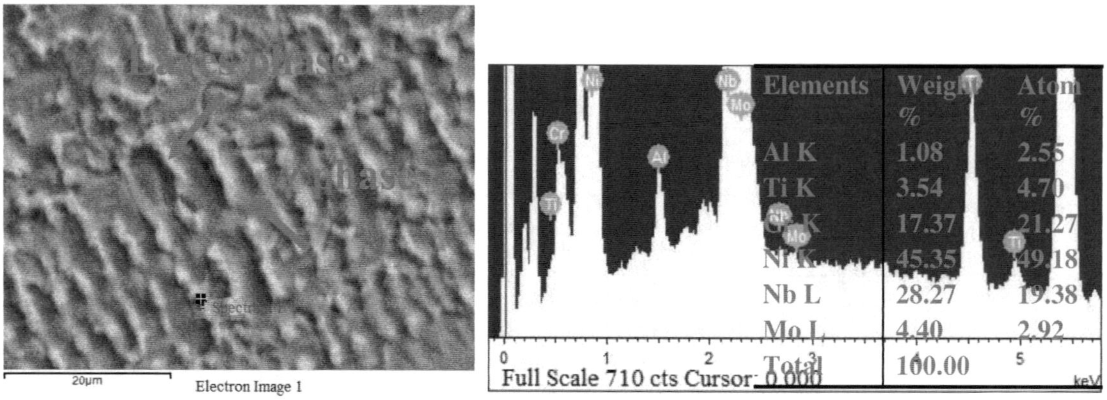

a) Morphology and size of intermetallic phase b) Phase chemistry microanalysis

Figure 4. Terminal solidification product of nonequilibrium phase transformation at dendrite interstice

Discontinuous bright part is Laves eutectic phase, and dark part is austenite γ phase. The former is easily spread in the interdendrite location of high-angle grain boundaries that is attributable to excess of alloying elements, selective partition and chemical interaction in the liquid to reach eutectic reaction concentration and drive primary solidification path and cooling sequence of solidification various stages. Supersaturation of residual interdendrite liquid is eventually relieved through intractable Laves eutectic reaction L→ Laves + γ to complete solidification process because of high affinity of Niobium, Molybdenum and Titanium for Nickel and low solid solubility in γ phase during continual enrichment ahead of solid/liquid interface, which are vital metallurgical factors to lead to superlattice reflection and mismatch, elastically and plastically accumulate stress concentration, and significantly embrittle grain boundaries. In other words, Niobium, Molybdenum and Titanium are consumed by low-temperature eutectic Laves phase formation. The nonequilibrium phase transformation is heterogeneously overwhelmed in the shoulder junction region between neck head and spike. It is emphasized that location-dependent rapid solidification conditions predominantly decrease solute back diffusion and severely increase proportion of nonequilibrium Nb-rich intermetallic phase. In order to minimize detrimental nonequilibrium eutectic-type reaction of welding metallurgy, combination of welding variables, i.e. peak temperature and cooling rate, are restricted, but it is fairly difficult to simultaneously meet grain refinement kinetics and complete eutectic-free solidification thermodynamics during weld cooling, result in dilemma for weldability improvement. The contradiction facilitates better comprehension of adverse eutectic-aided microstructure degradation without any tendency towards hot cracking during polycrystalline material welding.

Multiscale fractography of full penetration weld fusion zone is shown in figure 5. Nonequilibrium segregation and phase transformation thermodynamically produce a variety of fracture-susceptible intermetallic phase and chemical inhomogeneity with no distinct preferential orientation within the fusion zone that is detrimental to weld strength and ductility. Essentially, there is weak interface or incoherent phase boundaries between Laves phase and γ phase to potentially increase stress concentration, deformation and grain boundary decohesion in the vicinity of dendrite interstice, where crack is aggressively initiated and propagated at high-angle grain boundaries fracture region during tensile test to deteriorate variation of anisotropic mechanical properties. The detrimental intermetallic phase is particularly concentrated in the waist neck region. Intricate dendrite substructure with Laves/γ eutectic phase cluster delineates morphology undulation of intergranular crack path on the fracture surface. Moreover, dominant dimple fracture irregularly appears. Mechanical and metallurgical factors adversely contribute to not only nucleation, growth and prorogation of crack, but also losses of strength and ductility for fracture failure, however grain refinement of solid solution γ phase and γ′′ phase precipitation critically strengthen the weld through correct welding variables and thus provide sufficient ductility for stress relief with more available grain boundary surface area.

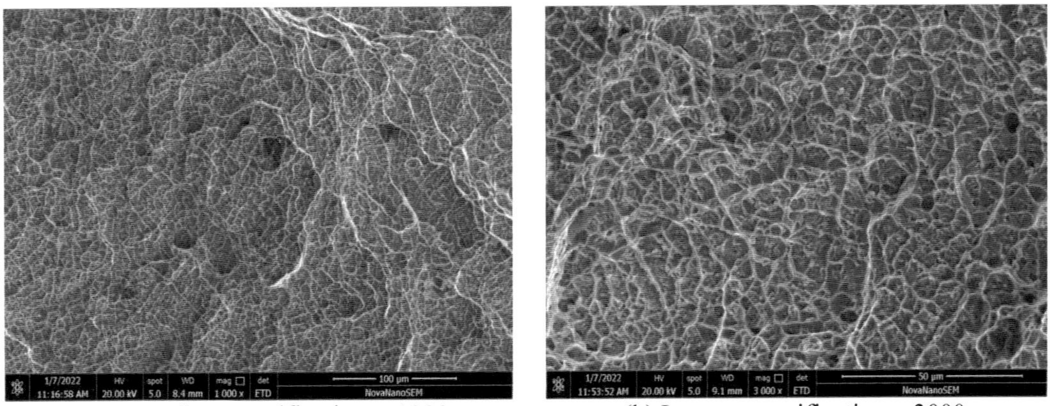

(a) Smaller magnification × 1000 (b) Larger magnification × 3000

Figure 5. Multiscale fractography of full penetration weld fusion zone

The weld suffers ductile dimple and brittle intergranular fracture failure together. Therefore, heat input is improvingly chosen to satisfy good metallurgical and mechanical properties. The higher amount of insidious nonequilibrium Laves/γ eutectic phase is dispersed alongsite dendrite boundaries, the larger size misfit of disregistry is spontaneously developed for stress buildup of polycrystalline weld and the smaller weld strength and ductility occur with irregular fracture surface because of dendrite boundary embrittlement and microstructure heterogeneity by various metallurgy reactions for inferior mechanical properties. Amount, distribution and morphology of intermetallic Laves phase are altered through appropriate welding procedure for thermodynamics reaction modification with good resistance to fracture so as to advertently increase attractive fine-grained single phase solid solution of fracture-insusceptible region and optimize mechanical properties.

Tensile properties comparison between substrate material and weld samples before and after laser welding is shown in figure 6. Samples #1-1-1,#1-1-2,#1-1-3 and #1-1-4 are welded by means of the same welding variables. Fine equiaxial grain of substrate material is more ductile and stronger than weld to decrease mechanical performance abnormality. Obvious differences of weld strength factors, such as tensile strength and yield strength, and ductility factor, especially elongation, are observed. Mechanical properties heterogeneity phenomena are ascribed to a series of grain growth and phase transformation variation in different regions to highly increase instability of fracture failure. As long as undesirable eutectic reaction is thermodynamically appalling, growth stress development of intermetallic phase with heterogeneous distribution plays an important role in the eventual variability of mechanical properties. There is a mismatch between ductile FCC γ phase, BCT γ′′strengthener phase and brittle HCP Laves eutectic phase for losses of coherency and homogeneity and degrades failure resistance during tensile test. Crack initiation and propagation as result of high stress concentration are very sensitive to nonequilibrium phase sites for interface decohesion of dendrite multistructure. Reduction of problematic metallurgical discontinuities, microstructure difference and eutectic-unresistant region size to hinder grain boundary sliding by proper processing parameters is essential to ameliorate weld properties and reduce fluctuation. Meanwhile, interplay between metallurgical and mechanical factors morphologically increase susceptibility to grain growth variation. Since weld is completely devoid of homologous preferential orientation in the γ phase grain and dendrite satisfactorily interlocks each other, the larger single γ phase region of austenite solid solution is produced, the smaller eutectic-driven mechanical anomalies of asymmetrical weld occurs and the more elasticity and plasticity are attained to abate mechanical heterogeneity and vice versa. No cracking is observed, nevertheless, conservative welding variables optimization for grain refinement is insufficient to compensate for weld mechanical impairment. Effect of deleterious intermetallic phase and abnormal multiscale grain growth on mechanical integrity of polycrystalline weld is elucidated to extensively comprehend response of metallurgical and mechanical behaviors to welding variables on both sides of weld pool.

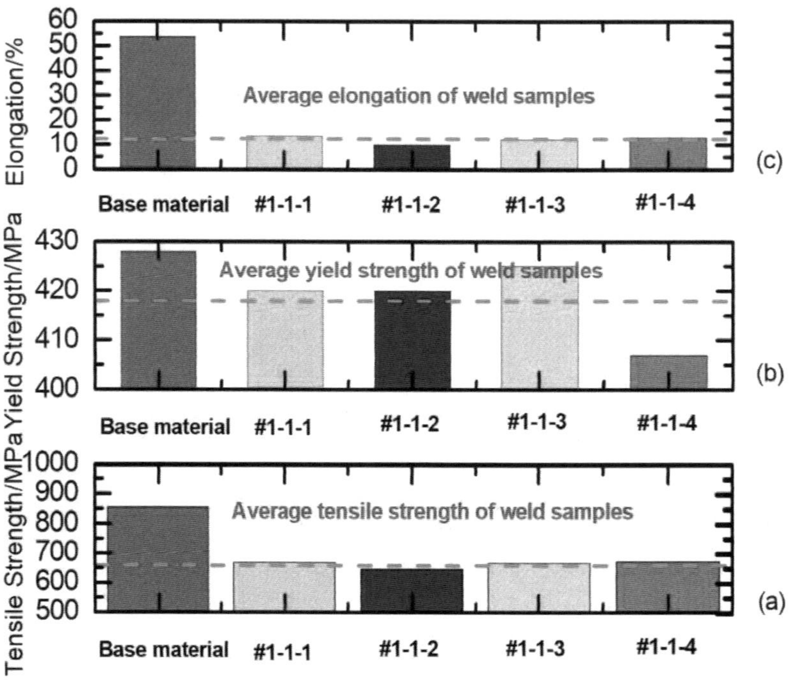

(a) Tensile strength, (b) Yield stress and (c) Elongation

Figure 6. Tensile properties comparison between substrate material and weld samples

4. Conclusions

1. Optimal strength and ductility of tensile properties are appropriately attained by amelioration of grain refinement kinetics and phase transformation thermodynamics to thermo-mechanically withstand grain growth abnormality and thermo-metallurgically suppress nonequilibrium brittle Laves/γ eutectic reaction as consequence of stringent microstructure and phase stability control.

2. No homologous crystallographic orientation is developed, nevertheless, numerous advantages of laser welding leads to final high-quality keyhole weld to minimize adverse effect of unsymmetrical weld pool shape on grain growth. Precise butt alignment and proper sample preparation are required.

3. Eutectic-susceptible portion of weld as result of supersaturation relief, which is entrapped within lateral dendrite rifts and interphase boundaries, drastically weakens grain substructure strength and ductility during weld cooling to embrittle incoherent dendrite with weak metallurgical bond, facilitate grain movement as well as separation and reduce resistance to fracture failure. Non-segregation solidification conditions are necessary for high strength and ductility of multicomponent weld.

4. Laser welding variables are rapid enough to initiate nucleation and growth of nonequilibrium intermetallic phase by various metallurgical reactions, and meanwhile favor flaw-free grain growth and tensile properties improvement. It is fairly difficult to achieve partition-resistant microstructure and superior mechanical properties of polycrystalline weld without detrimental nonequilibrium phase transformation, when either welding heat input or gain size of substrate material is critically as small as possible.

5. References

[1] Shuai Guan and Chuanyong Cui 2015 A newly developed wrought Ni-Fe-Cr-based superalloy for advanced ultra-supercritical power plant application beyond 700°C *Acta Metall. Sin. (Engl. Lett.)* **28(9)** 1083-1088

[2] Miao Wang and Xingwei Huang 2021 Effect of rotation rate on microstructure and mechanical properties of friction stir processed Ni-Fe-based superalloy *Acta Metall. Sin. (Engl. Lett.)* **34** 1407-1420

[3] Tianfu Wang and Xinjie Di 2019 Effect of δ phase on microstructure and hardness of heat-

affected zone in TIG-welded GH4169 superalloy *Acta Metall. Sin. (Engl. Lett.)* **32** 1041-1052

[4] Linxu Li and Xiufang Gong 2021 Correlation between phase stability and tensile properties of the Ni-based superalloy MAR-M247 *Acta Metall. Sin. (Engl. Lett.)* **34** 872-884

[5] L O Osoba and A K Khan 2013 Cracking susceptibility after post-weld heat treatment in Haynes 282 nickel-based superalloy *Acta Metall. Sin. (Engl. Lett.)* **26(6)** 747-753

[6] Xinjie Di and Xiaoqian Liu 2016 Effect of post-weld heat treatment on microstructure and corrosion resistance of deposited metal of high-chromium nickel-based alloy *Acta Metall. Sin. (Engl. Lett.)* **29(12)** 1136-1143

[7] Jun Zhang and Jide Liu 2021 Effect of high density current pulses on microstructure and mechanical properties of dual-phase wrought superalloy *Acta Metall. Sin. (Engl. Lett.)* **34** 1635-1644

[8] Zijian Zhou and Rui Zhang 2021 Effects of homogenization treatment on the microsegregation of a Ni-Co based superalloy produced by directional solidification *Acta Metall. Sin. (Engl. Lett.)* **34** 943-954

[9] Anwen Zhang and Yan Yang 2019 Distribution of phosphorus and its effects on precipitation behaviors and tensile properties of IN718C cast superalloy *Acta Metall. Sin. (Engl. Lett.)* **32** 887-899

[10] Dawei Han and Lianxu Yu 2018 Effect of heat treatment on the microstructure and mechanical properties of the Modified 718 alloy *Acta Metall. Sin. (Engl. Lett.)* **31** 1224-1232

[11] Yang Liu and Lei Wang 2021 Effects of thermally assisted warm laser shock processing on the microstructure and fatigue property of IN718 superalloy *Acta Metall. Sin. (Engl. Lett.)* **34** 1645-1656

[12] K D Ramkumar and R J Sai 2015 Influence of filler metals in the control of deleterious phases during the multi-pass welding of Inconel 718 plates *Acta Metall. Sin. (Engl. Lett.)* **28(2)** 196-207

[13] Yan Guo and Bohan Wang 2013 Aging precipitation behavior and mechanical properties of Inconel 617 superalloy *Acta Metall. Sin. (Engl. Lett.)* **26(3)** 307-312

[14] Xiangwei Li and Li Wang 2019 Effect of aging heat treatment on the microstructure and creep properties of the cast Ni-based superalloy at low temperature *Acta Metall. Sin. (Engl. Lett.)* **32** 651-658

[15] Tao Liu and Mei Yang 2019 Effect of heat treatment on microstructure and stress rupture properties of a Ni-Mo-Cr-Fe-based corrosion-resistant superalloy *Acta Metall. Sin. (Engl. Lett.)* **32** 116-126

[16] Enxiang Pu and Wenjie Zheng 2017 Effect of precipitation on hot deformation behavior and processing maps of nickel-based UNS N10276 alloy *Acta Metall. Sin. (Engl. Lett.)* **30(11)** 1119-1134

[17] Jinwen Zou and Xiaofeng Wang 2018 Characterization of interfacial bounding mechanism for graphene-modified powder metallurgy nickel-based superalloy *Acta Metall. Sin. (Engl. Lett.)* **31** 753-760

[18] Xu Zhang and Dianzhong Li 2016 Effect of Nb and Mo on the microstructure, mechanical properties and ductility-dip cracking of Ni-Cr-Fe weld metals *Acta Metall. Sin. (Engl. Lett.)* **29(10)** 928-939

[19] K S Prasad and C S Rao 2012 Effect of pulsed current microplasma arc welding process parameters on fusion zone grain size and ultimate tensile strength of Inconel 625 sheets *Acta Metall. Sin. (Engl. Lett.)* **25(3)** 179-198

[20] Shuang Gao and Jieshan Hou 2017 Influences of cooling rate after solution treatment on microstructural evolution and mechanical properties of superalloy Rene 80 *Acta Metall. Sin. (Engl. Lett.)* **30(3)** 261-271

Acknowledgement

The project is financially supported by the Anyang Institute of Technology (Grant No. BSJ2019031).

Thermal Decomposition of GAP Studied by Dynamic Pressure-measuring Thermal Analysis

Huang Keqi[1], Xia Lianghong[1], Long Maohao[1], Li Zhimin[1] and Zhang Tonglai[1,*]

[1]State Key Laboratory of Explosion Science and Technology, Beijing Institute of Technology, Haidian District, Beijing 100081, China

*ztlbit@bit.edu.cn

Abstract: The thermal decomposition process of GAP in a vacuum confined space was measured by Dynamic Pressure-measuring Thermal Analysis(DPTA). In order to avoid the test error, the test process was improved: the method of liquid nitrogen condensation was used to prevent GAP volatilization in the process of vacuum pumping in the reaction test tube, and the vapor pressure at different temperatures was measured by thermogravimetric analysis. The results show that at 80°C to 120°C, the mechanism function of GAP non-isothermal decomposition reaction is Jander equation, the apparent activation energy is 159.8kJ·mol-1, and the pre-exponential factor lnA is 25.8. At the GAP isothermal reaction stage, the mechanism function changes with the temperature. When the temperature increases, the thermal decomposition reaction changes from diffusion reaction to activation reaction. The decomposition rate constant K increases exponentially with increasing temperature. When T=80°C, K=1.27×10-6; when T=120°C, K=36.93×10-6. According to the data of the decomposition time of 0.1% in the temperature range of 80~120 °C, the Semenov equation is lntT = 19039.90/T- 43.012. The storage life is 35.96 years at 25°C.

1. Introduction

As the current research hot spot, Polyglycidyl azide (GAP) has high generate heat, large density, low pollution, no corrosive gas, low characteristic signal, low combustion temperature, and low sensitivity, which often considered making high energy solid propellant [1]. Kubota and Brill [2] et al studied the initial stage of GAP by T-jump and FTIR, then made a detailed review of GAP thermal decomposition. Chen Zhiqun et al [3] studied the process of GAP thermal decomposition by DSC,TG-FTIR and other technologies, and proposed the possible decomposition mechanism. Liang Lei et al [4] analyzed the structure of GAP by infrared spectroscopy and tested its thermal decomposition characteristics. They believed that GAP's decomposition peak temperature was high and its thermal stability was good, so it could be selected as the material of high-energy low characteristic signal propellant.

Although the above method can obtain the kinetic parameters of GAP thermal decomposition process, it cannot accurately obtain the thermal decomposition process of GAP in vacuum environment, and cannot analyze the slow decomposition process of GAP at low temperature, which has little significance for practical application. Therefore, the research method should be improved. Based on the principle of vacuum stability (VST), Zhang Tonglai et al [5] establish the dynamic vacuum stability test (DVST) with pressure and temperature sensors, which provides a new method for studying the stability of energetic materials and predicting their storage life. In 2010, Rui Liu [6] et al studied the thermal decomposition process of HMX by using dynamic vacuum stability experiment, and obtained the thermal stability of decomposition mechanism and storage life of

Journal of Physics: Conference Series **2361** (2022) 012002 doi:10.1088/1742-6596/2361/1/012002

HMX In 2015. Liu Jianchao et al [7] studied the thermal decomposition mechanism of lead azide (LA) by DVST method, and analyzed the influence of crystalline controller on the thermal stability and thermodynamic parameters of LA. In 2015, Rui Liu [8] established and improved the Dynamic Pressure-measuring Thermal Analysis (DPTA) technology, which was used to study the thermal properties of various types of energetic materials. In this study, DPTA) method was used to study the thermal decomposition process of GAP, and the boiling volatilization problem was improved for GAP in the process of external vacuumization, focusing on GAP heating at low temperature with high vacuum degree Long time storage process caused by heat and low pressure double factors (thermal pyrolysis production of gas product and the process of gas production rate, the thermal decomposition kinetic parameters were obtained through the quantitative analysis of reaction mechanism of the thermal stability and storage life, and the test results for direct production using the safe operation of storage and transportation process control storage conditions have important application value.

2. Experiment

2.1. Experimental materials

GAP, which is from Xi 'an Modern Chemical Institute; Non-volatile at room temperature, brown-yellow viscous liquid, vitrification temperature of -67°C, number average molecular weight Mn=5013, weight average molecular weight M33=6617, hydroxyl value 40.21mgKOH/g dispersion index 1.32, molecular formula is CH3[CH2CH(CH2N3)O]nCH3.

2.2. Test device

DPTA test system [9]: DPTA instrument was developed by the State Key Laboratory of Explosion Science and Technology, Beijing Institute of Technology. The instrument is composed of three parts: thermal decomposition reaction unit, temperature control program, heating unit, data acquisition and processing unit. The instrument is used for dynamic real-time monitoring of the pressure and temperature changes during the thermal decomposition of materials at a certain vacuum and temperature. The volume of the test tube is 25mL.

Thermogravimetric Analyzer: PerkinElmer Pyris type, the sample volume is about 1mg. Open platinum crucible, dynamic N_2 atmosphere flow rate 20mL/min.

2.3. Test process

The thermal decomposition test process includes two stages: the non-isothermal stage of uniform temperature rising to the set temperature and the isothermal process at the set temperature. In the process of vacuuming, the pressure in the sealed test tube is reduced, the boiling point of GAP is reduced, and volatilization is prone to increase in test error. Therefore, the reaction test tube is purged with nitrogen to dilute the air before the test, and the reaction test tube is cooled with liquid nitrogen in liquid nitrogen to cool down, which can prevent GAP boiling volatilization; When the sample is heated in the test tube, part of the sample will also volatilize and increase the dynamic pressure. Therefore, the vapor pressure should also be calibrated and deducted, The specific steps are as follows:

(1) The nitrogen replacement air process

As is shown in figure 1 first open the valve, then extend the rubber tube connected to the nitrogen gas tank into the bottom of the reaction test tube, and slowly inject nitrogen into the reaction test tube. After continuous 5 minutes, the air in the reaction test tube can be slowly removed.

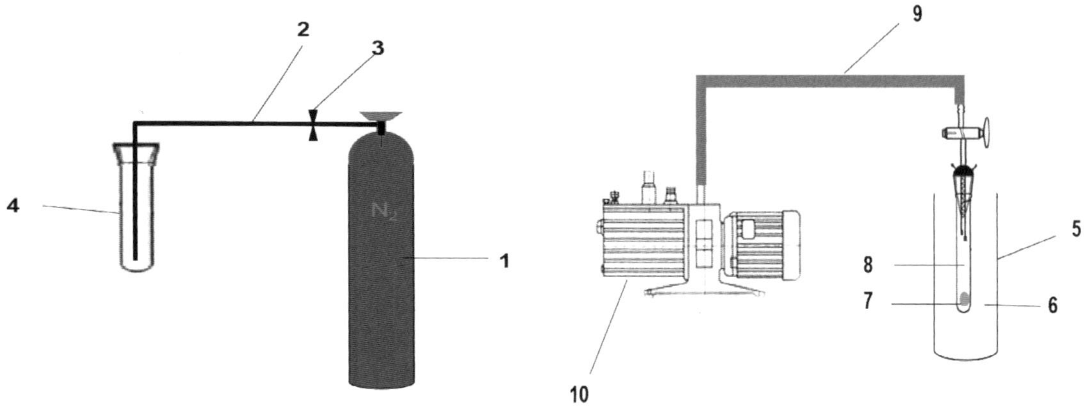

1- High pressure nitrogen cylinder; 2-rubber hose; 3- valve; 4- Partial reaction test tube; 5- Dewar bottles; 6- Liquid nitrogen; 7- Volatile energetic materials; 8- Nitrogen; 9- Rubber hose; 10- Vacuum pump

Figure 1. Nitrogen replacement device and liquid nitrogen cooling unit.

(2) Sample loading

Weigh about 0.5g of sample and load it into a reaction test tube with good air tightness, the reaction test tube is built with a miniature pressure sensor and a miniature temperature sensor, GAP does not contact with the pressure sensor and temperature sensor; The internal temperature T and pressure P of the viscous liquid are different from the temperature and pressure of the gas in the reaction test tube.

(3) Sample condensation process of liquid nitrogen

The reaction test tube with GAP is placed in a dewar bottle filled with liquid nitrogen for condensation treatment. After the condensation, the reaction test tube is vacuumized to a pressure less than 10Pa. After the pressure in the reaction test tube is stabilized, the pressure and temperature in the reaction test tube are recorded. When the sample is condensed, the external temperature is lower than the freezing point of GAP.

Liquid nitrogen is colorless, tasteless, non-corrosive and very low temperature liquid. The temperature of liquid nitrogen is between -196°C and -210°C, and the freezing point of nitrogen is -209.86°C, which can absorb a lot of heat during gasification. Put the test tube into liquid nitrogen, on the one hand, liquid nitrogen gasification will absorb the heat inside the test tube to reduce the temperature of GAP; On the other hand, as a result of the test tube inside and outside temperature difference will generate heat transfer, the temperature of the test pipe internal temperature is greater than the external cause heat transfer results in the decrease of internal temperature of the test tube. This methon is safe and simple, to solve the pressure in the closed container when volatile energetic materials volatilized caused by the reduction of quality affect DPTA test experimental data results.

(4) The dynamic stress test

The reaction test tube is removed from a low temperature environment and placed in a heating furnace for heating, Set the heating rate to 6 °C/min, constant temperature time records for 48 h, Record the pressure and temperature in the reaction test tube in real time online; The pressure is normalized and then standardized, and the pressure value P_1 is obtained.

(5) The vapor pressure test process

It is also weighed about 1mg sample, using thermogravimetric analysis (TGA) method to test GAP evaporation process at different temperatures and calculate the corresponding vapor pressure value P_0.

(6) Thermal decomposition pressure analysis

The pressure value P_1 after treatment at the same temperature is subtracted from the vapor pressure value P_0 to obtain the pressure increase value P caused by the decomposition of the energetic material, and the data is used for thermal analysis of the energetic material.

3. Results and discussion

3.1. Relationship between GAP vapor pressure and temperature

Assuming that the free surface area is constant, both evaporation and sublimation reactions are zero-order reactions, so in TG test, the sample is as flat as possible on the platinum crucible.

Under constant temperature, the mass loss rate caused by evaporation is constant. Evaporation under vacuum is described by Langmuir equation [10]:

$$-\frac{dm}{dt} = p\alpha\sqrt{\frac{M}{2\pi RT}} \quad (1)$$

-dm/dt-Evaporation rate per unit area, P-vapor pressure, M-molecular weight of vaporized gas, R - ideal gas constant, T - thermodynamic temperature, M - molecular weight of gas, T- temperature of gas, K;

The vapor pressure P of a standard substance can be calculated from its known Antoine equation [11]:

$$\lg P = A - B/(C+T) \quad (2)$$

Langmuir equation and Antoine equation were combined:

$$\ln P = A\ln(dm/dt) + B \quad (3)$$

Through experiments, Rui Liu obtained the corrected Antoine equation of standard substance benzoic acid [12]:

$$\ln P = 1.0964\ln\left({}^{dm}\!/_{dt}\right) + 1.5901 \quad (4)$$

The dm/dt value of GAP was measured by thermostatic TG method and substituted into equation (4). The corresponding relationship between dm/dt and temperature T is shown in the following table. The vapor pressure increases with the temperature.

Table 1. GAP vapor pressure at different temperatures.

T/°C	T/K	dm/dt /μg·min^{-1}	ln$P$	P
80	353.15	0.33	0.3745	1.454
90	363.15	0.81	1.345	3.838
100	373.15	3.50	2.964	19.37
110	383.15	9.01	3.999	54.54
120	393.15	12.1	4.314	74.74

3.2. Decomposing the corresponding relationship between pressure and time

Five test temperatures were set at 80, 90, 100, 110 and 120°C. respectively, to obtain the original curve of GAP thermal decomposition pressure measured with time in the temperature-programmed stage and isothermal stage, and to deduct the atmospheric pressure and initial value of the original curve. Due to the volatile nature of GAP, the vapor pressure of the material when heated to a set temperature in a vacuum is also subtracted. Then, the curve of standard pressure (P) of GAP thermal decomposition gas changing with time (t) is obtained, as shown in figure 2. Figure 2 shows that when the initial pressure is similar, the gas pressure of GAP thermal decomposition increases with the increase of heating time. GAP gas pressure at the same time and at different temperatures also increases with the increase of temperature. The results show that the storage temperature has a significant effect on the stability, storage safety and service life of GAP.

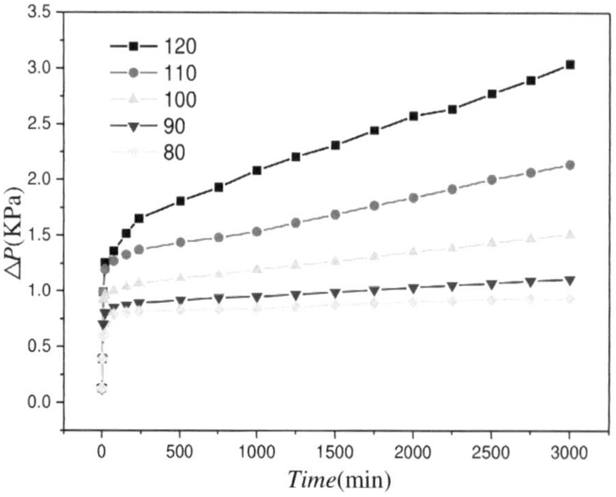

Figure 2. Variation of standard pressure of GAP net decomposed gas with time at different temperatures.

3.3. thermal decomposition kinetics research

Measured with the sample in a moment at the termination of the released gas volume V and test out the ratio of gas volume V_L said sample at the moment of conversion rate (or decomposition depth) alpha test the product gas pressure data manager wants to gas state equation transformation, said after the decomposition conversion for:

non-isothermal phase:

$$PT_L/P_LT=\alpha \tag{5}$$

Isothermal stage:

$$P/P_L=\alpha \tag{6}$$

P-the standard pressure of gas released at a certain measurement moment, kPa; P_L - the standard pressure of gas released at the time of test termination, KPa; T-the temperature at a measurement moment in the non-isothermal stage, K; T_L-the set test temperature, K.

3.3.1. Dynamic study at non-isothermal stage. The GAP was heated from 25 to 80,90,100,110,120°C. According to formula(5), the data of net separation liberated gas pressure (P) in five non-isothermal stages were converted to conversion rate (α), and the data of conversion rate (α) and its rate of change with temperature (dα/dT) were used for thermal analysis dynamics calculation. The general integral method [13]64 and differential equation method [13]80 were used to perform linear fitting for the known mechanism functions [13]151-160 of different types No.1-41. The optimal mechanism functions and corresponding dynamic parameters were screened by taking the highest linear correlation coefficient (R) as the criterion The results of dynamic parameters in the non-isothermal stage are shown in table 2. The results show that the thermal decomposition process in GAP non-isothermal stage at different test temperatures conforms to the No.6 mechanism function, namely Jander equation,n=2, which describes the three-dimensional diffusion, spherical symmetry and deceleration α -T curve. The activation energy of GAP at different temperatures is close to the pre-exponential factor value, which indicates that the activation energy of GAP thermal decomposition process is independent of heating temperature. The average E_a of GAP calculated by excluding errors generated by different test conditions is 159.8kJ·mol^{-1} and the average lnA is 25.8.

Table 2. Kinetic parameters of GAP thermal decomposition at non-isothermal stage.

T/°C	Function number	E_a/(kJ·mol^{-1})		ln(A·s^{-1})		standard deviation r	
		Universal integration	Differential equation method	Universal integration	Differential equation method	Universal integration	Differential equation method
80	6	158.9	159.1	25.15	24.14	-0.9938	-0.9927
90	6	159.3	160.2	26.35	25.13	-0.9945	-0.9819
100	6	157.8	160.0	25.18	27.12	-0.9967	-0.9925
110	6	158.2	161.2	24.95	26.17	-0.9939	-0.9945
120	6	160.2	163.2	26.45	27.19	-0.9956	-0.9813

As shown in figure 3,the thermal decomposition process of GAP is a complex process, it is generally believed that GAP first occurs the fracture rearrangement of azide group -CH_2-N_3,-CH_2-N_3 becomes -CH=NH, and then in the heating process, the main chain skeleton decomposes to generate CO、N_2、CO、HO_2. CH_4 and other substances. The GAP activation energy Ea measured in the experiment is close to the bond strength of RN=N_2 146~172, so it can be inferred that the GAP activation reaction is the fracture of azide group.

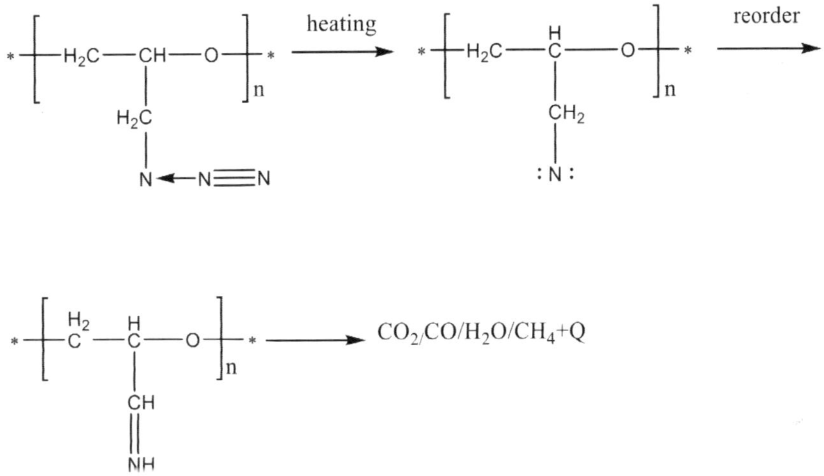

Figure 3. Thermal decomposition process of GAP under heating condition.

3.3.2. Kinetic study of thermal decomposition at isothermal stage. The GAP thermal decomposition gas pressure (*P*) at isothermal stage is converted into conversion rate (α) according to Formula (6). The data of conversion rate (α) and constant temperature time (t) are linearly fitted according to the kinetic equation of heterogeneous reaction G(α)= kt [14]. For different types of mechanism functions from 1 to 41, the least square method is used to make linear regression of G(α)-t relation y=Ax+B. The suitable mechanism function G(α) is selected with the maximum linear correlation coefficient R and the minimum intercept B as the comprehensive criterion. The optimal function is selected as the reaction mechanism function and the least square method is used to make linear regression y=Cx again The slope C of the fitted line is the reaction rate constant k at this temperature. The isothermal data of GAP at different temperatures from 80 to 120 are processed by this method. The results obtained are listed in table 3.

Table 3. Kinetic parameters of thermal decomposition at GAP isothermal stage.

T/°C	Function Number	Function name and mechanism	$K/(10^{-6}S^{-1})$	G(α)
80	9	Zhuralew-Lesokin-Tempelman Equation, three dimensional diffusion	1.27	$[(1-\alpha)^{-1/3}-1]^2$
90			2.79	
100	8	Anti-Jander Equation, three dimensional diffusion	6.45	$[(1+\alpha)^{1/3}-1]^2$
110			16.13	$[(1+\alpha)^{1/3}-1]^2$
120	5	Jander equation，n=1/2，three dimensional diffusion	36.93	$[1-(1-\alpha)^{1/3}]^{1/2}$

When the temperature increases, the intermolecular interaction of GAP is not enough to restrain the stronger intermolecular motion, and the intermolecular distance increases, and the internal friction decreases. For liquid viscosity mainly depends on intermolecular friction, when the temperature increases, the viscosity of GAP decreases, increasing the intermolecular diffusion For the GAP of single molecules, according to the lindemann monomolecular reaction activation mechanism: A molecular decomposition or isomerization reaction must go through with high-energy collisions among the molecules become activated molecules A*, if the energy is transferred to need breaking by impact the key of forming P, if A* internal energy is not fully transfer with low molecular collision energy, residual energy into A* will lost For ordinary molecule A, the reaction rate is the formation rate of product P. If the reaction activation energy is large, it is an activation controlled reaction. If the intermolecular viscosity of reactants is large, the reaction is mainly controlled by diffusion.

3.4. Study on thermal stability

The measured net decomposition standard pressure P data of GAP placed at five different temperatures at a constant temperature were converted into the gas discharge under standard condition by using the ideal gas equation according to the calculation method of gas discharge stipulated in military standard GJB772-97. The results are shown in table 4 below.

Table 4. Net gas emission of thermal decomposition obtained by DPTA at different temperatures.

t/°C	△*P*/KPa	*V*/（$ml\cdot g^{-1}$）
80	0.941	0.2324
90	1.064	0.2620
100	1.509	0.3731
110	2.140	0.5280
120	3.030	0.7484

As can be seen from table 4, with the increase of heating temperature, the net liberated gas volume increases. According to GJB772-97, if the gas volume of 1 g sample is less than 2.00mL at 48 h under 100, it indicates that the vacuum stability and the gas volume released by each GAP at 100 is 0.3731mL /g, and the thermal stability is qualified.

3.5. Research on storage

Semenov equation is used when GAP storage life is estimated at different temperatures [6]. Semenov equation is:

$$\ln t_T = a + b_T \quad (7)$$

T-the test temperature, K; tT-Reaction time at a certain depth of decomposition, s; a and b are undetermined coefficients;

The decomposition depth (α) is defined as the ratio of the volume of gas released by the sample at a certain measurement moment to that released by the sample after complete decomposition. The end point of the storage life of explosives is defined as the decomposition depth of 0.1% [15]. The pressure data of gas phase products are used to represent the complete decomposition depth:

$$Pa/P=\alpha \quad (8)$$

In the formula, P_a is the standard pressure KPa of gas released by thermal decomposition of the sample at a certain measurement moment; P is the standard pressure KPa of gas released by complete decomposition of the sample.

According to the law of mass conservation and the equation of state of ideal gas, the theoretical pressure of gas released when GAP is completely decomposed is 3465.5KPa. According to Equation(8), the decomposed gas pressure P of GAP at a measured moment in the temperature range of 80~120 is converted to the total decomposition depth α, and the time required for GAP decomposition depth to reach 0.1% at different temperatures can be calculated according to the relationship between α and *T*.

Table 5. Time t_T needed for GAP to decompose 0.1% at different temperature *T*.

T/°C	t_T/s
80	54295.7
90	12295.6
100	3016.9
110	797.1
120	224.9

The data in table 5 are substituted into the Semenov equation, and the linear fitting of lntT and 1/T is obtained as lntT = 19039.90/T- 43.012. The fitting curve of r=0.9916 is shown in figure 4. According to the equation, the effective storage life of GAP is 35.96 years when the storage temperature is 25°C.

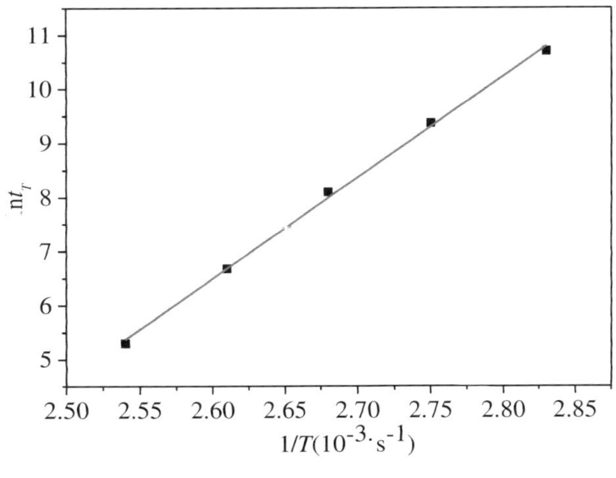

Figure 4. Correlation curve between lnt_T and 1/*T* of GAP.

4. Conclusion

(1) The thermal decomposition process of GAP at five different constant temperatures at 80~120 was studied by dynamic pressure measurement and thermal analysis. The thermal decomposition mechanism function and kinetic parameters of GAP were obtained according to the process of pressure and temperature changing with temperature.

(2) At these five set temperatures, the decomposition mechanism function of GAP at non-isothermal stage is Jander equation, n=2, and the integral form of the mechanism function is $G(\alpha)=[1-(1-\alpha)1/3]^2$. The activation energy E and pre-exponential factor A are close to each other at

different temperatures. The mechanism function of GAP is different at different test temperatures. The value of reaction rate constant K increases with the increase of test temperature.

(3) At five test temperatures, the net decomposed gas volume of GAP calculated by the net decomposition pressure is less than 1mL/g, which can be judged that GAP has good thermal stability.

(4) Semenov equation fitted by DPTA method is $\ln t_T = 19039.90/T - 43.012$, and storage life is 35.96A when extrinsic 25.

5. References

[1] Wang Yawei, Yang Huiqun, Han Jinchao, et al. Research progress on GAP [J]. Applied Chemical Industry, 2017, 46(12): 2459-2462.

[2] Kubota N, Sonobe. T. Combustion mechanism of azide polymer[J]. Propellants, Explosives, Pyrote chnic, 1988, 13: 172-177.

[3] Chen Zhiqun, Liu Yan, Liu Ziru. Study on kinetics and mechanism of GAP thermal decomposition [J].Solid rocket technology,2003,26(4):52-54.

[4] Liang Lei, Yun Ni, Geng Xiaoheng, Lin Shaoting. Journal of north university of China (natural science edition), 2014, 35(02): 177-181.

[5] Zhang Tonglai, Hu Xiaochun, Yang Li, Li Kunyuan, Zhang Jianguo, Wang Wenjie, Wang Liqiong. Study on dynamic vacuum stability test method [J]. Energetic materials, 2009, 17(05): 549-553.

[6] Rui Liu, Yin Yanli, Zhang Tonglai, Yang Li, Zhang Jianguo, Zhou Zunning, Qiao Xiaojing, Wang Wenjie, Wang Liqiong. Study on dynamic vacuum stability test method (IV): thermal decomposition of HMX [J]. Energetic materials, 2011, 19(06):650-655.

[7] Liu Jianchao, Jiang Yutong, Zhang Tonglai, Yang Li. Evaluation of thermal stability of LA and CMC-LA by dynamic vacuum stability method [J]. Energetic materials, 2015, 23(10): 1020-1023.

[8] Liu Rui. Research on Dynamic Pressure Measurement and Thermal Analysis Technology and Its Application[D]. Beijing Institute of Technology, 2015.

[9] Zhang Tonglai, Hu Xiaochun, Yang Li et al. Study on Dynamic Vacuum Stability Test (DVST) Method [J]. Energetic Materials, 2009 17(5):549-553.

[10] J.P.Elder. Sublimation measurements of pharmaceutical compounds by isothermal thermogravimetry[J]. Journal of Thermal Analysis. 1997, 49 (2): 897-905.

[11] J. Oxley, J.L. Smith, J. Brady, S. Naik. Determination of urea nitrate and guanidine nitrate vapor pressures by isothermal thermogravimetry[J]. Propellants Explosives Pyrotechnics. 2010, 35 (3): 278-283.

[12] K. Chatterjee, D. Dollimore, K. Alexander. A new application for the Antoine equation in formulation development[J]. International Journal of Pharmaceutics. 2001, 213 (1-2): 31-44.

[13] Hu Rongzu, Gao Shengli, Zhao Fengqi et al. Dynamics of thermal Analysis [M]. 2nd edition. Beijing: Science Press, 2008.64.

[14] ChovancovaM, ZemanS. Study of initiation reactivity of some plastic explosives by vacuum stability test and non-isothermal differential thermal analysis[J].Thermo-chimica Acta 2007.460: 67-7620.

[15] Chu Shijin. Thermal Analysis of Explosive[M]. Beijing: Science Press 1994:226-230.

Weldability of Nickel/Aluminum Dissimilar Materials Laser Welding Part I: Microstructure Development, Solidification Behavior and Mechanical Properties of Polycrystalline Nickel-based Superalloy

Zhiguo Gao

School of Mechanical Engineering, Anyang Institute of Technology, Huang He Da Dao (West Side), Anyang City, Henan Province, 455000, China

E-mail: gao_zhiguo@ayit.edu.cn

Abstract: Weldability-related issues, including asymmetrical weld pool shape, γ phase multicomponent microstructure development, solidification behavior and mechanical properties, of γ″ precipitation-strengthened polycrystalline nickel-based superalloy are experimentally analyzed during keyhole laser welding repair. The crucial relationship between them provides weldability improvement opportunities and defect-free high quality weld for feasible aerospace materials laser processing. Attractive neck transition region of weld is more vulnerable to coarser dendrite than bottom part in the vicinity of fusion boundary, and this location is also liable to abundant Laves/γ eutectic phase formation because of heat and solute accumulation. The dominant chemical inhomogeneity brings about microstructure anomalies and irregular morphology to deteriorate hot cracking resistance. Because of thermometallurgical factors of selective alloying partition in the molten pool, partition-resistant Cr is chemically accumulated in dendrite core, while partition-susceptible Ti, Al, Mo and Nb are enriched in high-supersaturation interdendrite region, which consequently result in serious segregation and extensive Nb-rich intermetallic phase formation across solidification interface. Dendrite refinement and aggressive Laves/γ eutectic phase formation suppression are not simultaneously satisfied by low heat input under nonequilibrium solidification conditions. Therefore, mechanical properties are only partially improved by diffusion-limited microstructure refinement as result of phase instability. In other words, reductions of tensile strength and ductility are attributed to brittle Laves/γ eutectic phase formation during terminal stage of solidification. The severity of eutectic reaction thermodynamically depends on nail-shaped keyhole profile, solidification conditions and solute redistribution ahead of solid/liquid interface. Additionally, viable control of weld metallurgical and mechanical properties progressively encourages optimal combination of laser power and welding speed to interdendritically minimize disadvantages of eutectic phase formation by a variety of welding conditions optimization.

1. Introduction

Because of good performance at high service temperature, corrosion-, oxidation- and heat-resistant nickel-based supralloys are applied in aircrafts, gas turbines, rock engines and petrochemical equipments. Defect-free repair of damaged and degraded aerospace components is preferred by useful laser welding to economically reduce manufacture cost and time. The research and methodology contributions of nickel-based superalloy weldability problems and welding metallurgy in the published literature are briefly introduced next. Guanfei Xiao *et al.* and Yongquan Ning *et al.* [1,2] discussed effect of processing parameters on microstructure development of Ni-based superalloys GH 4037 and GH 4133B by semisolid metal formation and electron beam welding. Shengchao Duan *et al.* and Guo Wei *et al.* [3,4] analyzed effect of thermodynamic properties of liquid Ni/Ti, Ni/Al and Ti/Al binary alloys and nickel-rich Ni-Al-Ti ternary alloy on phase transformation and mechanical properties through atom and molecule theory during laser welding. Qiu Hu *et al.* and Liu Enze *et al.* [5,6] analyzed effect of carbon, boron and rhenium elements on phase transformation, segregation behavior and microstructure morphology of nickel-based superalloys. C Wang *et al.* and Mahmood Moradi *et al.* [7,8] discussed effect of multiple processing parameters on weld profile, grain refinement and mechanical properties of nickel-based alloys K403 and Rene 80 by laser shock processing and laser welding. S F Gao *et al.* and Xinbao Zhao *et al.* [9,10] discussed grain selection, crystal orientation and grain competitive growth of Ni-based single-crystal superalloys DD3 and DD407 during directional solidification casting. Yong Yaowei *et al.* and H Xiao *et al.* [11,12] metallurgically discussed effect of processing parameters on alloying segregation, phase transformation, microstructure development and mechanical properties of WC/TaC nickel-based composition coating on carbon steel by laser additive manufacturing, such as laser cladding. O A Ojo *et al.*, G H Chen *et al.*, P A Molian *et al.* and Dongyun Zhang *et al.* [13-16] discussed effect of welding thermal cycle and postweld heat treatment on heat-affected zone hot cracking susceptibility, grain boundary liquation, phase transformation and microstructure refinement of Ni_3Al-based superalloy by laser welding. T Rojhirunsakool *et al.* and Daeyoung Kim *et al.* [17,18] metallurgically discussed effect of welding parameters on segregation, dendrite growth and hot cracking susceptibility of precipitation-hardened Ni-based superalloy IN738 by several fusion repair processes, such as laser cladding, laser welding and gas tungsten arc welding. K Shinozaki *et al.* [19,20] discussed relationship between process parameters, material-related factors, weld geometry and heat-affected zone hot cracking susceptibility of heat- and corrosion-resistant Inconel 718 superalloy by laser welding. In essence, in order to improve weld integrity and mechanical properties, motivation of this paper is to thermodynamically and kinetically mitigate thermal and metallurgical factors to withstand insidious morphology, size and distribution of microstructure degradation under nonequilibrium solidification conditions.

2. Experiment procedure and material

Butt joint of wrought precipitation-strengthened Ni-based superalloy Inconel 718 is welded by optimal fiber laser processing parameters to produce good weld quality. The chemical compositions of commercial Inconel 718 are listed in Table 1. The dimensions of sample are 300 (length)×100 (width)×5 (thickness) mm. Phase transformation temperatures and welding parameters are listed in Table 2. The size, morphology, distribution and chemistry of microstructure development, especially secondary dendrite arm spacing, and rapid solidification behavior, especially alloying element partition and nonequilibrium phase, inside transverse cross section of weld are sequently characterized by optical microscopy (OM), scanning electron microscopy(SEM) and energy dispersive spectroscopy(EDS) after a series of polishing and etching preparation to metallographically analyze kinetic and thermodynamic properties and metallurgically clarify the important relationship between macroscale weld geometry and microscale microstructure development. Ten cross sections of weld are observed and measured. Furthermore, tensile test of mechanical properties is applied, and stress-strain data as well as fractograph are also discussed.

Table 1. Chemical composition of wrought Inconel 718 (weight percent, wt%).

Ni	Fe	Cr	Nb+Ta	Mo	Ti	Al	Si	Mn	C
Bal.	18.49	19.3	5.05	3.12	1.02	0.54	0.04	0.02	0.02

Table 2. Summary of metallurgical reaction temperature and welding parameters.

Liquidus, °C	Solidus, °C	Eutectic Laves/γ, °C	Laser power, kW	Welding speed, m/min	Beam focus, mm	Shielding gas flow rate, L/min	Heat input, J/mm
1360	1320	1200	2.4	0.6	0.4	20	240

3. Results and discussion

3.1. Microstructure development

3.1.1. Transverse cross section of weld. Metallograph of weld shape and size with partial and complete penetration is shown in figure 1. Different appearances of weld geometry are particularly attributable to unexpected factors, especially gap misalignment, incorrect sample preparation and incidence angle of laser beam inclination, with wide choice of laser welding application, and these phenomena frequently occur. Promising laser welding is capable of full penetration and narrower heat-affected zone (HAZ) to facilitate advantageous weld geometry development with high energy input efficiency and sufficient productivity. The fluctuation of weld penetration is closely related to keyhole instability. Nail-shaped keyhole is an essential metallurgy-related phenomenon. Inclination of keyhole weld shape complicates fluid flow, heat transfer, crystal orientation and dendrite growth from top, middle to bottom of weld. Heat stagnation exists in the neck transition region of keyhole to decrease temperature gradient and increase solute transport.

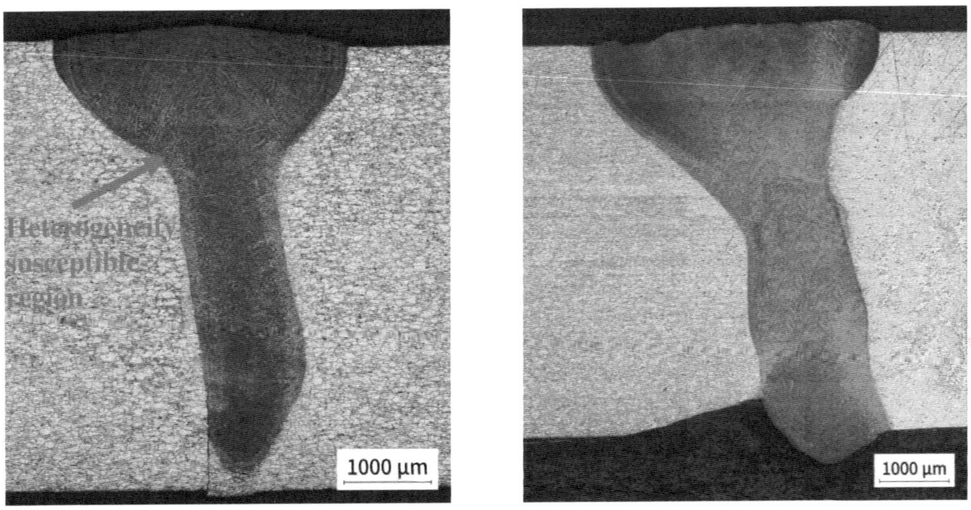

(a) Partial penetration (b) Full penetration

Figure 1. Weld shape inside transverse cross section.

The weld shape deviates from symmetrical centerline. Dendrite on left and right sides grows under different rapid solidification conditions. The asymmetrical weld is free of underfill, porosity and hot cracking. Defectless full penetration is successfully obtained with high aspect ratio. Interaction between laser radiation and material thermally deepens weld penetration and gradually broadens weld width. Weld width of upper portion is fairly larger than bottom part, where more laser energy is consumed and dissipated. Laser energy is proportionately attenuated by increase of weld penetration. Confinement of predominant heat flow accumulation at intermediate neck region with longer molten weld pool existence inevitably splits the weld pool into two parts, i.e. wider upper side and narrower bottom side to initiate subsequent microstructure inhomogeneity. High quality weld without defect is an important prerequisite for microstructure control. Comparison of polycrystalline weld morphology development at (a) left side along fusion boundary, (b) weld center, (c) right side along fusion boundary and (d) HAZ are shown in figure 2.

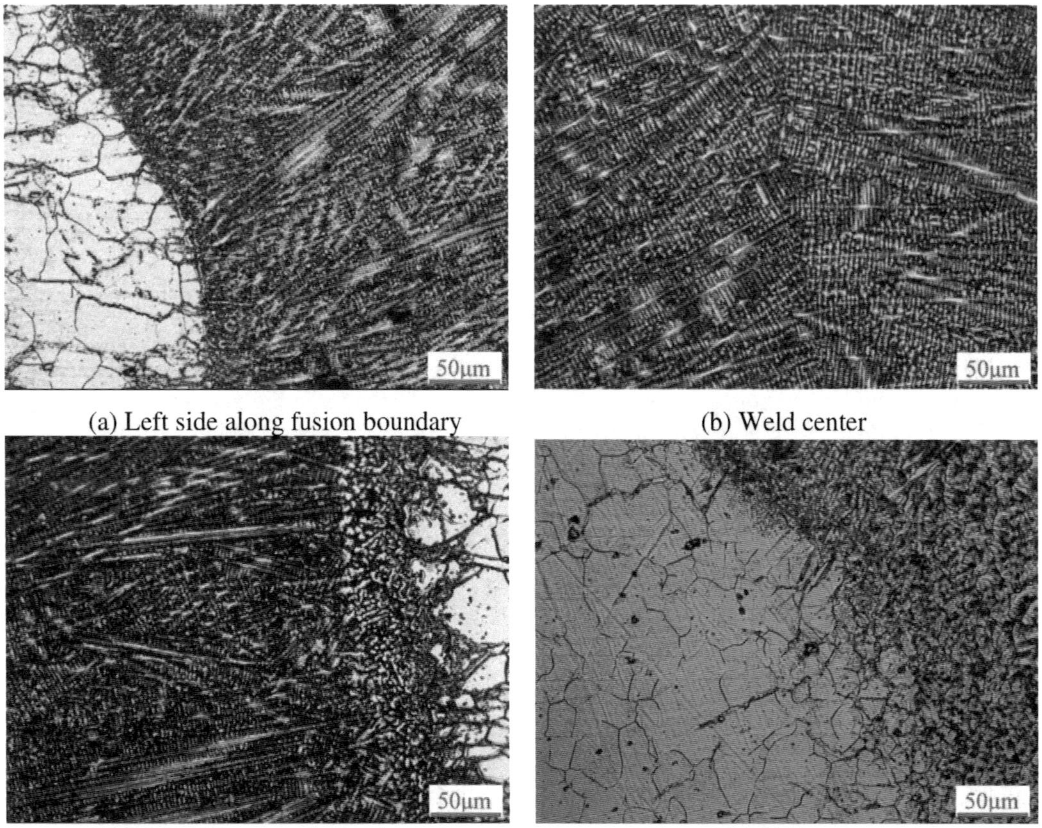

(a) Left side along fusion boundary (b) Weld center

(c) Right side along fusion boundary (d) HAZ

Figure 2. Polycrystalline weld morphology development at different locations.

In (a) and (c), for face-centered cubic (FCC) materials, crystallography-dependent columnar dendrite heterogeneously nucleates and epitaxially grows at fusion boundary. Primary columnar dendrite morphology along fusion boundary is prominent, and preferential crystallographic orientations of nuclei are vertical to fusion boundary with favorable solidification conditions, especially steep temperature gradient and slow solidification rate. Heterogeneous microstructure is severely located near neck transition region and microstructure is incompletely transformed into columnar dendrite because of insufficient solute diffusion and rapid solidification across solid/liquid interface. Intergranular grain boundary liquation occurs. Both increase susceptibility to hot cracking initiation and propagation along HAZ grain boundary. Semi-solid transition interface of metallurgical bonding exists between heat-affected zone and weld fusion region. Thereinto, microstructure is metallurgically divided into three different areas, i.e. HAZ, mushy zone and dendrite growth region. Disoriented dendrite side-branch, lateral growth, deflection angle and competitive growth are unfavorably developed in the molten pool. In (b), naturally, primary and secondary dendrites are mostly developed. Misalignment of polymorphic dendrite branches is thermally activated into the central molten pool with secondary arm. Tertiary arm is scarce. There are divergence angle and boundary misorientation between preferential crystallographic orientations during weld pool rapid solidification. Competitive growth is distinct. Abundant dendrite growth intersections and high-angle grain boundaries occur. Dendrite tip undercooling is low enough to suppress equiaxial dendrite growth and crystallization. The complexity of undercooling-driven dendrite morphology and size is determined by multiple metallurgical factors, such as thermal and solute distributions. Finer dendrite size improves weld integrity through optimization of laser processing parameters. Microstructure refinement potentially avoids hot cracking. In (d), in order to highlight microstructure variation and interface morphology at periphery of neck region, which is partially melted, nonhomogeneous recrystallization in heat-affected zone along fusion boundary is clearly observed without intergranular liquation cracking. Recrystallization grain excessively coarsens with grain boundary liquation as consequence of less heat dissipation and more heat accumulation. Crack-free HAZ is an essential part for high quality weld requirement.

3.1.2 Secondary dendrite arm spacing. The microscale measurement results emphasize effect of asymmetrical weld shape on nonlinear distribution of secondary dendrite arm spacing in the weld pool, as shown in figure 3.

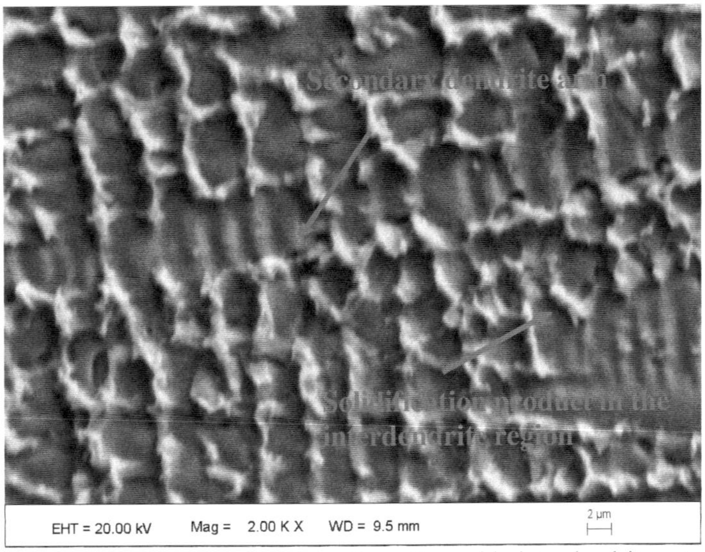

(a) Secondary dendrite arm morphology with unfavorable interdendrite eutectic phases

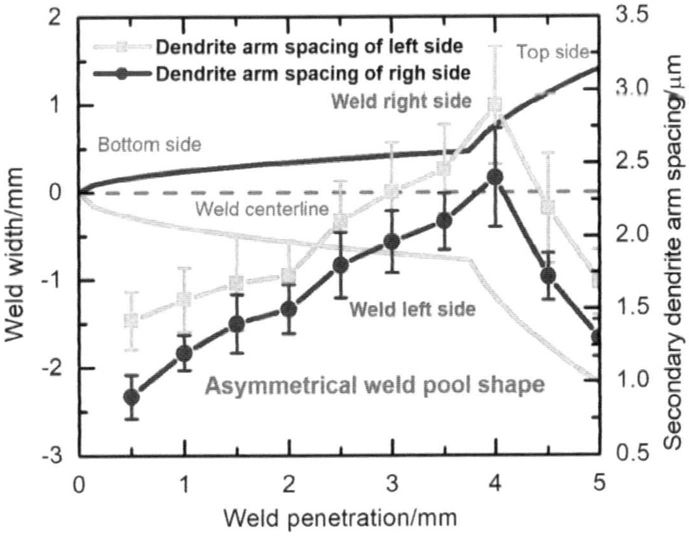

(b) Superimposition of average secondary dendrite arm spacing along fusion boundaries

Figure 3. Effect of asymmetrical weld pool shape on secondary dendrite arm spacing.

In (a), in accordance with solidification path sequence, the microstructure wholly consists of γ solid solution of columnar dendrite and interdendrite Laves eutectic phase, which detrimentally contributes to microstructure heterogeneities and seriously deteriorates weld integrity. In (b), welding thermal cycles and solute concentration profile are inhomogeneously distributed at different locations. Location-dependent secondary dendrite arm spacing of primary dendrite branches initially increases and then decreases around fusion boundary from weld top surface, middle to bottom sides. The coarsest secondary dendrite arm spacing is mainly located at neck transition region due to heat flow stagnation. In other words, dendrite grows faster in this location. Keyhole weld fusion zone is preferentially deeper and narrower at root part and molten pool solidifies quickly with steep temperature gradient to simultaneously facilitate nucleation and growth of finer dendrite as well as preferential crystallographic orientation. Dendrite growth significantly differs on both sides. Overall, the diffusion-driven multicomponent secondary dendrite arm spacing of left side fusion boundary is coarser than right side throughout keyhole penetration, because weld pool is asymmetrically larger with significant heat and mass transport phenomena. Besides, it is worth noting that laser-induced keyhole shape plays an important role in thermodynamics and kinetics of dendrite growth, and is one of the most critical factors to determine final weld quality. The mechanism of keyhole-induced dendrite growth variation is thermally and metallurgically proposed to improve better understanding of microstructure development and solidification behavior. More importantly, this mechanism provides valuable information of weld interior beneath the top surface for accurate and proper alteration of weld geometry with feasibility of high energy intensity laser welding.

3.2. Solidification behavior

3.2.1 Alloying partition coefficient. The measurement results of solute redistribution in dendrite core and interdendrite region within acceptable limits is shown in figure 4. Comparatively, morphology and concentration of dendrite core and interdendrite region around fusion boundary are straightforwardly observed and measured by SEM and EDS under nonequilibrium solidification conditions to chemically provide opportunities for necessary reduction of segregation region during dendrite growing.

The diffusion-controlled interaction between laser and material is thermometallurgically analyzed to kinetically elucidate contribution of solute redistribution to well-developed nucleation and growth of multicomponent dendrite growth through chemical analysis and utterly reveal interrelationship between dendrite shape, size, solute redistribution, selective partition, dendrite tip undercooling and segregation in the internal and external dendrite of favorable crystallographic orientation, especially a wide range of Ni, Cr, Ti, Al, Mo and Nb concentrations. Alloying elements laterally diffuse towards primary γ dendrite side-branch below solid solubility within saturation according to Ni-Cr, Ni-Ti, Ni-Al, Ni-Mo and Ni-Nb binary phase diagrams in all cases and this microstructure is free of deleterious eutectic phases. A variety of alloying concentrations occur. Solute accumulation, concentration gradient and mutual interaction are imposed by thermal and solute diffusions across dendrite interface to develop new dendrite branch with high dendrite tip undercooling as result of heat and mass transfer during solidification. Ni and Cr are enriched at central of dendrite. These elements strongly partition to dendrite core. Copious Ti, Al, Mo and Nb are preferentially concentrated at dendrite interface. These elements substantially partition to interdendrite regions to satisfy dendrite growth kinetics and restrict solidification behavior. Favorable alloying partition is a contributing factor to withstand dendrite anomalies and thereby ameliorate diffusion kinetics.

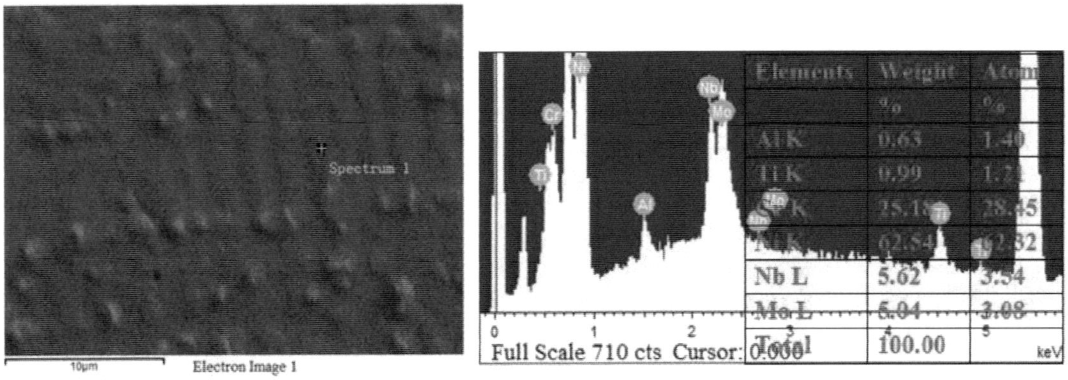

(a) Concentration in the dendrite core

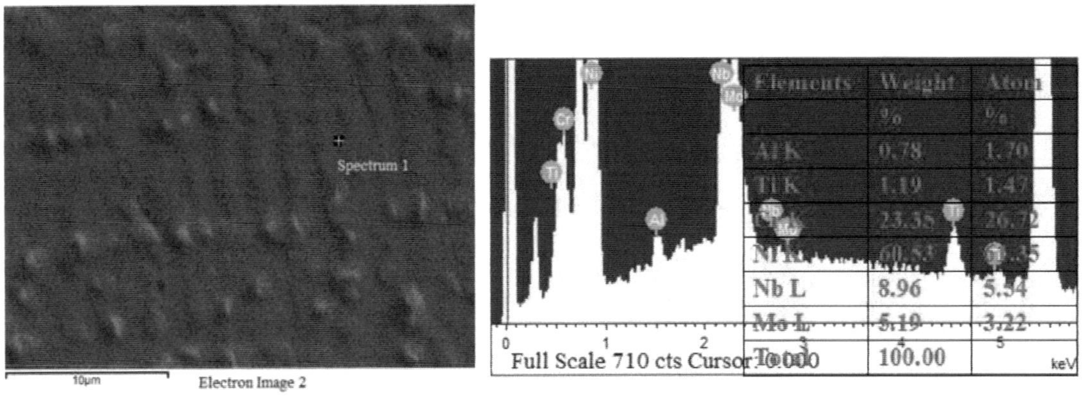

(b) Concentration in the interdendrite region

Figure 4. Measurement results of solute redistribution in the dendrite core and interdendrite region.

The variation of γ phase solute partition in the secondary dendrite arm with asymmetrical weld pool shape is separately shown in figure 5. Significant alloying partition discrepancy between dendrite core and interdendrite region are thermodynamically and kinetically developed. Microstructure essentially experiences different welding thermal cycle at various locations. It is noteworthy that solute partition initially decreases and then increases from top side to bottom side of weld because of heat stagnation in the neck transition region. In other words, alloying elements are primarily enriched in this location with disadvantageous solidification conditions to coarsen dendrite growth and weaken hot cracking resistance. Stage of increment is attributed to nonequilibrium partitionless solidification, where alloying segregation is deficient, to refine dendrite growth. In contrast, there is inverse relationship between aforementioned secondary dendrite arm spacing and alloying partition coefficient across dendrite branches through solute atoms outward diffusion. Epitaxial growth in close proximity to fusion boundary prevails, and solute partition coefficient on the right side fusion boundary is increasingly larger than left side, and the latter is completely larger than center part to suppress infelicitous nucleation, growth and distribution of eutectic phase as result of asymmetrical weld pool shape under nonequilibrium solidification conditions. Similar partition behavior is nearly consistent for all cases. Overall, the partition coefficient of Cr, which illustrates value of around 0.95-1 and are close to 1 to impede interdendrite segregation, are chemically larger than Ti, Al and Mo, which illustrate value of around lower and upper bounds 0.8-0.9, and the latter are significantly larger than Nb within solubility limits, which illustrates value of around 0.7-0.8. Nb is of particular interest, since Nb is the most potent element to easily concentrate at interdendrite region with strong preference to partition in residual liquid and increase propensity for detrimental segregation and Laves eutectic phase formation. Therefore, limitation of diffusion-driven alloying partition through nonequilibrium solidification is provided by welding metallurgy with high priority to improve chemical homogeneity. Cr-enriched and Nb-depleted dendrite core inside γ phase advances supersaturation- and undercooling-dependent dendrite size and morphology development due to numerous thermal and chemical contributing factors at solid/liquid interface.

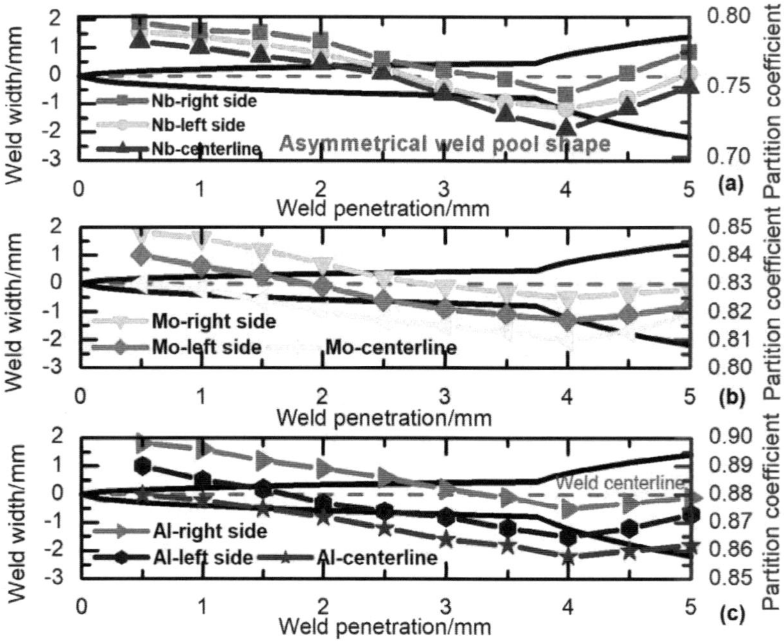

(a)-(c) Nb, Mo and Al partition coefficients

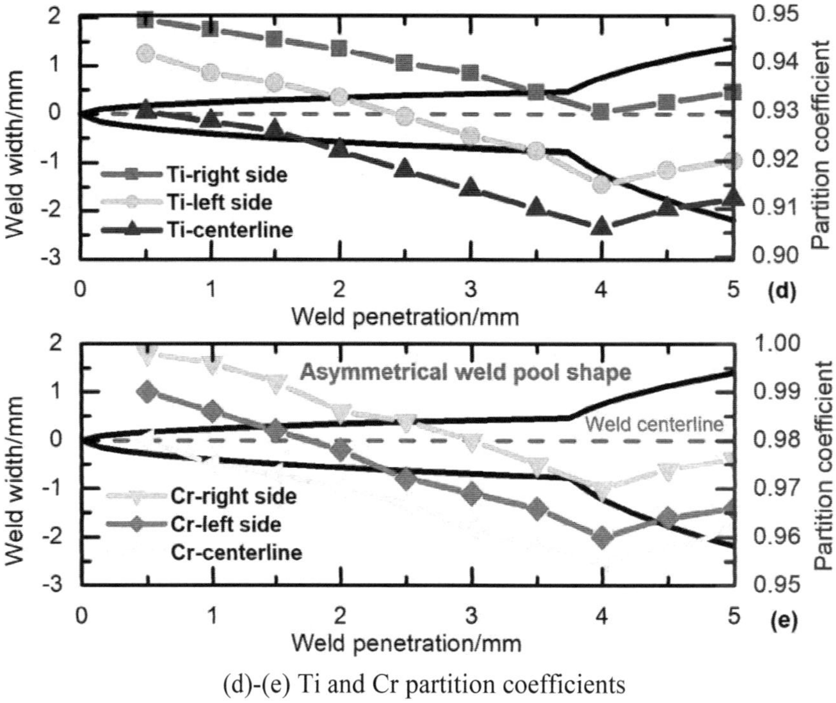

(d)-(e) Ti and Cr partition coefficients

Figure 5. Effect of weld pool shape on alloying partition coefficients through chemical analysis.

3.2.2 Phase transformation. The Laves/γ eutectic phase morphology, size, distribution and chemistry in the center part of weld neck transition region are shown in figure 6, where heat and solute accumulation are intractable and conspicuous. This place is more vulnerable to eutectic reaction. It is energetically more favorable for Nb to segregate at dendrite interfaces for Laves/γ eutectic phase formation.

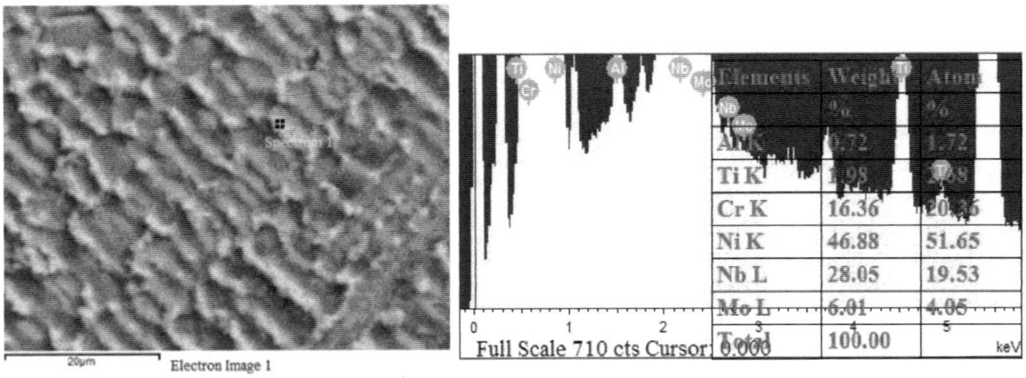

(a) Laves/γ eutectic phase morphology (b) Chemical analysis

Figure 6. Laves/γ eutectic phase morphology, size, distribution and chemistry.

Besides high-supersaturation primary γ phase solidification, phase transformation of solidification product, such as Laves phase, emerges alongside interdendrite network region in the weld pool under nonequilibrium solidification conditions, and significantly deteriorates weld quality, integrity and weldability [21,22]. Bright phase is Laves phase, while dark phase is γ phase. Laves/γ eutectic phases are heterogeneously distributed in (a). Interdendrite region is favored as nucleation site of eutectic phase formation, as solidification proceeds towards hypereutectic reaction. Despite Nb composition is quite low in the material, nonequilibrium solidification is another reason for a large amount of interdendrite Laves/γ eutectic phase formation. Nb-rich Laves phase is thermodynamically favorable, and morphology and size are sensitive to welding thermal cycle and weld pool shape. Nb concentration is increased in the interdendrite region in (b). Nb severely segregates into remnant interdendrite liquid rather than dendrite core as result of solute redistribution and diffusion ahead of solid/liquid interface. Solidification initially advances by primary reaction L→γ. Subsequently, residual liquid is consumed by L→γ+Laves eutectic reaction, because concentration predominantly exceeds solid solubility at terminal stage of solidification in accordance with solidification sequence of Ni-Nb binary phase diagram. Continuous Laves/γ eutectic phase development is one of critical metallurgical driving forces for microstructure and mechanical properties degradation. Therefore, weld pool shape, chemical inhomogeneity, Laves/γ eutectic phase formation, alloying solute partition, segregation and dendrite growth are interrelated. In order to thermometallurgically mitigate eutectic phase formation disadvantages of polycrystalline superalloy, alleviate solute accumulation and improve microstructure and phase stability with high feasibility and flexibility of laser welding, it is imperative to modify thermodynamics and kinetics of dendrite growth, and suppress secondary phase nucleation and growth conditions with aid of solute concentration control ahead of solidification interface by optimal combination of laser power and welding speed.

3.3. Mechanical Properties

3.3.1. Fractography of weld. Morphology analyses of tensile properties are shown in figure 7. All samples involve centerline fracture during tensile test. Inhomogeneous and discontinuous metallurgy-related defects in the central of polycrystalline weld are evidently detrimental to mechanical properties, especially low strength and low ductility. Significant prevalence brittle and ductile fracture morphologies are observed. With closer observation of fracture surface, internal secondary phase abundance of dimple is contributing factor for microstructure irregularities and insufficient mechanical properties, and predominantly decreases dendrite boundaries resistance. Brittle nonequilibrium Laves/γ eutectic phases inside the interdendrite region are intermixed together, thus result in the potential site for interface decohesion of fracture failure. Phase interface of terminal solidification product is weak to promote crack nucleation, growth and propagation. Aberrant eutectic behavior with ability of dendrite boundaries embrittlement is attributable to nonequilibrium solidification of weld pool, notwithstanding pertinent low heat input is conducive to weld quality and weldability. There is variability in microstructure development that metallurgically alters mechanical properties. Corroboratively, fractography of fracture-susceptible microstructure phenomenally elucidates good consistency with aforementioned analysis. Interaction controllability between dendrite and eutectic regions through optimum heat input microstructurally and morphologically mitigates risk of strength, ductility and toughness losses. Hence, it is imperative to steer diffusion-driven segregation behavior and dendrite network permeability, thermodynamically abate morphology, size and distribution of disadvantageous phase transformation product and lessen severity of mechanical properties fluctuation through achievable chemistry and solidification conditions of disoriented and interlocked substructure during primary phase solidification. Acceptable defect-free deep penetration keyhole weld is the most critical step to decrease mechanical heterogeneity. Amelioration of phase boundaries or dendrite boundaries strength is progressively driven by metallurgical and mechanical factors, which is prerequisite to warrant ascendancy of microstructure integrity and extend usage of faster laser welding repair.

(a) Fracture interior at lower magnification (b) Fracture interior at higher magnification

Figure 7. Morphology analyses of tensile properties.

3.3.2 Stress-strain curves. Tensile test results comparison between base material and welds are shown in figure 8. Tensile tests of weld under optimum welding conditions are applied. Weld samples facture in the fusion zone. This location is prone to failure. Numerous stress-stain data are obtained. The tensile strength and elongation of base material are 853MPa and 54%. Comparatively, the tensile strength and elongation of weld sample #1-1-1 are 669MPa and 13.5%. these results are 647MPa and 10% for weld sample # 1-1-2, 667MPa and 12% for weld sample #1-1-3, 672MPa and 13% for weld sample # 1-1-4, separately. Base material significantly possesses higher tensile strength and elongation than four sets of weld sample. Equiaxial grain of austenite γ phase base material contributes to superior mechanical properties. Based on previous macro- and micro-scale analyses, mechanical properties of weld strongly depend on multiple factors, such as multicomponent microstructure development and solidification behavior. The inevitable difference between weld and base material is primarily attributed to detrimental dendrite morphology and Laves/γ eutectic phase formation. Undesirable microstructure, especially low melting point eutectic phases, leads to serious degradation of mechanical properties, such as reductions of tensile strength and ductility, during laser welding. Intermetallic phase weakens dendrite boundary strength to initiate crack during tensile test due to lack of homogeneity and hence exacerbate weldability and repairability. Because of unfavorable brittle eutectic phase in the weld interior, desirable microstructure refinement partially compensates for strength and ductility reductions through solidification conditions optimization. It is therefore necessary to appropriately balance between mechanical properties, microstructure refinement, segregation behavior and brittle phase formation with low heat input under nonequilibrium solidification conditions. Defect-free weld and acceptable mechanical properties are simultaneously satisfied through viable experiment design methodology. Innovative keyhole laser processing is an integral part of polycrystalline weld anomalies suppression for aerospace material welding and repair with capability of high energy efficiency and flexible process versatility to overcome inadequate mechanical properties.

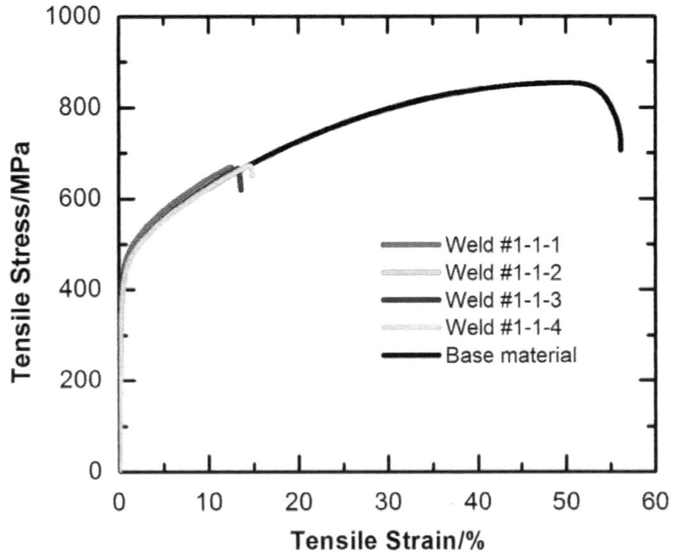

Figure 8. Tensile properties comparison between base material and welds.

4. Summary

1. Diffusion-limited microstructure development, segregation-driven solidification behavior and eutectic-induced mechanical properties degradation are properly interrelated under optimum welding conditions to thoroughly advance understanding of weldability through important thermal, metallurgical and mechanical determinants during full penetration keyhole laser welding polycrystalline nickel-based superalloy.

2. Disadvantageous dendrite growth kinetics along fusion boundaries is essentially ascribed to asymmetrical molten pool development. Neck transition region of weld is particularly susceptible to coarser dendrite arm spacing as consequence of heat and solute accumulation, while finer dendrite arm spacing is located in the bottom of weld.

3. Alloying solute partition between γ phase dendrite core and high-supersaturation interdendrite region is chemically observed in the neck transition region of weld. The chemical inhomogeneity and solute redistribution are capable of thermodynamic solidification behavior and kinetic dendrite growth modification. Ni and Cr are accumulated in the dendrite core, and Ti, Al, Mo and Nb are selectively rejected into interdendrite liquid.

4. In spite of low Nb initial composition in the material, detrimental Laves/γ hypereutectic reaction occurs in the neck transition region, which is preferable site for microstructure anomalies. Enormous amount of brittle Nb-rich eutectic phases are heterogeneously distributed, morphologically augments susceptibility to intergranular fracture and worsens tensile properties.

Acknowledgement
The project is financially supported by the Anyang Institute of Technology (Grant No. BSJ2019031).

References
[1] Guanfei Xiao and Jufu Jiang 2019 Comparison of semisolid microstructure evolution of wrought nickel-based superalloy GH4037 with different solid fraction *Solid State Phenomena* **285** 105-110
[2] Yongquan Ning and Zekun Yao 2014 Hot deformation behavior and hot working characteristic of nickel-based electron beam weldments *J. Alloys Compd.* **584** 494-502
[3] Shengchao Duan and Xiao Shi 2019 Determination of the thermodynamic properties of Ni-Ti, Ni-Al, Ti-Al and nickel-rich Ni-Al-Ti melts based on the atom and molecule coexistence theory *J. Mol. Liq.* **294** 111462
[4] Guo Wei and Song Peng 2013 Effect of laser welding on microstructure and mechanical behavior of Ni-49 at% Ti *Rare Met. Mater. Eng.* **42** 78-82
[5] Qin Hu and Lin Liu 2013 Effect of carbon and boron additions on segregation behavior of directionally solidified nickel-based superalloys with rhenium *Trans. Nonferrous Met. Soc. China* **23** 3257-3264
[6] Liu Enze and Guan Xiurong 2011 Effect of rhenium on solidification segregation of nickel-based superalloy *Rare Met.* **30** 320-322
[7] C Wang and X J Shen 2016 Effects of laser shock processing on microstructure and mechanical properties of K403 nickel alloy *Mater. Des.* **89** 582-588
[8] Mahmood Moradi and Majid Ghoreishi 2011 Influences of laser welding parameters on the geometric profile of Ni-based superalloy Rene 80 weld bead *Int. J. Adv. Manuf. Technol.* **55** 205-215
[9] S F Gao and L Liu 2012 Grain selection during casting Ni-based single-crystal superalloys with spiral grain selector *Metall. Mater. Trans.* A **43** 3767-3775
[10] Xinbao Zhao and Lin Liu 2015 Investigation of grain competitive growth during directional solidification of single-crystal nickel-based superalloys *Appl. Phys.* A **120** 793-800
[11] Yong Yaowei and Fu Wei 2017 In-situ synthesis of WC/TaC reinforced nickel-based composite alloy coating by laser cladding *Rare Met. Mater. Eng.* **46(11)** 3176-3181
[12] H Xiao and S M Li 2017 Effects of laser modes on Nb segregation and Laves phase formation during laser additive manufacturing of nickel-based superalloy *Mater. Lett.* **188** 260-262
[13] O A Ojo and R G Ding 2006 Heat-affected zone microfissuring in a laser beam welded directionally solidified Ni_3Al-based alloy *Scr. Mater.* **54(12)** 2131-2136
[14] G H Chen and C Chen 1992 Investigation of the heat-affected zone fracture in Ni_3Al welds *J. Mater. Res.* **7** 1076-1082
[15] P A Molian and Y M Yang 1992 Laser welding behavior of cast Ni_3Al intermetallic alloy *J. Mater. Sci.* **27** 1857-1868
[16] Dongyun Zhang and Haiyun Gao 2012 Ni_3Al-based alloy of laser welding behavior research *Adv. Mater. Res.* **418-420** 1531-1537
[17] T Rojhirunsakool and K Sirivedin 2020 Investigation of welding speed parameters on Ni-based superalloy by laser welding process *Key Eng. Mater.* **856** 85-91
[18] Daeyoung Kim and Jonghyun Hwang 2001 A study on fusion repair process for a precipitation-hardened IN738 Ni-based superalloy *J. Eng. Gas Turbines Power* **122(3)** 457-461
[19] K Shinozaki and H Kuroki 2000 Effects of welding parameters on laser weldability of Inconel 718. Study of laser weldability of Ni-based heat-resistant superalloys (1st Report) *Welding Research Abroad* **46(1)** 3-9
[20] K Shinozaki and H Kuroki 2000 Comparison of hot cracking susceptibilities of various Ni-based heat-resistant susperalloys by U-type hot cracking test. Study of laser weldability of Ni-based heat-resistant superalloys (2nd Report) *Welding Research Abroad* **46(2)** 10-17
[21] G A Knorovsky 1989 Inconel 718: a solidification diagram *Metall. Trans.* A **20** 2149-2158
[22] Akin Odabasi 2010 A study on laser beam welding (LBW) technique: effect of heat input on the microstructural evolution of superalloy Inconel 718 *Metall. Mater. Trans.* A **41** 2357-2365

Kevlar 49 yarn pull-out test and numerical analysis

Xiang Hong[1], Zhenkun Lei[1*], Yu Ma[2,3], Ruixiang Bai[1#], Weikang Li[1]

[1]State Key Laboratory of Structural Analysis for Industrial Equipment, Department of Engineering Mechanics, Dalian University of Technology, Dalian 116024, China
[2]State Key Laboratory of Explosion Science and Technology, Beijing Institute of Technology, Beijing 100081, China
[3]Beijing Institute of Technology Chongqing Innovation Center, Chongqing, China

Email: *leizk@dlut.edu.cn, #bairx@dlut.edu.cn

Abstract. The yarn pull-out tests are of great importance to study the frictional energy dissipation mechanism between yarns and the overall deformation behavior of fabrics. In this study, the stress transfer mechanism of the yarn pull-out process of Kevlar 49 woven fabric is investigated by the digital image correlation (DIC) marker method, considering both the warp waves and transverse preload variables, and have analyzed the shear deformation behavior of textile by the change history of fabric shear angle. The results show that the peak load of the yarn pull-out process increases nonlinearly with the increase of the transverse preload force and the number of warp waves, the maximum pull-out displacement of the yarn is sensitive to the preload force only and is largely independent of warp waves. Finally, the finite element (FE) calculation of the single yarn pull-out process was carried out by using the mixed element method with the restart analysis technique, and the numerical simulation results achieved a good match with the peak load of the experimental data.

1. Introduction

At present, aramid fabrics are widely used in the field of high-speed impact protection for their light weight, high specific strength and good impact resistance. Ballistic penetration experiments are the main method to study the mechanical properties of fabric impact, but as the penetration experiments can only analyse the kinetic energy change of the bullet and the change of fabric morphology before and after the impact, the stress transfer process between fabric yarns and the energy dissipation mechanism cannot be observed, some scholars propose to model the ballistic impact experiments as quasi-static experiments such as yarn pulling, so as to realise the analysis of the impact process and help to conduct an in-depth study of the fabric deformation mechanism.

Zhu et al[1]. divided the yarn pull-out process into two stages: curvilinear elongation and frictional slip, and defined the interfacial shear stress transfer state respectively. Qin et al[2]. applied Raman optical measurements to investigate the interfacial stress transfer state between yarns during yarn pull-out. Dong et al[3]. conducted yarn pull-out experiments on five fabrics, including K310 and K706, and the results showed that Tapie et al[4]. designed a new method of yarn pull-out test, and the experimental and finite element simulation results showed that during the yarn pull-out process, most of the energy was dissipated by friction and the fabric stored strain energy. Pre-stretching has a significant effect on both frictional and strain energy. SiO2 impregnated fabrics with mass fractions of 62%, 65% and 67% were tested in a pull-out test by Bai et al[5]. and a new frictional energy dissipation mechanism was proposed. The results showed that the work done by external forces in yarn pulling tests could be equated to the energy dissipation by inter-yarn friction. Further tests on the

pulling behaviour of single yarns of aramid plain fabrics with different preloads and different yarn pulling rates were carried out, and a bilinear analytical model of inter-yarn stress transfer was proposed, which was in good agreement with the experimental load displacement curves[6].
This study uses the DIC marker method to analyse the deformation process of the fabric during yarn pulling and to explore the link between the shear deformation of the fabric and the static and dynamic friction process. A FE analysis of the fabric pull-out process is carried out.

2. Yarn material properties

The material used in this study is Kevlar 49 plain woven fabric produced by DuPont, and its material properties are shown in Table 1. The basic geometric properties of the fabric are shown in Figure1, which consists of warp and weft yarns orthogonally woven. Both warp and weft yarns can be regarded as consisting of several woven sinusoidal waves, and there are two warp and weft yarn intersections in the woven waves. After polishing the fabric cured with resin, the cross-sectional shape of the yarn was determined by optical microscopy to be elliptical, and the long-axis dimensions of the weft and warp yarns are $W_a = 1$ mm and $W_b = 1.1$ mm, the yarn gaps are $G_a = 0.1$ mm and $G_b = 0.05$ mm, the short-axis of the yarn was $T_a = 0.23$ mm and $T_b = 0.24$ mm.

Table 1. Material properties of the fabric

Type	Dtex	Density [g/cm^3]	Number of fibers	Fiber diameter [μm]
Kevlar 49	1580	1.44	1000	12

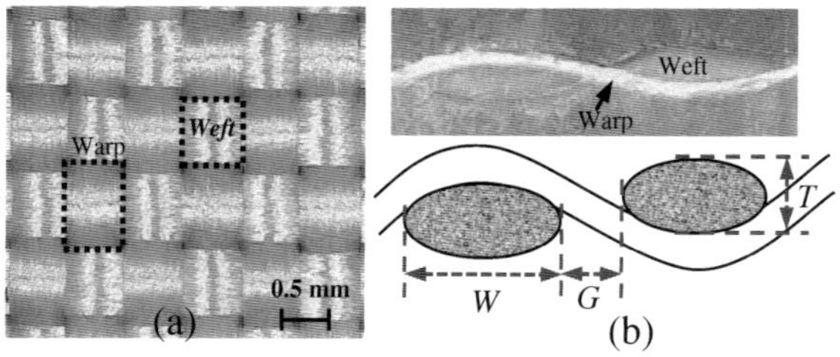

Figure 1. (a) Fabric geometry dimensions and (b) yarn cross-section diagram

The axis distribution of Kevlar 49 warp and weft yarns shows a sine wave variation pattern, and the neutral axis distribution equation of the yarn is established as follows.

$$\begin{cases} y = \dfrac{T_a}{2}\sin\left(\dfrac{\pi}{W_a + G_a}x\right) & Warp \\ y = \dfrac{T_b}{2}\sin\left(\dfrac{\pi}{W_b + G_b}x\right) & Weft \end{cases} \quad (1)$$

Where x represents the displacement along the yarn direction and y represents the initial buckling shape perpendicular to the yarn direction. With the difference in the choice of reference position, the sine function can also be replaced by the cosine function.

3. Single yarn pull-out experiment

The single yarn pull-out test is a quasi-static modeling test method for the behavior of yarn flexural elongation and slips pull-out in the ballistic impact test of flexible fabrics. This study helps to analyze systematically the stress transfer mechanism between yarns and the overall deformation behavior of fabrics during the high-speed impact of fabrics.
The Instron 3345 tension machine was used for the single yarn pull-out test. The yarn other than the pulled yarn was cut out before the experiment, as shown in Figure 2(a). A clamping area of 5 cm in width was reserved on both sides of the test piece. The fabric specimens were sized to 65, 85, and 105 woven waves in the warp direction and 65 woven waves in the weft direction, respectively, and then

clamped with a special fixture (Figure 2(b)), which could apply transverse preload to the fabric specimens through a slide rail. The marker method is used to measure the shear deformation state of the fabric during yarn pull-out, and the markers are made at equal intervals on the right side of the fabric specimen. By tracking the displacement of the markers during yarn pull-out, the change of the shear angle θ of the fabric specimen can be obtained, as shown in Figure 2(c).

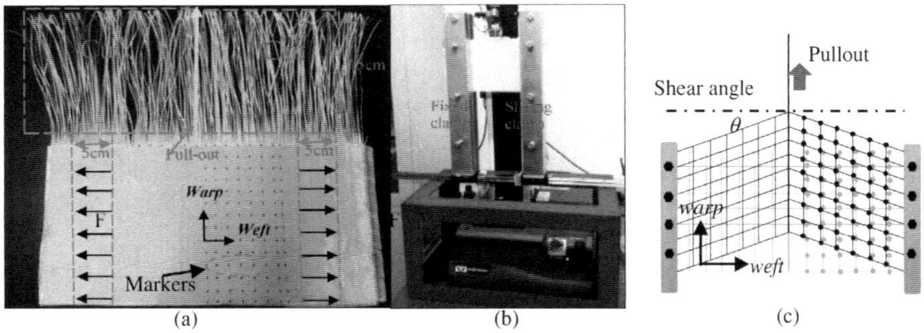

Figure 2. Schematic of (a) fabric specimen, (b) single yarn pull-out test system and (c) overall deformation state of fabric pull-out

4. Results and Discussion

4.1. Effect of wave numbers

For the load-displacement curve case of the yarn pull-out process, the ultimate pull-out load increases with the increase of wave number, as shown in Figure 3(a). The ultimate load of the 65-wave specimen is 8.8 N, while the ultimate loads of the 85-wave and 105-wave specimens increase to 10.1 N and 14.9 N, respectively. This is because the increase in the number of waves increases the contact area between the warp and weft yarns, and the maximum pull-out force increases with the same contact relationship. This shows that the frictional properties between warp and weft yarns are not a simple linear superposition relationship.

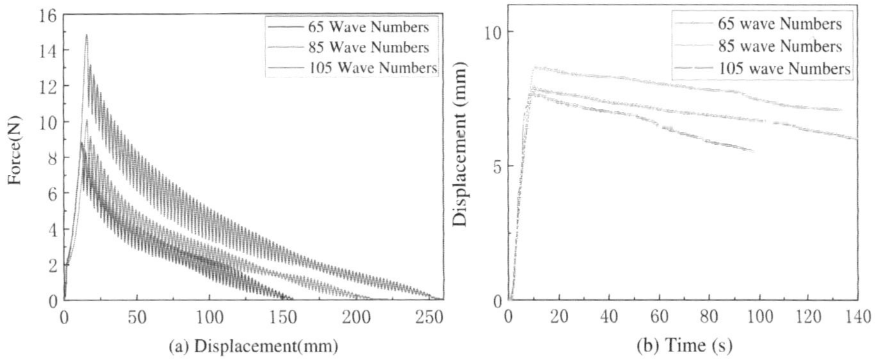

Figure 3. (a) Load-displacement curve of yarn pull-out process with different wave numbers and (b) displacement-time curve of markers in the 5th row and the 2nd column

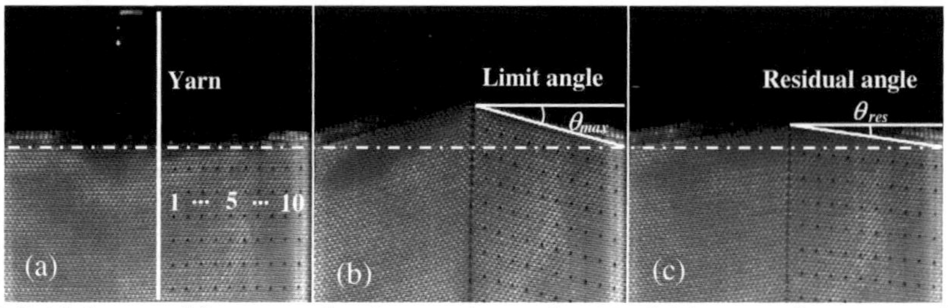

Figure 4. (a) Initial state, (b) state under the maximum load and (c) state after yarn pull-out for the 85-wave specimen

The overall deformation of the fabric specimen during yarn pull-out undergoes three typical shear deformation states, which are the initial state (θ = 0), the ultimate deformation state (θ_{max}), and the residual deformation state (θ_{res}). The displacement changes of the marker points with time were analyzed using the DIC method (Figure 4(b)), and the maximum displacements of the three were 7.7 mm, 7.9 mm, and 8.7 mm, and the displacement trends along the warp direction were all sharply rising and then gradually decreasing, with a certain deformation remaining at the end.

4.2. Effect of transverse preload force

The yarn pull-out load-displacement results for fabric specimens with wave numbers 65 at 25, 65, and 100 N transverse preloads, respectively, are represented in, Figure 5(a). The ultimate yarn pull-out load increases with increasing transverse preload, with a peak load of 8.59 N based on 25 N preload force, and the corresponding maximum peak loads increase by 124.4% (10.5 N) and 133.3% (11.0 N) when the preload force increases by 160% (65 N) and 400% (100 N), respectively. This is because the preload makes the weft straighten and the weft fiber bundle spread and extend, which expands the contact area between warp and weft yarns; at the same time, the preload also makes the weft loaded, which increases the contact pressure at the intersection of warp and weft yarns, thus increasing the friction force on the contact surface, which increases the peak load. As the preload increases, the maximum displacement at the marker becomes smaller and the residual deformation after the yarn pull-out is also smaller (Figure 5(b)), which shows that the increase of preload limits the overall deformation of the fabric specimen, and the larger ratio of maximum displacement/ residual displacement indicates that the fabric deformation recovery ability is stronger.

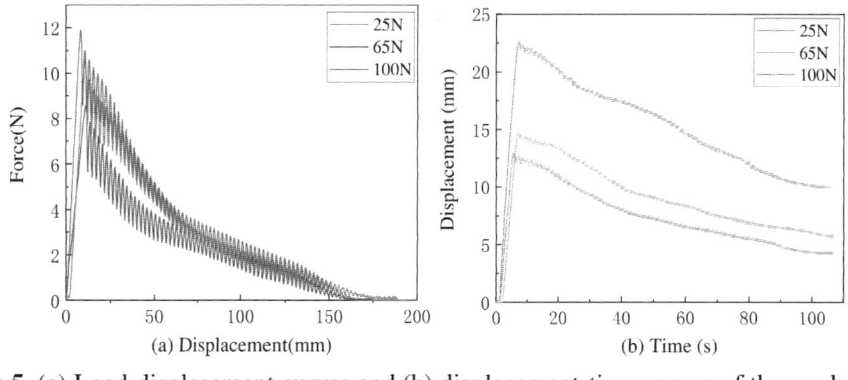

Figure 5. (a) Load-displacement curves and (b) displacement-time curves of the marker in the 5th row and the 2nd column for fabrics with different transverse preloads

5. Finite element analysis

5.1. The selection of elements

The commercial software ABAQUS was used to perform the FE analysis of the yarn pull-out test. According to the actual deformation process of yarn pull-out, the mixed elements modeling method of membrane elements combined with shell elements is choose. For warp yarns under tension, the shell elements and set only three integration points in the thickness direction are used, while for other yarns, the membrane elements are used.

A restart analysis was used to introduce the effect of preload force. First, a uniformly distributed preload force load is applied in the horizontal direction to the fabric in the static analysis step (Figure 6(a)) and its stress-strain state is calculated. Then, a dynamic analysis step is re-established by applying a boundary condition with both fixed ends and the loaded yarn in Figure 6(b). Finally, the stress-strain field from the static analysis is introduced into the dynamic analysis step for calculation.

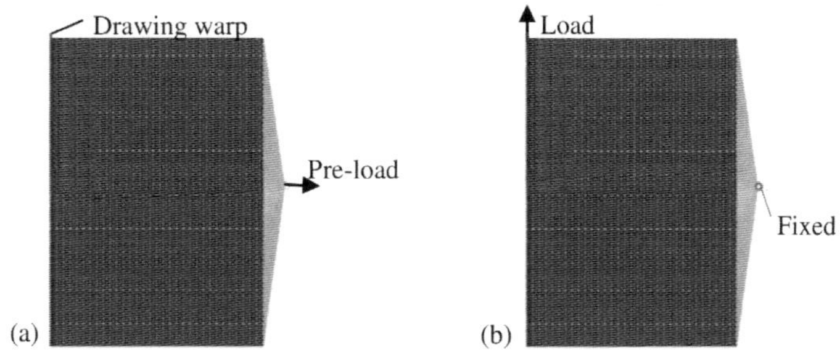

Figure 6. Boundary conditions of (a) applied preload force and (b) single yarn pull-out

5.2. Simulation results and discussion

The displacement diagram of the yarn pull-out process is shown in Figure 7(a-d). The pulled yarn first produces a semi-elliptical deformation area on the pulled side, where the warp yarn starts to straighten gradually from its initial flexural state and the shear angle at this point is θ_1 as the pull-out process continues, the semi-elliptical area becomes larger in the direction of the warp yarn and the shear angle θ_2 increases during this process; eventually, the deformation area spreads to the end of the yarn, when the yarn pull-out load reaches its maximum and the shear angle reaches a maximum of θ_3; thereafter, the drawn warp yarn as a whole starts to slip and the shear angle starts to decrease gradually, eventually retaining a certain residual deformation degree of θ_4.

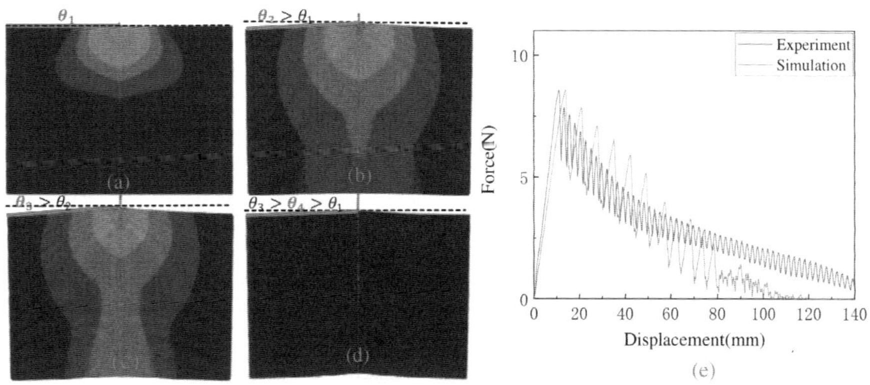

Figure7. (a-d) Deformation of the fabric at different stages of the drawing process and (e) comparison of experimental and simulated load-displacement curves for 65-waves specimen under 25 N preload

A comparison of the experimental and simulated load-displacement curves for a preload force of 25 N at 65-waves is shown in Figure 7(e). The peak loads of both are nearly the same, the experimental result is 8.58488 N the simulated result is 8.58504 N, with only a difference of 0.00186%, the slope of the linear rise phase is relatively close, the main difference is reflected in the oscillating fall phase.

In the interval from 15 mm to 80 mm during the oscillatory descent phase, it can be observed that the number of peak-valleys in the simulated curve is less than that of the experimental curve. On the one hand, this study adopts a Dynamic Explicit solution. The introduction of acceleration makes the yarn buckling and straightening process that should have been completed in the static friction stage enter the dynamic friction stage before it has been fully straightened, which is equivalent to adding a yarn buckling and straightening displacement to the yarn sliding in the dynamic friction stage. The phenomenon of the peak-valleys mismatch between simulation and experiment is caused. On the other hand, the introduction of inertial acceleration makes the yarn that would not deform produce an elastic deformation. With the increase of displacement and pulling force, the elastic deformation caused by inertial acceleration also increases, which is equivalent to adding a yarn elastic deformation displacement to the yarn slip in the dynamic friction stage.

6. Conclusion

In this study ,the yarn pull-out test was carried out and the shear deformation state of the fabric during the yarn pull-out process was measured using the DIC marker method to analyze the frictional properties between the yarns and the shear deformation of the fabric. FE analysis of the single yarn pull-out process was carried out using mixed elements modeling combined with restart analysis, and a good match was obtained with the experimental results.

1) The yarn pulling process can be divided into two stages: static and dynamic friction. In the static friction phase, the pull-out load and shear deformation of the fabric specimen both increase sharply to the extreme, after which the pulled yarn slips into the dynamic friction phase, and the load and shear deformation gradually decrease as the yarn slips out.

2) The ultimate load pulled out by the yarn increases non-linearly with the increased waves of the fabric specimen and the level of transverse preload, which indicating that inter-yarn friction is not simply linearly related to external factors; fabric shear deformation is only related to transverse preload, not to the number of waves of the fabric.

3) A mixed element modeling method combined with restart analysis was used to analyze the shear deformation of fabric and load-displacement during yarn pull-out, improving the calculation efficiency and achieving good matching results.

Acknowledgments

The authors thank the Postdoctoral Science Foundation of China (2020TQ0033), National Natural Science Foundation of China (Nos. 11972106, 11772081)

References

[1] Zhu D, Soranakom C, Mobasher B, et al. Experimental study and modeling of single yarn pull-out behavior of kevlar 49 fabric. Composites Part A: Applied Science and Manufacturing 2011, 42(7): 868-879.
[2] Qin F, Lei Z, Ma Y, et al. Stress transfer of single yarn drawing in soft fabric studied by micro Raman spectroscopy. Composites Part A: Applied Science and Manufacturing 2018, 112: 134-141.
[3] Dong Z, Sun C T. Testing and modeling of yarn pull-out in plain woven Kevlar fabrics. Composites Part A: Applied Science & Manufacturing 2009, 40(12):1863-1869.
[4] Tapie E, Guo Y B, Shim V P W. Yarn mobility in woven fabrics – a computational and experimental study. International Journal of Solids & Structures 2016, 80:212-226.
[5] Bai R, Ma Y, Lei Z, et al. Energy analysis of fabric impregnated by shear thickening fluid in yarn pullout test. Composites Part B: Engineering 2019, 174:106901.
[6] Bai R, Ma Y, Lei Z, et al. Shear deformation and energy absorption analysis of flexible fabric in yarn pullout test. Composites Part A: Applied Science and Manufacturing 2020, 128:105678.

Fatigue Damage Characteristics and Morphology of Fiber Braided Laminated Flexible Composites

Yuqi Yang[1,a], Longbin Liu[1,b,*] and Shaozhe Ding[1,c]

[1]College of Aerospace Science and Engineering, National University of Defense Technology, Changsha City， Hunan Province, 410073, China

*E-mail: longbuaa@163.com

Abstract. In this paper, the fatigue damage evolution law and damage failure mode of flexible braided thin film composites are studied by stress fatigue damage test and scanning electron microscopy morphology test of uniaxial 45° off-axis sample. The results show that the failure mode of the specimens with 45° off-axis angle is characterized by fatigue damage of both the matrix and the functional film and the fracture of the fiber bundle has little influence. In addition, the pattern fatigue stress and fatigue life data of the fatigue test of viscose matrix and functional film under different fatigue loads have the rule of linear equation, which has predictive significance on the safety of aircraft.

1. Introduction

With the continuous development of technology, aerospace structures are developing towards the direction of lightweight, high reliability, long life and high efficiency and flexible fiber braided thin film structure has very light weight and strong toughness [1]. As a result, flexible fiber-braided thin films are playing an increasingly important role in many areas of aerospace. In the field of aircraft, flexible fiber composite materials are often used as the basic materials for flexible inflatable airbags of high-altitude balloons and airships [2]. But in aerospace, flexible fiber woven type membrane structure working environment is very tough and need to suffer from electromagnetic effect, ionizing radiation, and thermal layer issues which need flexible fiber composite materials to maintain the bearing capacity of the aircraft so it is of great significance to accurately measure the strength and tear properties of fiber-reinforced composites [3].

Cox studied the internal stress of the damaged composite laminate and established the stiffness degradation mode of the material [4]. Sheng Zhao et al. proposed a normalized derived fatigue damage model based on the residual strength of composites and described the three-stage development law of the strength of fiber-reinforced resin composites in the fatigue loading process [5]. Maoxu Lu et al. proposed the modified Hoffman strength theory and established a fatigue life prediction model under complex in-plane loads based on this theory [6].

In this article, uniaxial fatigue loading damage method was used to study the damage law and failure morphology of flexible fiber braided materials. The fatigue damage test data of laminate fiber composite materials were analyzed and the morphology observation of microscopic components was conducted by scanning electron microscopy. After the test, the fatigue damage characteristics, failure morphology and failure mode of the fiber braided laminated flexible composite were obtained. The research results have important significance for the safety, flight performance and service life of flexible inflatable airbag of near space vehicle and can support the design and engineering optimization of flexible thin film fiber reinforced composite materials.

2. The experimental process

2.1. Off-axis angle 45° fatigue damage test material

Fiber reinforced flexible composite materials are mostly high performance composite materials, laminated base films composed of film materials with various properties, which are composed of load-beari

toget

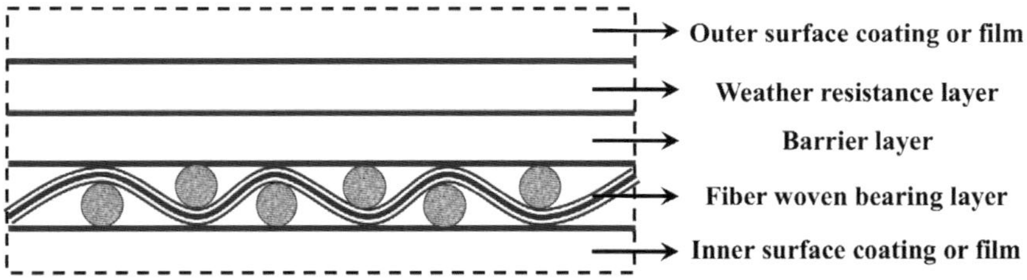

Figure 1. Section structrue of fiber reinforced flexible compesites.

Among them, the preparation method of the bearing layer is plain weave, which vertically intersects the yarns in the horizontal and vertical directions [8].

2.2. Uniaxial stress fatigue test

According to standard GJB 2637-96 and ASTM D1004-09, the size of sample stress fatigue test is shown in Fig. 2[9,10]. In the figure, x direction represents the weft tow direction and y direction represents the warp tow dire

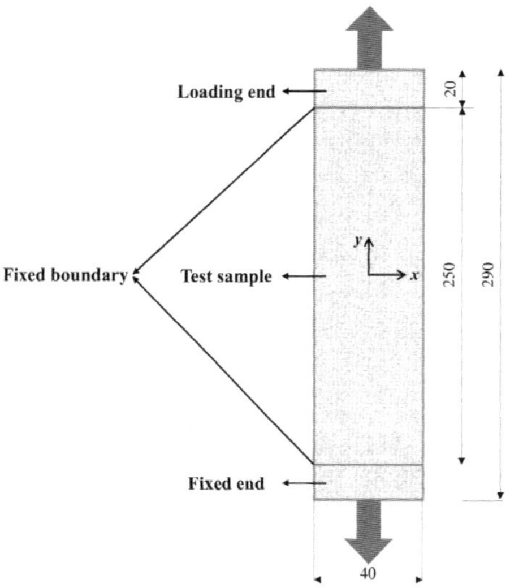

Figure 2. Fiber reinforced flexible composite sample geometry.

The experimental device is a stress alternating fatigue testing machine as shown in Fig 3. Limited by the clamping head of the stress alternating fatigue testing machine, the total length and width of the flexible fiber braided thin film sample are 290mm and 40mm with 20mm fixed clamping ends being reserved at both ends of the sample. The constant strain control mode is adopted on the stress alternating fatigue testing machine. The lower end of the sample was fixed, and the upper end was sinusoidal cyclic loading according to the fatigue cycle. The off-axis angle θ=45°was observed under different stress loads. Adjust the instrument according to the quality of the expansion from the electron microscope until the morphology of fiber bundles is clear and the display ratio is shown in Fig 4.

It should be noted that the loading test environment is room temperature air and the temperature control range is 25 ± 5°C, relative humidity is (50 ± 10)% RH.

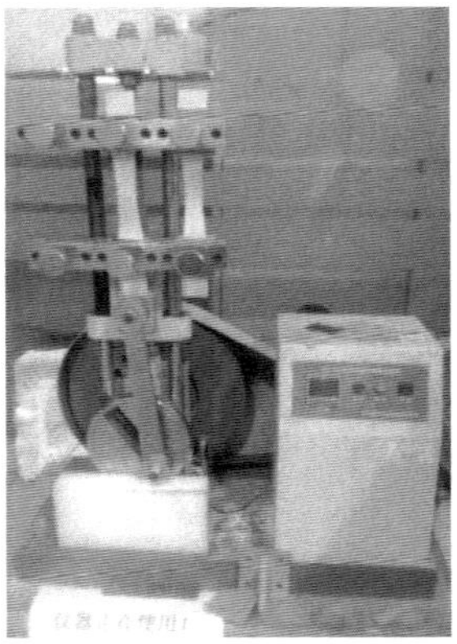

Figure 3. Uniaxial elongation fatigue testing machine for fiber reinforced flexible composite materials.

2.3. Failure mode analysis of fatigue damage at 45° off-axis angle.

As can be seen from the macroscopic deformation of the fatigue sample with 45° off-axis Angle, the transverse deformation of the large range of tensile deformation has obvious Poisson Contraction Effect because the fiber yarn bundle is a material with high elastic modulus, it is impossible to have significant variation. Therefore, the failure mode of the 45° sample is characterized by the joint fatigue damage of the matrix and the functional film, while the fatigue bearing effect of the fiber bundle is ineffective.

Macroscopic fracture morphology of sample fatigue damage is shown in Fig. 5. Under tensile and bending fatigue load, the failure fracture of sample is matrix degumming and fracture and the fracture direction is perpendicular to the long side direction of the sample. It can be seen from scanning electron microscopy that obvious cracking of the matrix and the functional film leads to free buckling of the braided fiber bundle, as shown in Fig. 4. At the same time, a small amount of fatigue fracture appeared in the fiber bundle near the fracture attachment area and a fraction of degumming phenomenon occurring at the interface and the fracture began to expand from the transverse truncation plane with the minimum shrinkage deformation. With the increase of fatigue damage times, the area of failure fracture area expands but the fracture morphology is almost similar.

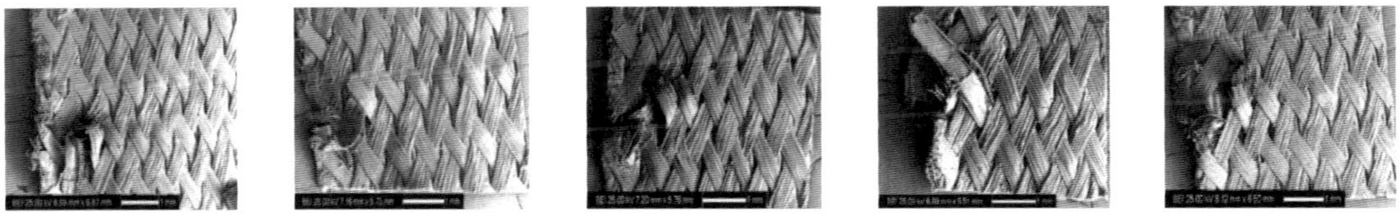

(a) F_m**=625N** **(b)** F_m**=605N** **(c)** F_m**=585N** **(d)** F_m**=565N** **(e)** F_m**=545N**

Figure 4. SEM images of fatigue damage fracture of specimens with 45° off-axis angle.

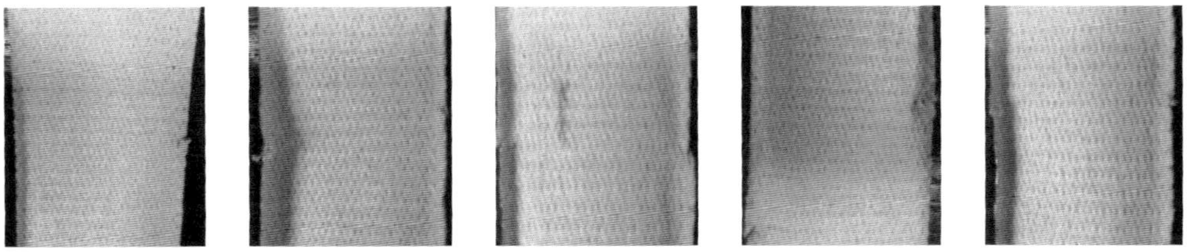

(a) F_m**=625N** **(b)** F_m**=605N** **(c)** F_m**=585N** **(d)** F_m**=565N** **(e)** F_m**=545N**

Figure 5. Local morphology of fatigue fracture of specimens with 45° off-axis angle.

Under the cyclic fatigue load, the minimum cross section of samples with 45° off-axis angle has a high stress concentration effect, which leads to the micro-cracks in the matrix and the debonding of the functional film. With the increase of the number of cycles, the cracks will gradually spread forward and then lead to the cracking failure of the matrix and the functional film and encountering the fiber also can cause fiber fracture damage. The cracking failure of the matrix and the functional film leads to the deterioration of the air tightness and environmental resistance of the fiber reinforced flexible composite, which leads to the destruction of the material. Therefore, the damage mechanism of the 45° sample is controlled by both the matrix and the functional film, and its fatigue properties directly affect the fatigue characteristics of the sample while the fracture of the fiber bundle has little effect.

2.4. Data processing of 45° off-axis angle fatigue damage test

Macroscopic fracture morphology of samples with off-axis angle of 45° under different fatigue alternating loads is shown in Fig. 6, in which (a)~(e) respectively represent samples under different fatigue loading levels. It can be seen from the diagram that the fatigue under the action of alternating load are present in the side viscose matrix and function of membrane layer damage fracture damage failure mode at the same time, the lateral Poisson Effect obviously began to sprout from the edge of the macroscopic fracture specimens from the middle area, expanding gradually to sample, and sample didn't momentarily complete rupture (membrane as matrix and function layer appeared fatigue crack, the airtightness of the skin is damaged. In this paper, the specimen is defined as damage failure when the fatigue crack length reaches 2mm).

In the experiment, the test data of samples with 45° off-axis Angle under different fatigue loads were recorded and tabulated in Table 1:

Table 1. Fatigue test parameters of specimens with off-axis angle θ=45°.

The serial number	The sample number	The width[mm]	Maximum load[N]	The corresponding stress [MPa]	Destroy number N
1	F-45-01	40	625	260.417	4673
2	F-45-02	40	625	260.417	4115
3	F-45-03	40	625	260.417	4964
4	F-45-04	40	605	252.083	6055
5	F-45-05	40	605	252.083	5813
6	F-45-06	40	605	252.083	5362
7	F-45-07	40	585	243.750	7343
8	F-45-08	40	585	243.750	8011
9	F-45-09	40	585	243.750	7178
10	F-45-10	40	565	235.417	9710
11	F-45-11	40	565	235.417	8935
12	F-45-12	40	565	235.417	9296
13	F-45-13	40	545	227.083	11285
14	F-45-14	40	545	227.083	12457
15	F-45-15	40	545	227.083	10179

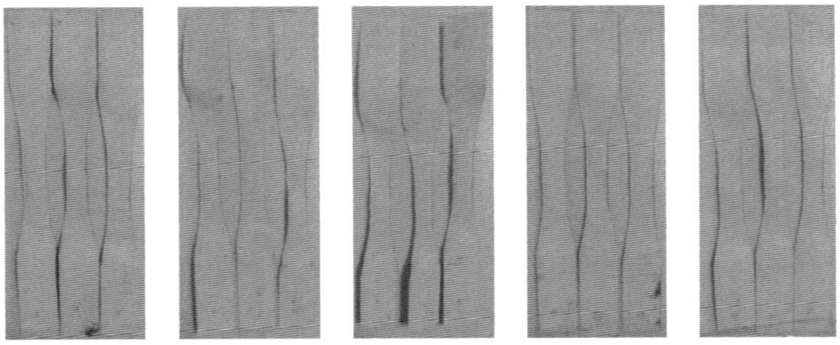

Figure 6. Fatigue fracture morphology of specimens with 45° off-axis angle under different stress levels.

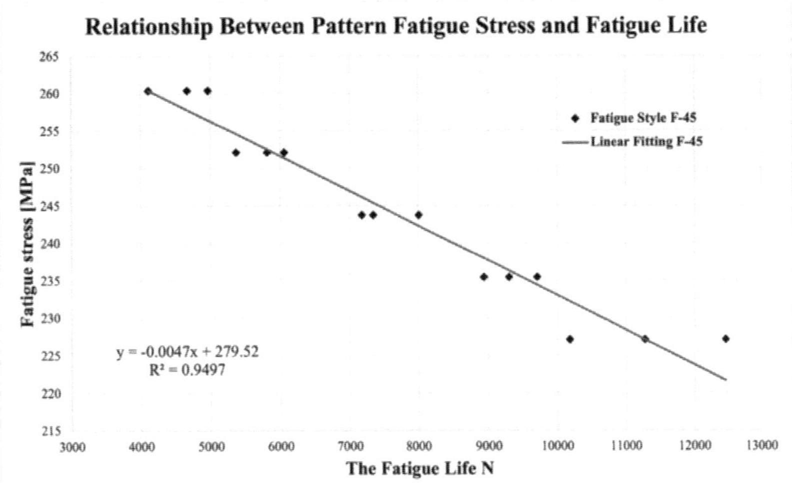

Figure 7. Relationship between pattern fatigue stress and fatigue life.

It can be seen from the test data and fitted curves in the Fig. 7 that the S-N curves of samples with 45° off-axis angle under different fatigue loads are approximately linear and the fitted S-N curves are expressed as follows:

$$N = -0.0047\sigma_{max} + 279.52 \tag{1}$$

The linear correlation coefficient (mean variance of fitting) for all test data is $R^2 = 0.9497$.

Therefore, the fatigue test data of the specimen with an off-axis angle of 45°, that is, the viscose matrix and the functional film have a good linear relationship under different fatigue loads and the S-N data of the specimen with an off-axis angle of 45° can be fitted by the linear equation.

3. Conclusion

The linear relationship between the fatigue times and the stress of the viscose matrix and the functional film of the flexible fiber braided film was obtained by the off-axis fatigue test and SEM morphology observation. The fatigue damage morphology and failure mode of 45° direction flexible thin film were obtained. The failure mode of the material is characterized by fatigue damage of both matrix and functional film. The matrix of failure fracture degumming and fracture, resulting in the cracking failure of matrix and functional film. At the same time, a small amount of fatigue fracture occurs in the fiber bundle near the fracture attachment area. Therefore, the damage mechanism of the

45° sample is controlled by both the matrix and the functional film layer and is little affected by the fracture of the fiber bundle.

Acknowledgements
This work was sponsored by the National Natural Science Foundation of China (No.11802334).

References

[1] Zhao. Z, Wang. H, Hu. L. Research status and development trend of aerospace advanced Composites. No.2 (2021), p. 24-29. (in Chinese)
[2] Ma. Z, Hou. Z, Yang. X. Structural characteristics analysis of large flexible airbag in adjacent space. Journal of National University of Defense Technology. Vol. 37 (2015), p. 25-30. (in Chinese)
[3] Information on https://baike.baidu.com/item/Spacecraft environmental engineering /5028500?fr=aladdin. (in Chinese)
[4] Cox H L. The elasticity and strength of paper and other fibrous materials. Vol. 3 (1952), p. 72-79.
[5] Zhao. S, Zhang. J. A derivative of fatigue damage model based on the residual strength of composites. Journal of composite materials. Vol. 5 (2020), p. 2473-2481. (in Chinese)
[6] Lu. M, Ji. X, Hao. Z, Zhang. L, Liu. L. Strength and Fatigue Life Prediction method of Planar Braided High-Aluminum Fiber Reinforced Alumina Matrix Composites under Complex In-plane Stress State. Journal of composite materials. Vol. 38 (2021), p. 3785-3798. (in Chinese)
[7] Gu. Z. Study on the skin material of stratospheric airship. Spaceflight return and remote sensing. Vol. 28 (2007), p. 62-66. (in Chinese)
[8] Longbin L, Zheng Y, Jiang Z. Study on evolution mechanism of uniaxial fatigue damage of fiber woven reinforced laminated composites. Materials Research Express. Vol. 6 (2019), p. 126407.
[9] Information on https://www.doc88.com/p-88161132800839.html
[10] Information on https://www.astm.org/Standards/D1004.htm

Evaluation interfacial properties of resin with bond exchange reaction / T700 carbon fiber by micro-drop test and FEM

Heshan Bai, Tianyu Zhao, Ruixiang Bai#, Zhenkun Lei*, Chen Liu

State Key Laboratory of Structural Analysis for Industrial Equipment, Department of Engineering Mechanics, Dalian University of Technology, Dalian 116024, China

Correspondent author's e-mail: #bairx@dlut.edu.cn, *leizk@dlut.edu.cn

Abstract. Vitrimer epoxy resin is a newly developed resin with bond exchange reaction, which has potential applications in the field of recyclable fiber composites. The force-displacement curve of T700 carbon fiber monofilament pulled out from Vitrimer epoxy microdroplet was obtained by the microdrop test, and the interfacial shear strength of the composite system was obtained. Based on the cohesive contact theory, the finite element model (FEM) of the micro-drop test was set up. The calculated force-displacement curve is consistent with experimental data. The fracture strength and parameter characterization of the interface between Vitrimer epoxy and carbon fiber composites are further discussed.

1. Introduction

Fiber-reinforced composites formed by thermosetting resin and carbon fiber are widely used in transportation, aerospace, wind turbine blades, and various industrial fields because of their excellent mechanical and physical properties[1]. Recently, Leibler's group presented a new material category called Vitrimer epoxy resin based on the classical transesterification reaction in epoxy network. Vitrimer epoxy resin has dynamic covalent bond networks, namely dynamic covalent bond networks or covalent bond adaptive networks, which makes it have amazing properties just like thermoplastic with recyclability and repairability, and makes it have great application potential[2].

Interfacial shear strength (ISS) is the preferred index to characterize the bonding property in different phases, and the stress transfer behaviors, and the result of fiber surface modification. The corresponding interface parameters can be included by a single fiber mechanical test. At present, the quantitative measurement techniques to evaluate the bonding properties of the interface between matrix and fiber mainly include fiber pull-out test[3], fiber extrusion test[4], microdrop test[5], and fiber fragmentation test[6]. The microdrop test is widely used because of its simplicity and effectiveness.

In terms of finite element (FE) simulation, for the glass fiber/thermosetting resin, Minnicino et al.[7] used a simplified finite element model which is a quarter symmetrical model to analyze the stress distribution during interface fracture of the fiber/matrix. The influence of cohesion model strength and toughness parameters on its macro properties is analyzed. Sockalingam et al.[8] evaluated the effects of material properties and geometry in Glass fiber/ epoxy on ISS and failure mechanism in micro-drop experiment using FE method.

In order to characterize the interfacial properties of a new resin composite with bond exchange reaction, the interfacial bond strength of Vitrimer epoxy /T700 carbon fiber was evaluated by microdroplet test. By reasonably setting the experimental load and boundary conditions, the FE numerical model is established, and then the interface damage process of a new resin composite with bond exchange reaction in the microdrop test is discussed. The cohesive zone model (CZM) of mixed mode is used to simulate the interface failure, and the interface parameters are discussed and analyzed.

2. Microdrop testing

2.1. Specimen preparation

The T700 carbon fiber used in this study is produced by Turing Evolution company in China Shenzhen. The resin composition and ratio of resin are showed in Tab 1. The epoxy monomer, curing agent and catalyst are produced by the Shanghai McLean Co. The purpose of zinc acetylacetonate catalyst is to activate the bond exchange reaction in the matrix, to achieve the effect of repairable and recyclable.

Table 1. The ratio and composition of the resin

Type	Ingredients	Proportioning
Vitirmer resin	Bisphenol A (epoxy monomer), zinc acetylacetonate (catalyst), glutaric anhydride (curing agent)	1:1:0.05

The sample of T700/Vitrimer epoxy resin microdroplet is shown in Fig 1. The preparation process is described as follows. At first, a monofilament from the purchased Carbon fiber bundle is selected and straighten on the hard cardboard. Then, the resin is prepared according to Tab 1 and mixed at 80 °C using a collector-type constant temperature heating machine. Dip a small amount of resin with the tip of a needle under a hand-held microscope, apply epoxy microdroplets on each fiber, and control the microdroplet size to be close to 200 μm. Put the microdroplet sample into a vacuum drying chamber (DZF-6050) for curing. The temperature conditions are 130 °C for 4 h and 160 °C for 2 h. Finally, the size of the solidified microdroplet sample was measured by a microscope.

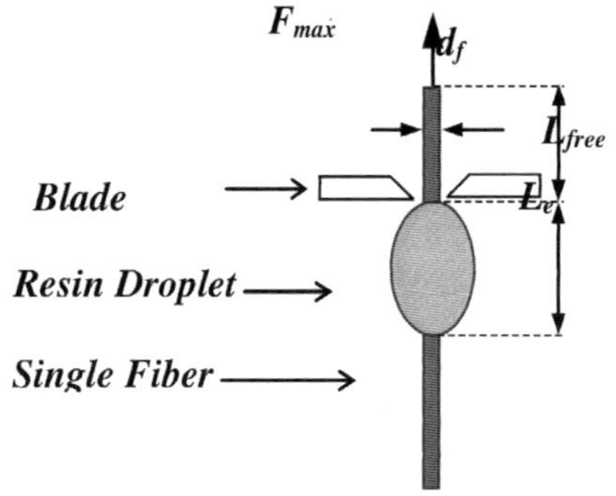

Figure 1. Microdroplet test principle

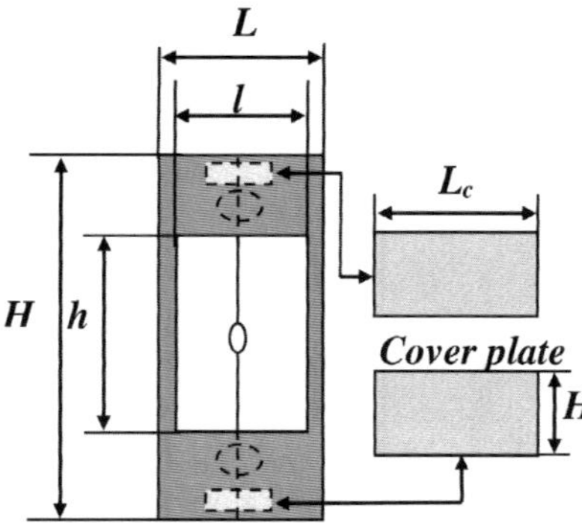

Figure 2. Microdroplet sample

2.2. Test procedure

The principle of the micro-drop test is shown in Fig 2, For the outer frame size, H = 65 mm, L = 30 mm; for the inner frame size, h = 35 mm, l = 24 mm; for the cover piece size, L_c = 15 mm, H_c = 30 mm. Before the test, secure the upper part of hard cardboard microdroplet sample in the fixture on the experimental machine, then place the lower part of the sample in self-made knife-edge fixture. Then, cut off the hard cardboard on two sides of the test piece, regulate the distance between the blades by swiveling the fastening bolt, and clamp microdroplets. Fine-tune the beam of the testing machine to make the force sensor exactly zero loads, set the tensile rate at 0.002 mm/s during a formal test, at the same time the force displacement curve was recorded by the machine. The ISS is calculated by combined with the load-displacement curve, that is

$$\tau = \frac{F_{\max}}{\pi d_f L_e} \tag{1}$$

where τ indicates the ISS, F_{max} is the peak load, d_f is fiber diameter and L_e is the length of the fiber embedded in the resin.

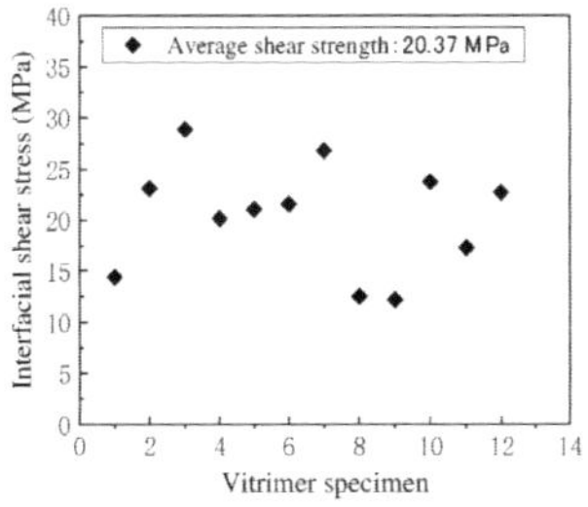

Figure 3. ISS distribution for the T700 carbon fiber/Vitrimer

In the test, 12 groups of effective data were extracted from 100 groups of the tested microdroplet samples. It can be seen from Figure 3 that the data of the microdrop test is very discrete. The average ISS is used to characterize the interfacial bonding performance between resin and fiber, about 20.37 MPa, for the T700 Carbon Fiber / vitrimer epoxy resin system.

3. FE Simulation

3.1. Establishment of model

The microdroplet shape is simplified into a plane sphere, and the established FE model of symmetrical loading is shown in Figure 4. The diameter of carbon fiber is 7 μm, so the fiber width of the axisymmetric model is 3.5 μm. The embedding length L_e of resin droplet is 0.2 mm (the embedding length corresponds to the specimen of the typical load-displacement curve). Mesh refinement is performed in the area where the fiber is embedded in the resin, and the mesh size is set to 0.35 μm, the refinement area is shown in the thumbnail figures. The aspect ratio of the grid is close to 1. The carbon fiber and epoxy element types use 4-node / bilinear / axisymmetric / quadrilateral elements (CAX4).
Due to the high requirements of interface contact convergence on the mesh, the element size of fiber and resin matrix should be small enough and uniform. The mesh convergence was analyzed as the mesh size from 1 μm to 0.1 μm, and assessed by the difference of the maximum force less than 2%. Both the fiber and matrix are regarded as isotropic elastic materials, and their properties are listed in Table 2.

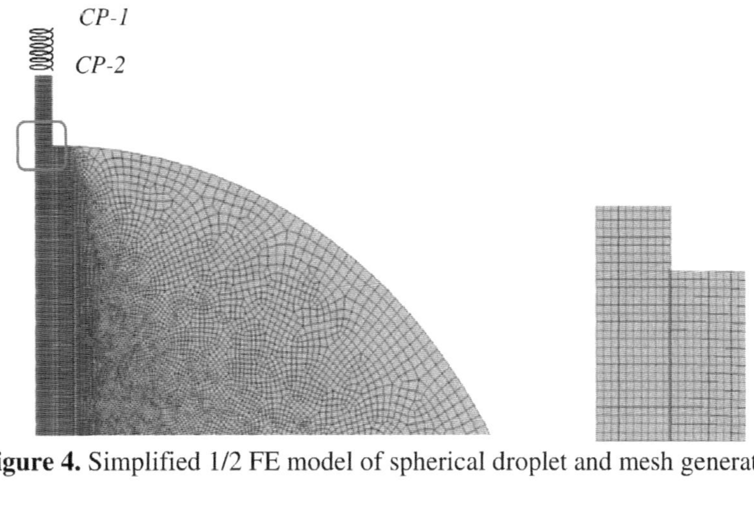

Figure 4. Simplified 1/2 FE model of spherical droplet and mesh generation

Table 2. Properties in FEM

Property	Fiber	Resin
Modulu (GPa)	236	2.78
Poisson's ratio	0.21	0.31

The boundary conditions were imposed by the FE model. The top part of carbon fiber is fixed, the spring element k is created on the upper surface of fiber, coupled to the upper surface of the fiber through the control point CP-1 at the lower part of the spring, and then the load is added to the control point CP-2 at the upper part of the spring. All simulations are performed using the commercial FE program ABAQUS with implicit solver standard.

3.2. Cohesion law at the interface

To effectively simulate the interfacial debonding between droplets and fibers, a cohesive force model with traction-separation constitutive relationship is implemented by the finite element method. The core of CZM is to simulate the fracture process zone before the crack, which can eliminate the singularity of the material crack tip. The damage-free and uncoupled linear elastic traction-separation surface behavior is defined as follows, and the area under the curve represents the fracture energy under the corresponding fracture mode.

$$\sigma = \begin{Bmatrix} \sigma_n \\ \sigma_s \end{Bmatrix} = \begin{bmatrix} K_n & 0 \\ 0 & K_s \end{bmatrix} \begin{Bmatrix} \delta_n \\ \delta_s \end{Bmatrix} = K\delta \tag{2}$$

where σ_i^0 $(i = n, s)$ is the cohesive traction force, K_i is the cohesive stiffness, δ_i is the cohesive separation distance, the subscripts n and s are the normal and tangential direction components respectively, and the initiation damage separation δ_i^0 is $\delta_n^0 = \sigma_n^0 / K_n, \delta_s^0 = \sigma_s^0 / K_s$.

The CZM achieves cohesive behavior by cohesive contact method by ABAQUS. The fiber-matrix interface adopts the CZM with simplify traction-separation law such as bilinear cohesive model, adopts the quadratic stress criterion to define the damage initiation criterion, adopts the power criterion to carry out the damage evolution, and the power exponent is set as $\alpha = 2$. Eqs. (3) and (4) are given respectively.

$$\left\{\frac{\sigma_n}{\sigma_n^0}\right\}^2 + \left\{\frac{\sigma_s}{\sigma_s^0}\right\}^2 = 1 \tag{3}$$

where σ_i^0 (i = n, s) is the cohesive normal traction strength and shear traction strength at the onset of damage.

$$\left(\frac{G_I}{G_{IC}}\right)^\alpha + \left(\frac{G_{II}}{G_{IIC}}\right)^\alpha = 1 \tag{4}$$

where G_I and G_{II} express the fracture energy release rate in normal and shear modes, and the subscript "C" express the critical fracture energy release rate required for complete damage in corresponding mode.

4. Results and discussion

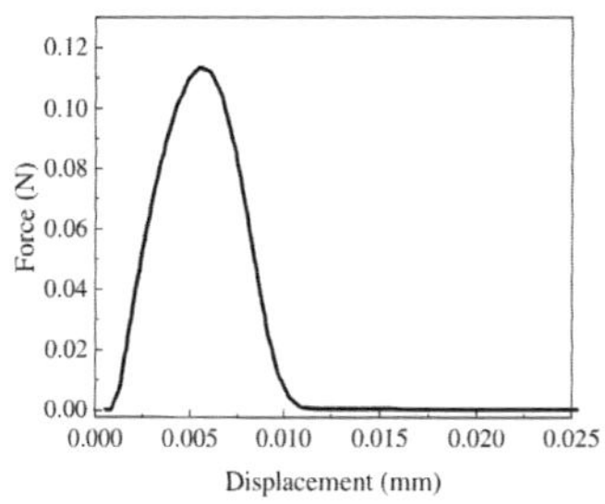

Figure 5. Force-displacement curve of Microdrop simulated by FE method (free section of fiber is not considered)

It can be seen from Fig 5 that the displacement shown by the FE simulation results is too small compared with the experiment. It can be seen from reference [9] that the slope of the force-displacement curve of the micro-drop test is greatly affected by the free fiber length L_{free}, so to match the initial slope of the load-displacement curve measured in the experiment, the free length of the top edge fiber is simulated between two nodes in a model with stiffness of 0.18 kN/m including a spring element. The equivalent relationship between the fiber-free length L_{free} and the spring element stiffness k is

$$k = \frac{\pi E_{fiber} d_f^{\ 2}}{4L_{free}} \quad (5)$$

where free fiber stiffness is k, E_{fiber} is the elasticity modulus of fiber, d_f is the diameter of fiber and L_{free} is the free fiber length.

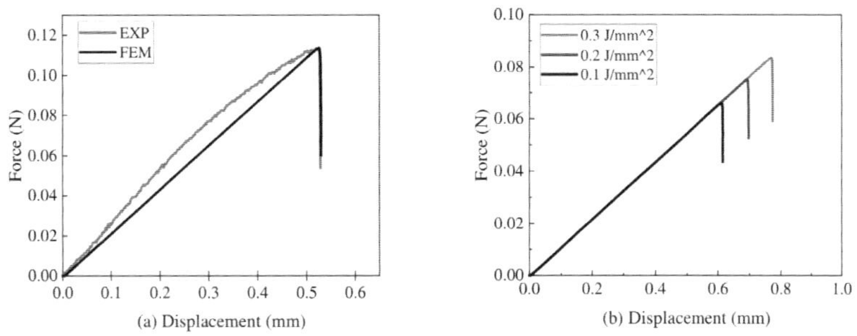

Figure 6. (a) Comparison of load-displacement curves of microdrop test (b) the effect of fracture energy on a load-displacement curve under the same ISS

Figure 6(a) shows the comparison of load-displacement curves of Virimer resin and carbon fiber microdrop test considering the free-section fiber. The FEM force-displacement curve is in good agreement with the tested force-displacement curve in terms of the peak load and displacement, indicating that the can reflect the damage evolution process during droplet pullout. At the same time, the determined cohesive model parameters are effective.

Ignoring the interference signal of the tested curve caused by the vibration of testing machine, it can be found from Fig 6(a) that the CZM simulates the whole process curve of microdrop test from the beginning to complete debonding, which is almost linear. The simulated debonding process can be divided into three main stages. Firstly, the deformation only occurred the free fiber L_{free} and the elastic deformation of the interface. At this time, the cohesive elements of the interface are not broken, and the damage factor is zero. In the second stage, the interface element reaches the given ISS resulting in initial damage according to the secondary stress failure criterion. In the third stage, when the stress of the interface element is close to the ISS, the interface exceeds the carrying capacity and quickly drops the load to zero, which shows that the cohesive element is destroyed and the crack propagates along with the interface. The nonlinearity of the curve in the microdrop test may be caused by the plasticity of Virimer resin.

The influence of cohesive parameters on interface failure behavior is further discussed. The cohesive constitutive relationship of the interface unit includes three critical parameters, such as the ISS, the max separation distant, and the fracture energy release rate. In this study, because the ISS has been determined, the interface parameter values that can be highly consistent with the experimental results can be selected by a trial method, considering the different combinations of interface cohesion strength and fracture energy. The results show that the interface fracture energy release rate has a significant effect on the load and displacement at complete failure, as shown in Fig 6(b). When the ISS at a certain time, the debonding force and corresponding displacement increase with the increase of fracture energy. Finally, the type II fracture energy release rate of Vitrimer interface is determined to be 20 J/m^2 by the trial method.

5. Conclusion

In this study, the interfacial properties of Virimer resin and T700 Carbon fiber were studied by the microdrop test. The deformation and debonding process of the T700 carbon fiber/resin polymer interface were simulated by the FE model based on cohesive contact method. The numerical results are in good agreement with the tested data. The failure mechanism of the interface is discussed and the effective ISS of virimer/T700 Carbon fiber is 20.37 MPa, and the interfacial fracture energy obtained by the FE inversion analysis is 20 J/m^2. The displacement of the interfacial force-displacement curve is closely related to the free-end fiber length, while the ISS is only related to the fiber/polymer system used.

Acknowledgments

The authors thank the National Key Research and Development Program (No. 2019YFA0706803).

References

[1] Chung DDL. A review of multifunctional polymer-matrix structural composites. Composites Part B: Engineering. 2019; 160:644-660.
[2] Montarnal D, Capelot M, Tournilhac F, Leibler L. Silica-like malleable materials from permanent organic networks. Science. 2011; 334(6058):965-968.
[3] Chandra N, Ghonem H. Interfacial mechanics of push-out tests: theory and experiments. Composites Part A: Applied Science and Manufacturing. 2001; 32(3):575-584.
[4] Tandon GP, Pagano NJ. Micromechanical analysis of the fiber push-out and re-push test. Composites Science and Technology. 1998; 58(11):1709-1725.
[5] Day RJ, Rodrigez JVC. Investigation of the micromechanics of the microbond test. Composites Science and Technology. 1998; 58(6):907-914.
[6] Deng S, Ye L, Mai Y-W, Liu H-Y. Evaluation of fibre tensile strength and fibre/matrix adhesion using single fibre fragmentation tests. Composites Part A: Applied Science and Manufacturing. 1998; 29(4):423-434.
[7] Minnicino MA, Santare MH. Modeling the progressive damage of the microdroplet test using contact surfaces with cohesive behavior. Composites Science and Technology. 2012; 72(16): 2024-2031.
[8] Sockalingam S, Dey M, Gillespie JW, Keefe M. Finite element analysis of the microdroplet test method using cohesive zone model of the fiber/matrix interface. Composites Part A: Applied Science and Manufacturing. 2014; 56:239-247.
[9] Minnicino MA, Santare MH. Predicting microdroplet force response using a multiscale modeling approach. Composites Part A: Applied Science and Manufacturing. 2016; 80:39-52.

Effects of Vanadium on the Kinetics of Austenitization from Pearlite in Fe–V–C Alloys

Shihu Hu and Lingnan Kong*

Department of Mechanical and Electrical Engineering, Hunan University of Humanities, Science and Technology, Loudi 417000, China

The corresponding author's e-mail: unhyde@163.com

Abstract. Vanadium effects on the kinetic characteristics of the austenit transformation in Fe-V-C ternary alloys were investigated by microstructure analysis, differential scanning calorimetry (DSC) and the Johnson-Mehl-Avrami-Kolmogorov model (JMAK). The results show that the vanadium gives rise to the refinement of pearlite lamellar spacing before transformation, the original austenite grain size decreases after transformation, and and the increase of the amount of carbide at the grain boundarsy. Studies on kinetics showed that the austenization of the Fe-V-C ternary alloys is characterized by JMAK model involving site saturation nucleation, diffusion-controlled growth, and impingement correction. V increases the diffusion activation energy and decreases the pre-exponential factor for diffusion.

1. Introduction

It is well known that the performance of the material is controlled by the microstructure. For the steel, most of the microstructure are transformation products from austenite, such as ferrite, pearlite, bainite and martensite. Thus, a thorough research on the phase transformation kinetics, especially the kinetics of austenitization, is very necessary for steel. Although the formation of austenite plays an important role in the heat treatment of steel, the austenite formation process has not attracted enough attention, which has been reported only in limited literature. Miyamoto et al. studied alloying effects on the kinetics of austenitic phase transition or tempered martensite in Fe-C-M alloys (M=Mn, Si, Cr). They found that Mn or Si obstacles the phase transition due to reducing the carbon activity gradient, and the diffusion of Cr leads to a magnitude slowly austenitization process. Following closely, Li et al. reported that a similar result was obtained in the study of the effect of alloying on kinetics of austenitic phase transition of Fe-C-M alloy (M=Mn, Si, Cr). Zhang et al. studied the nucleation process of strain-induced austenitizing of low alloy hypereutectoid steel, and found that the fine microstructure and strain are conducive to the nucleation of austenite. Ma et al. analyzed that the austenization behavior from martensite of heat-resistant steel, they found that the diffusion of elements and the precipitation of carbides have affected on the austenitic phase transition. Recently, the author analyzed the kinetics characteristics of the austenitic phase transition in Fe-C-M alloys (M=Mn, Mo, Cr). The results show that the pearlite structure is refined, and the diffusion distance of the elements is reduced due to the addition of alloying elements. The amount of austenite along the cementite lamellae increases with the decrease of the interlayer spacing. In addition, the alloying elements combine with carbon and form carbides, which not only limit the diffusion of carbon, but also play a pinning role to hinder the growth of austenite. In conclusion, the existing research results show that the austenitization process is affected by many factors, such as the microstructure, alloying elements, stress and so on.

More and more attention has been paid to the kinetics of austenitization, but the affection of alloying elements on the kinetics of transition from pearlite to austenite is still insufficient. Therefore, The purpose of this paper is to prove the effect of V addition to the Fe-V-C alloys on the austenitization kinetics experimentally and theoretically based on the microstructure, alloying element content and carbon distribution and so on.

2. Experimental

2.1. The sample source

All samples were produced by the casting. A small amount of pure aluminum is added during smelting to remove oxygen from the liquid steel. When the temperature of the molten steel reaches 1853 K, which was injected into a sand mould and maintained for 24 h. The composition exhibited in Table 1.

Table 1. The compositional of the samples. (wt. %)

	C	V	S	P	Fe
1#	1.06	0.57	0.03	0.02	Bal.
2#	1.05	0.85	0.03	0.02	Bal.
3#	1.06	1.54	0.04	0.02	Bal.

2.2. DSC

All samples came from the same position of the Y-type ingot and Its size is approximately Φ 2.0–0.5 mm. All cylindrical shape samples were analyzed using a Netzsch DSC 404C. The DSC analysis was performed from room temperature up to1373 K at 10 K/min and then cooled down to room temperature at 40 K/min. In addition, pure argon was introduced into the DSC equipment to prevent oxidation.

2.3 Microstructural characterization

The as-cast samples were obtained at the same position as the sampling position of the DSC sample. And the samples were machined, polished, and etched with the 4 vol.% Nital solution and observed by scanning electron microscopy (SEM).

In order to obtain the original austenite grain size after the DSC analysis, the samples with the size of 10×10×10 mm was heat treated under the same heating process as the DSC analysis, and it is immediately cooled in water when the temperature reached 1373 K. In order to obtain the austenite grain size, the samples was machined, polished, and etched with a solution of picric acid and sodium dodecyl benzene sulfonate, and then analyzed with optical microscopy.

3. Results and discussion

The SEM morphology of the as-cast microstructure of Fe-V-C alloy steel is shown in Figure 1. The as-cast microstructure of all samples is fine lamellar pearlite. Moreover, the lamellar spacing (S_0) of pearlite was measured and listed in Table 2. During the measurement process, each sample was measured more than 25 times, and the average value of all the measurement results was invoked as the final lamellar spacing of pearlite. It was found that the lamellar spacing of pearlite decreases from 352.1 nm to 233.7 nm with the increase of V content. In this case, V was increased from 0.57 wt.% to 1.54 wt.%, since V can significantly increase the hardenability of steel, the hardenability of the sample is improved as the V content increases. And then, the degree of undercooling increases with the improvement of hardenability under the same cooling conditions. As we all know, the lamellar spacing of pearlite is inversely proportional to the degree of supercooling[1]. Therefore, the lamellar spacing decreases with the increase of V.

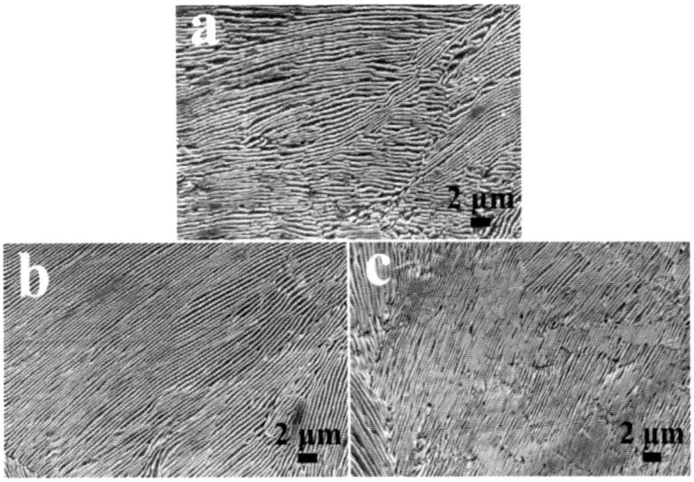

Figure 1. The SEM micrograph of the casting microstructure of the samples

Table 2. The lamellar spacing of pearlite.

	1#	2#	3#
S_0 (nm)	352.1	317.6	233.7

The OM morphology of the austenite are shown in Figure 2. It can be observed that the original austenite grains can be clearly identified, and the grain boundaries are smooth and clear. In addition, undissolved carbides (red circles) are found, and the amount increases with the increase of V content. The carbides play a pinning role during the formation process of austenite and hinders the growth of austenite grains.

The size of the original austenite grains was measured. More than 300 sets were collected, and the average of all data was used as the final grain size (d). Since the austenitic phase transition can be attributed to the site saturation model, the number of nucleation sites is constant during the phase transformation. To calculate the nucleation density (N^*), the geometric factor (g) is equivalent to 1 when assuming that the austenite is considered cubic[2]. Thus, the number of nucleation density (N^*) was calculated by $N^* = 1/d^3$. The original austenite grains size (d) and the nucleation density (N^*) are listed in Table 3.

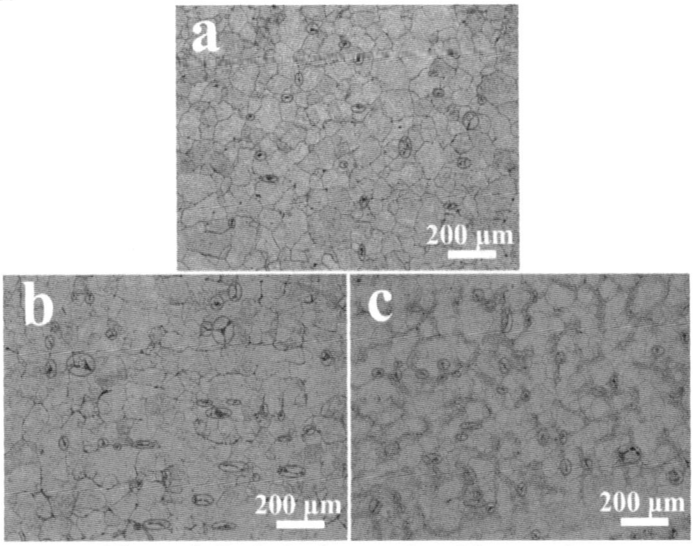

Figure 2. Optical micrographs of the samples

Table 3. The original austenite grains size (d) and the nucleation density. (N^*)

	1#	2#	3#
d (μm)	49.26	39.64	15.78
N^* (m^{-3})	8.37×10^{12}	1.61×10^{13}	2.54×10^{14}

In Figure 2 and Table 3, The original austenite grains size reduces from 49.26 μm to 15.78 μm with the increase of V content. There are fluctuations of energy, composition and microstructure at the grain boundary, so austenite grains tend to nucleate at the interface. As described above, as the content of V introduced gradually increases, the lamellar spacing of pearlite decreases from 352.1 nm to 233.7 nm. The amount of austenite nuclear sites is enhanced with the decrease of pearlite lamellar spacing, which means that the austenite nucleation density increases and the original austenite grains become more refined. Besides, V is a strong carbide forming element whose carbides are difficult to dissolve, just as carbides exist on the austenite grain boundariy. These carbides not only increase the number of the interface, but also play a pinning role to hinder the growth of austenite during the formation process of austenite. Under the combined effect of the above factors, the austenite grain size shows a decreasing trend with the introduction of V.

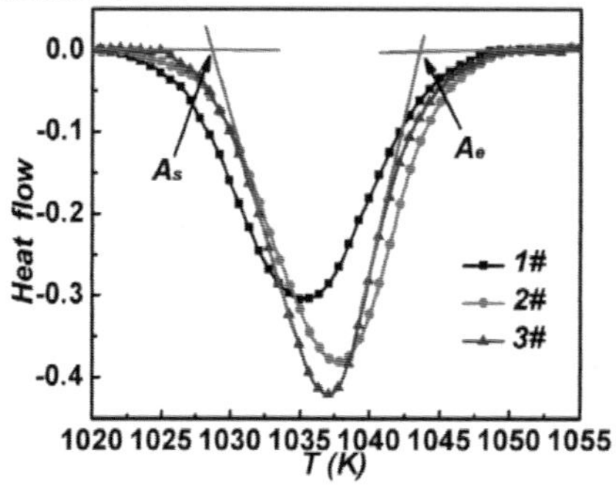

Figure 3. DSC curves of the samples

All samples were analyzed by DSC experiment, and the results are presented in Figure 3. It is shown that an endothermic peak, which belongs to the transformation from pearlite to austenite. Besides, the intersection point of the endothermic peak and the baseline is taken as the start temperature (A_s) and end temperature (A_e) of austenitization, respectively. The corresponding values of each temperature are given in Table 4.

As shown in Figure 3 and Table 4, the endothermic peak shows a tendency to move to a high temperature, and the start temperature increases as the V content increases, but the end temperature is reversed, the difference between the start and the end temperature ($A_e - A_s$) gradually decreases at the same time, which means that the time necessary to complete austenitization is shortened. In other words, the austenitization process can be completed more quickly as the V content is increased.

Table 4. The start and the end temperature of the austenization transformation.

	1#	2#	3#
A_s (K)	1025.49	1026.93	1029.53
A_e (K)	1044.38	1045.13	1043.8
$A_e - A_s$ (K)	18.89	18.2	14.27

In the case, since the heating process is isochronal experiment, the austenitization process can be characterized by the JMAK models. According to the JMAK models, the austenite fraction (f_γ), the total enthalpy changes of austenitization process (ΔH_A) and the heat flux (F_u)can be described by the following equation[3]:

$$f_\gamma(t) = (\frac{1}{\Delta H_A})\int_0^T F_u(t)dt \quad (1)$$

Therefore, The relationship curve between austenite fractions and temperature for various V content is obtained from the DSC curves as shown in Figure 4. It can be found that the austenite fraction changes with a S-shaped curve with temperature for all samples, which indicate that all austenitization processes have the same kinetics characteristics and phase transformation type.

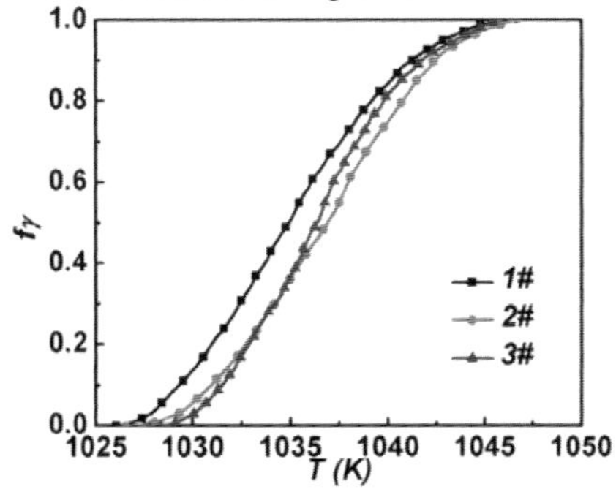

Figure 4. The relationship curve between austenite fractions and temperature

The austenite formation rate (df_γ/dt) can be obtained from the time derivative of the austenite fraction, and the relationship curve between the transformation rate and temperature or the austenite fraction is shown in Figure 5.
As shown in Figure 5, the start temperature of austenitization enhances significantly, and the end temperature decreases slightly. And then the overall time of austenitization reduces with the increase of V content. Furthermore, there is just one maximum value for the relationship curve between the transformation rate and temperature or the austenite fraction. At the same time, the maximum value increases as the V content is increased. The increase of the austenite formation rate shortens the overall austenitization time, which corresponds to the previous conclusion in Table 4.

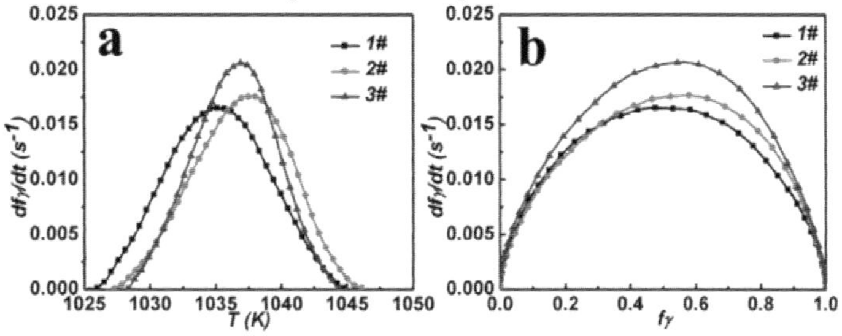

Figure 5. The curve between the austenite formation rate and temperature or the austenite fraction

It is a known fact that the formation rate of austenite is affected by by the nucleation rate and the diffusion rate of elements[4]. In the case, the influence of pearlite is very important. As mentioned earlier, the lamellar spacing of pearlite decreases from 352.1 nm to 233.7 nm when V content increases from 0.57wt. % to 1.54 wt. %. The nucleation rate of austenite can be significantly improved by shorten the lamellar spacing of pearlite. Moreover, element diffusion is accompanied at the interface of ferrite and austenite, cementite and austenite, ferrite and cementite and inside austenite during austenite formation. Therefore, the decrease of the pearlite lamellar spacing not only increases the nucleation rate of austenite, but also greatly reduces the effective diffusion distance of elements, and then accelerates the diffusion rate of elements. Otherwise, the number of carbides gradually increases as the content of V increases, which once again improve the nucleation rate of austenite. In summary, the transformation rate of austenite increases as the V content increases.
In the JMAK model, austenitizing is decomposed into three relatively independent parts: nucleation, growth and impingement. The impingement between particles are ignored when the particle volume is computed during nucleation and growth. Thus, the extended volume does not correspond to the actual volume, which must be corrected.Thus, for the modified JMAK model, the austenite fraction (f_γ) can be given as[5]:

$$f_\gamma = 1 - \exp\left\langle -N^* g \left(\frac{D_0}{\phi}\right)^{\frac{3}{2}} \left\langle \exp\left[-\frac{Q_d}{RT(t)}\right] \frac{RT(t)^2}{Q_d} \left\langle 1 - \exp\left\{\frac{Q_d[T_0 - T(t)]}{RT(t)^2}\right\}\right\rangle\right\rangle^{\frac{3}{2}}\right\rangle \quad (2)$$

where g is the geometry factor, which is equal to 1, ϕ is the heating rate, R is the gas constant.

Table 5. Kinetic parameters of the austenization process.

	1#	2#	3#
Q_d (kJ mol^{-1})	9.38	10.16	10.74
D_0 (m^2 s^{-1})	1.12×10^{-11}	7.59×10^{-11}	1.45×10^{-10}

According to the nucleation density (N^*) in Table 3 and the austenite fraction measured by the experiment in Figure 4, the values of the pre-exponential factor (D_0) and the activation energy (Q_d) for diffusion are obtained by fitting with Eq. (2). The results of the fitted curve and the corresponding kinetic parameters are presented in Figure 6 and Table 5.

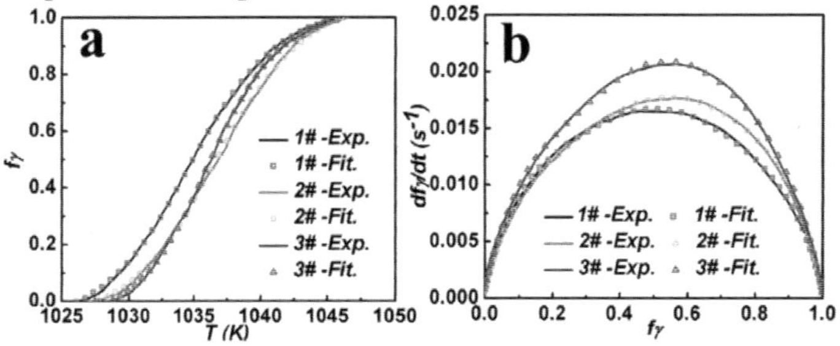

Figure 6. Comparison of the JMAK model and the experimental

As shown in Figure 6, there is a broad consistency between the JMAK model and the experimental, and the fitted values of the kinetic parameters are rough agreement with the pre-exponential factor and the activation energy of the diffusion of carbon.

Austenite transformation can be divided into two processes: nucleation and growth, which depends on the initial microstructure, composition and heating temperature. In the case, all samples were treated at the same heating rate and temperature. Therefore, the affect of heating temperature on the austenitization process is not considered. The only difference in the composition of all samples is that the V content gradually increases. V combines with carbon to form carbides, which is difficult to dissolve even at austenitizing temperature, as shown in the red circle in Figure 2. It means that the addition of V reduces the carbon activity gradient, and thus delays the austenitization process. Therefore, the diffusion activation energy (Q_d) increases as the V content increases, which is in line with the gradual increase of the austenitization start temperature (A_s) in Table 4. In addition, the initial microstructure of all samples is lamellar pearlite, and the introduction of V effectively reduces the lamellar spacing of pearlite. As mentioned above, the reduction of pearlite lamellar spacing not only improves the nucleation rate, but also reduces the effective diffusion distance of element. For the austenitization process, the austenite growth is still controlled by carbon diffusion. Hence, the finer lamellar spacing of pearlite contributes to the austenite growth.

4. Conclusions

By means of microstructure analysis, DSC experiment and JMAK model, the influence of V on the austenization kinetics was systematically studied, The results were as follows:

(1) The initial microstructure of all samples is lamellar pearlite and the lamellar spacing decreases gradually. After austenization treatment, the original austenite grains are refined, and carbides are present at grain boundaries.

(2) With the increase of V content, the start temperature of the austenization slightly increases, and the total time of the austenization process decreases.

(3) The austenitization process can be described adopting site saturation nucleation,

diffusion-controlled growth, and impingement correction.

(4) V increases the nucleation rate of austenite, reduces the diffusion distance and the activity gradient of carbon.

Acknowledgements

This work is supported by the double first-class discipline construction program and the scientific research project of education department of Hunan province (18A411; 18B452).

References

[1] C. Liu, Y. Liu, D. Zhang, Z. Yan, Kinetics of isochronal austenization in modified high Cr ferritic heat-resistant steel, Applied Physics A, 105 (2011) 949-957.
[2] S.C. Hong, S.H. Lim, H.S. Hong, K.J. Lee, D.H. Shin, K.S. Lee, Effects of Nb on strain induced ferrite transformation in C–Mn steel, Materials ence & Engineering A, 355 (2003) 241-248.
[3] N. Zotov, V. Marzynkevitsch, E.J. Mittemeijer, Evaluation of kinetic equations describing the martensite–austenite phase transformation in NiTi shape memory alloys, Journal of Alloys & Compounds, 616 (2014) 385-393.
[4] T. Zhou, H.S. Zurob, E. Essadiqi, B. Voyzelle, Kinetics of Delta-Ferrite to Austenite Phase Transformation in a Two-Phase Fe-Al-C Alloy, Metallurgical & Materials Transactions A, 42 (2011) 3349-3357.
[5] D. Wang, Y. Liu, Y. Zhang, Improved analytical model for isochronal transformation kinetics, Journal of Materials Science, 43 (2008) 4876-4885.

Residual Stress Simulation of Heterogeneous Cladding Layer Between SiC/Ti6Al4V

Junjun Liu[1], Hao Jiang[2,*], Shen Yu[1], Zhenkun Lei[1]

[1]State Key Laboratory of Structural Analysis for Industrial Equipment Department of Engineering Mechanics, Dalian University of Technology, Dalian 116024, China
[2]School of Materials Science and Engineering, Dalian University of Technology, Dalian 116024, China
Corresponding author: *jaymath@126.com

Abstract. SiC cladding on the alloy surface can significantly improve the high temperature resistance and oxidation resistance of the material. However, heterogeneous cladding usually produces large residual stress and affects the service life of the material. In the process of laser cladding simulated by finite element method, the mechanism of residual stress during SiC deposition on alloy surface is studied by using moving double ellipsoid heat source model. The simulation results show that the residual stress is unevenly distributed near the cladding layer, and the residual stress presents a compression-tension-compression distribution in the thickness direction due to the mixing law of material properties, which provides a reliable numerical prediction method for the residual stress of SiC cladding layer cladding on the alloy base.

1. Introduction

Laser cladding technology, as a kind of automatic rapid additive manufacturing and surface modification technology for metal parts, has received extensive attention in the field of mechanical manufacturing. The laser heated powder material and the base material are melted and solidified at a moment to form a good metallurgical compound interface, which can repair the nearly damaged mechanical parts and make them regain their service performance [1-3].

As a hot-melt deposition material, ceramic cladding on the surface of alloy material can significantly improve the high-temperature resistance and oxidation resistance of alloy. If the bonding between cladding layer and matrix is metallurgical bonding, it also has the advantages of high bonding strength. However, due to the rapid heating and cooling effect in the laser cladding area and the difference of thermal expansion and shrinkage coefficients of different materials between the deposition layer and the substrate layer, great residual stress will be generated near the hot melt deposition area, resulting in the generation and growth of thermal cracks and affecting the bearing capacity and service life of the material [4-5].

With the continuous development of simulation technology, more and more domestic and foreign scholars began to study laser cladding technology through numerical simulation. Khamidullin et al. [6] established a two-dimensional laser cladding model to simulate the macro morphology change, crystallization process and temperature field evolution of the cladding layer. Gu et al. [7] studied the numerical simulation of laser cladding. By analyzing the process of laser cladding, they found that complex thermal cycles and temperature distribution occurred in the deposition layer, which is the main reason for large tensile stress. Bruckner et al. [8] pointed out that compression plastic strain and

tensile residual stress are easily generated during laser cladding cooling, but they only studied the case of one cladding layer.
In this study, the finite element thermal mechanical sequential coupling method is used to simulate the residual stress distribution after laser cladding SiC particles on titanium alloy (Ti6Al4V or TC4) substrate to form alloy composite layer under the assumption of double ellipsoidal heat source, and the generation mechanism of residual stress at the fusion interface is revealed.

2. Basic principles

2.1. Phase transition equation

Due to the phase transformation during laser cladding, in addition to the conventional elastic-plastic deformation, it also produces thermal expansion strain and phase transformation expansion strain. The common non isothermal phase transformation includes diffusion phase transformation and non diffusion phase transformation [9]. This study simulates the phase transformation process of TC4 based on the former. In the case of diffusion phase transition, Johnson-Mehl-Avrami equation is usually used to describe the phase transition dynamics during cladding, that is

$$t_i^* = \left(-\ln\left(1 - f_{i-1}\right)/b_i\right)^{1/n_i} \tag{1}$$

$$f_i^* = 1 - \exp\left[b_i\left(t_i^* + \Delta t\right)^{n_i}\right] \tag{2}$$

$$f' = f_i^* f_{\max} \tag{3}$$

Where f' is the volume fraction of the new phase, f_i^* is the virtual fraction transformed, $f_{\max}$ is the maximum volume fraction, t is the isothermal duration, k is the coefficient depending on the temperature, parent phase composition and grain size, n is also the coefficient depending on the phase transformation type and grain growth, and the temperature dependence is included in the coefficients $b(t)$ and $n(t)$.

2.2. Mechanical principle

In solid-state phase transition, the total strain increment can be expressed as:

$$\mathrm{d}\boldsymbol{\varepsilon} = \mathrm{d}\boldsymbol{\varepsilon}_{\mathrm{e}} + \mathrm{d}\boldsymbol{\varepsilon}_{\mathrm{p}} + \mathrm{d}\boldsymbol{\varepsilon}_{\mathrm{T}} + \mathrm{d}\boldsymbol{\varepsilon}_{\mathrm{VOL}} + \mathrm{d}\boldsymbol{\varepsilon}_{\mathrm{trip}} \tag{4}$$

Where the each strain component in the right side of equation represents the strain increase caused by elastic load, plastic load and thermal load respectively, and the volume expansion caused by solid-state phase transformation is expressed as [10, 11]

$$\mathrm{d}\boldsymbol{\varepsilon}_{\mathrm{T}} = \alpha_{\mathrm{T}}\mathbf{I}\mathrm{d}T \tag{5}$$

$$\mathrm{d}\boldsymbol{\varepsilon}_{\mathrm{VOL}} = \mathrm{d}f \cdot \mathrm{d}\boldsymbol{\varepsilon}^{\Delta\mathrm{V}} \tag{6}$$

Where f is the phase transition volume fraction, $\mathrm{d}f$ is the increment of phase volume fraction, α_{T} is the mixed expansion coefficient, **I** is the third-order unit tensor, $\mathrm{d}T$ is the temperature difference, which is affected by factors such as phase transformation latent heat, $\mathrm{d}\boldsymbol{\varepsilon}_{\mathrm{trip}}$ is transformation induced plasticity strain, without considering in this study.

3. Numerical simulation

3.1. FEM model and boundary conditions

Since the laser micro melting process involves various physical and chemical changes, and there are still complex phenomena such as heat transfer and fluid flow in the formed molten pool. Therefore, in the finite element simulation of the laser micro melting temperature field, the following assumptions need to be put forward accordingly:
1) The cladding layer and matrix are isotropic materials, and the plastic yield of the material follows von Mises yield criterion, and the flow rule is described by bilinear strengthening behavior;

2) The temperature change caused by the liquefied flow of metal in the molten pool is not considered, and its influence on the heat transfer coefficient is ignored;
3) Ignoring the heat generated by material plastic deformation, the weak thermal mechanical coupling is adopted; The creep effect is ignored.
In order to deeply study the distribution and formation mechanism of residual stress in laser cladding functional material layer, a thermal mechanical coupling FEM is established by using the birth and death element algorithm. Fig. 1 is a FEM of single-layer cladding TC4 established with reference to the actual contour of the cladding. In order to improve the simulation accuracy and save the calculation cost, the symmetrical model is adopted, and the Y-Z plane is selected as the symmetrical plane of the model. Due to the different laser action positions, the temperature gradients in different regions of the micro melting model will be greatly different, so appropriate mesh generation is more important. Because the size of molten pool in laser cladding deposition process is generally 2~3mm, the precision and accuracy of cladding can be accurately controlled while improving the focusing performance of beam. In this case, the size of molten pool is small, the thermal load is small, the cooling speed is fast, and the area of final cladding structure is small. In order to analyze the temperature field, phase composition and stress field of laser cladding, a more accurate model is to be established.
High quality mesh can not only accurately complete the simulation work and ensure the accuracy of the results, but also effectively save the calculation time and space and improve the simulation efficiency. In this study, the transition grid division method is adopted. As shown in Fig. 1, the minimum grid size set at the laser action position is about 0.00167 mm, and the grid far away from the laser action position gradually increases from small to large according to a certain proportion, and the model size is 10×12×10 mm^3. Finally, the FEM of single-layer cladding of TC4 plate is composed of 13830 elements and 15686 nodes, as shown in Fig. 1.

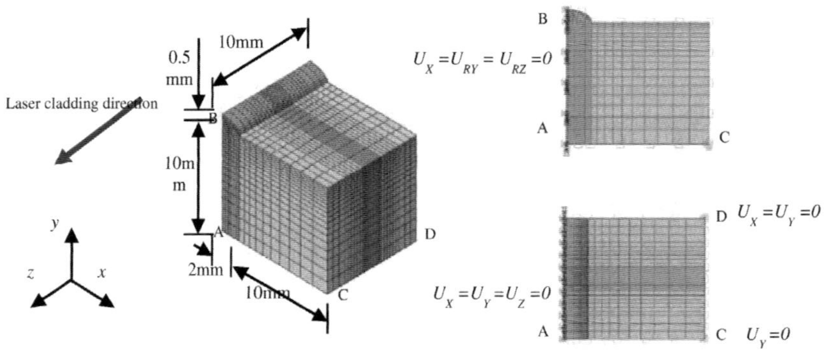

Figure 1. Schematic diagram of the FEM and boundary conditions

In simulating the thermos-mechanical coupling process of laser cladding, a three-dimensional thermo coupled finite element method was employed to simulate the thermo mechanical coupling process in laser cladding using various user-defined sub programs in ABAQUS software. The algorithmic flow is shown in Figure 2. In the software ABAQUS, user-defined subroutines were employed to simulate the complex interactions among thermal, mechanical and metallurgical phenomena, as shown in Fig.2. At each time point, When defining the heat source flow, the DFLUX subroutine is written, which can update the local coordinates of the heat source definition center to simulate the effect of moving forward in the deposition process. The user defined subroutine USDFLD is used to simulate the addition of filler metal in the cladding process, and the plasticity related material properties are defined as field variables. The life and death cell function is used. Before the cladding process, all the cells of the cladding track are killed. After the cladding starts, the cells on each cladding layer are activated in turn with the movement of the laser heat source, and it also can calculate the phase transition fraction according to Eq. 3. In the UEXPAN subroutine, the phase fraction transmitted is used to calculate the expansion coefficient and strain corresponding to each identified region. In order to incorporate the strain due to metallurgical transformation, the user subroutine UEXPAN is used to define the dynamic model to convert the phase volume fraction into strain, Through this method, the heat, metal and mechanical changes in the process of the laser melting were established. UEXPAN could not only calculate the strain caused by heat expansion, but also the strain and volume expansion caused by the metal transformation. The UHARD program was mainly used to define the mechanical construction behavior of materials, and it describes the elastic-plastic behavior of materials at different temperatures to calculate the elastic strain $\boldsymbol{\varepsilon}_e$, Plastic strain $\boldsymbol{\varepsilon}_p$ and stress $\boldsymbol{\sigma}$.

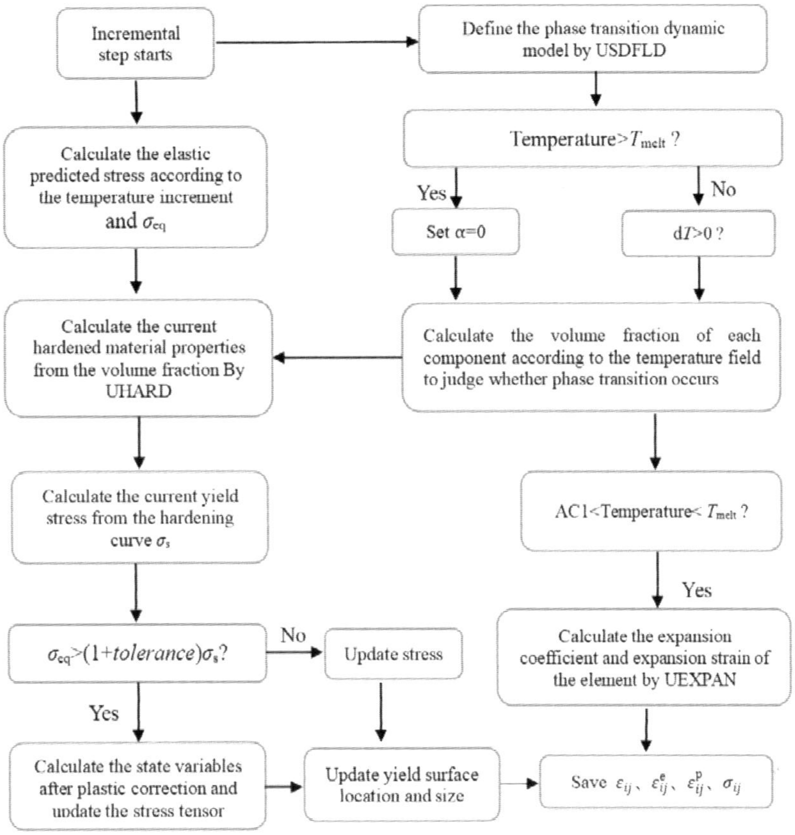

Figure 2. Finite element calculation flow chart considering phase transformation and temperature coupling

3.2. Heat source model

In the process of the melting, the heat source on the surface of Gauss couldn't describe the penetrating effect of the laser very well, so the distribution characteristic of the heat flow in the depth of the melting pool couldn't be taken into account. Therefore, Goldak and others [12] developed a distributed power density double oval heat source model to overcome these problems, so that they could imitate the type of melting with complex furnace shapes, as shown in Fig. 3. The laser power density of the model is controlled by the piecewise ellipsoid function described in Eq. 7.

$$\begin{cases} q_f(x,y,z,t) = \dfrac{6\sqrt{3} f_f Q}{abc_f \pi\sqrt{\pi}} \exp\left(-\dfrac{3x^2}{a^2} - \dfrac{3y^2}{b^2} - \dfrac{3(z+vt)^2}{c_f^2}\right), & z \geq 0 \\ q_r(x,y,z,t) = \dfrac{6\sqrt{3} f_f Q}{abc_r \pi\sqrt{\pi}} \exp\left(-\dfrac{3x^2}{a^2} - \dfrac{3y^2}{b^2} - \dfrac{3(z+vt)^2}{c_r^2}\right), & z \geq 0 \end{cases} \tag{7}$$

Where, q is the total input energy, in [W]; Q is the power density of the laser, in [W·m^{-3}]; f_f, f_r are the percentages of the anterior and posterior hemispheres in the total input energy; a, b, c_f and c_r are the shape parameters of ellipsoid, which determine the shape of double ellipsoid, and the unit is [m]; v is the moving speed of the heat source, in [m/s]; T is the welding time, in [s].

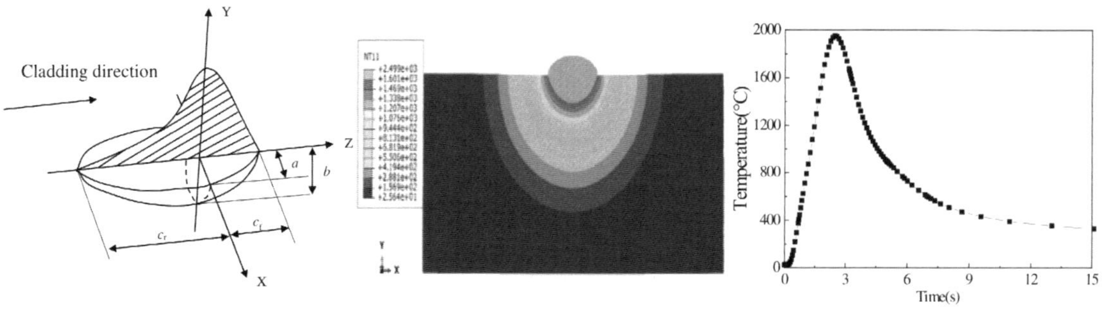

(a) (b) (c)

Figure 3. (a) Double ellipsoid heat source model, (b) the temperature distribution in FEM and (c) the temperature change history of an element on the cladding path

According to the linear mixing law [13], the mixed material properties of SiC and TC4 can be determined, including Specific heat coefficient, heat dissipation coefficient, Poisson's ratio, yield strength and so on, as follows:

$$\varphi=\varphi_{TC4}(1-V_{SiC})+\varphi_{SiC}V_{SiC} \quad (8)$$

Where represents the specific material properties of the first or second layer (e.g., Poisson's ratio and yield strength); φ_{TC4} and φ_{SiC} represents the specific material properties of TC4 and SiC respectively; V_{SiC} represents the volume fraction of SiC in the corresponding layer. In addition, their yield strength is calculated by the following formula (Li [14]):

$$\sigma = \sigma_{TC4}[V_{SiC}(s+2)/2+V_{TC4}] \quad (9)$$

Among them σ and σ_{TC4} is the yield strength of the corresponding layer and TC4 respectively; V_{SiC} and V_{TC4} are the volume percentages of SiC and TC4 in the corresponding layer, respectively; And s is a constant related to SiC size. The thermal and physical properties of TC4 and SiC are referred to the literatures [15].

4. Discussion

From the simulated result in Fig. 4, it can be seen that the residual stress in the cladding direction obeys the distribution of compression-tension-compression from the cladding area to the base metal area of the substrate. However, the von Mises equivalent stress is the largest at the interface between cladding layer and substrate, which explains the mechanical mechanism that the cladding interface is easy to be damaged. As can be seen from Fig. 5, the maximum tensile stress occurs 4.5mm from the top of cladding, which is roughly the heat affected zone outside the molten pool.

Fig. 4 (a) and (b) show the residual stress distribution in a single layer laser molten, and it can be seen that peak tensile stress is obtained in the position close to the melt interface. This is because a large amount of heat is transferred to the material below, the melt interface is solidifying and starts shrinkage. Since the shrinkage around the melt interface is resisted by the surrounding material, the surrounding material causes tensile stress near the melt interface. In addition, the residual stress in the cladding layer and the substrate pool area is negative, the value is close, the area is most affected by the heat source, and the phase transition begins at first because of the rapid temperature decrease and the non-diffusion phase transition finally takes place.

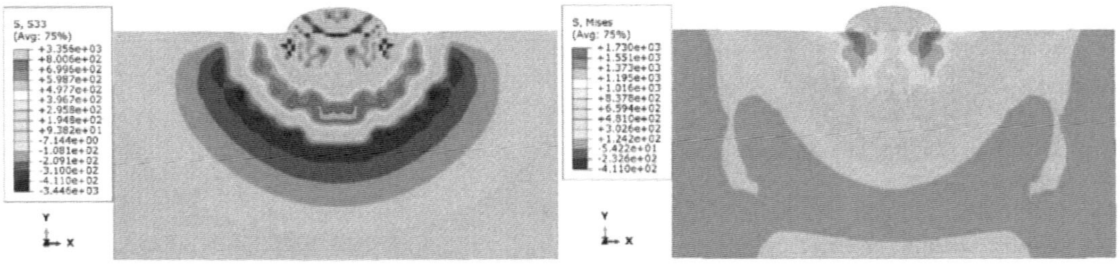

(a) (b)

Figure 4. Distributions of (a) S_{33} and (b) von Mises residual stress field after cooling

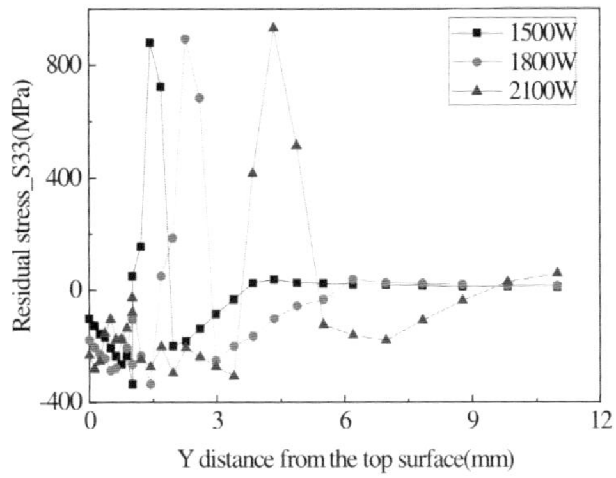

Figure 5. Residual stress distribution in thickness direction

As shown in Fig. 5, the residual stress exhibits different distributions at different heat source power. The peak of the residual stress is offset backwards as the power rises, and the maximum residual stress gradually increases with the rise of the power. However, the change in heat source power does not affect the distribution trend of 'compression-tension-compression' of residual stress.

This numerical model contains many coupling parameters between heat, machinery, and metallurgical, which are usually related to experimental conditions. That is, under a certain experimental conditions, the final simulated result of simulation can be consistent with the experimental results by adjusting these coupling parameters. This model is intended to upgrade in the later studies, using displacement field and strain field of the cladding layer to reverse phase change parameters, which can be used to analyze the residual stress and temperature of samples with multilayer overlapping coating and cladding path.

5. Conclusion

In this study, a micro melting model for simulating SiC cladding on the Ti6Al4V alloy surface is established by using the transient thermal coupling finite element method. Several user-defined subroutines are used to describe the phase transition of materials in the coupling solution process. The model simulates the diffusion and non diffusion phase transition during heating and cooling. Then, through this method, the temperature field and residual stress distribution of single path cladding are analyzed, and the main conclusions are as follows:

(1) The residual stress exhibits different distributions at different heat source power. The peak of residual stress is backwards offset as the power rises.

(2) The residual stress in the cladding direction obeys the distribution of compression - tension - compression from the cladding area to the base metal area of the substrate.

(3) The peak tensile stress is obtained in the position close to the melt interface. The maximum tensile stress occurs at 4.5mm from the top of cladding, which is roughly the heat affected zone outside the molten pool.

Acknowledgments
This work was financially supported by the National Natural Science Foundation of China (Nos. 12002078, 11772081, 11972106).

References

[1] F. Vollertsen, K. Partes and J. Meijer: State of The Art of Laser Hardening and Cladding. International Journal of Human Resources Development and Management, Vol. 70(4) (2005), p. 116
[2] W. Wang, A.J. Pinkerton, L.M. Wee and L. Li: Component Repair Using Laser Direct Metal Deposition. SAGE Publications (2007), p. 38-39
[3] P. Kattire, S. Paul, R. Singh and W. Yan: Experimental Characterization of Laser Cladding of CPM9V on H13 Tool Steel for Die Repair Applications. Journal of Manufacturing Processes, Vol. 20(3) (2015), p. 492-499
[4] J. Leunda, C. Soriano, C. Sanz, and V.G. Navas: Laser Cladding of Vanadium-carbide Tool Steels for Die Repair. Physics Procedia, Vol. 12 (2011), p. 345–352
[5] H. Paydas, A. Mertens, R. Carrus, J. Lecomte-Beckers and T. J. Tchoufang: Laser Cladding as Repair Technology for Ti-6Al-4V Alloy: Infuence of Building Strategy on Microstructure and Hardness. Materials and Design, Vol. 85 (2015), p. 497–510
[6] B. A. Khamidullin, I. V. Tsivilskiy, A. I. Gorunov, et al. Modeling of The Effect of Powder Parameters on Laser Cladding Using Coaxial Nozzle[J]. Surface and Coatings Technology, 2019, 364: 430-443.
[7] D. Gu, B. He. Finite element simulation and experimental investigation of residual stresses in selective laser melted Ti–Ni shape memory alloy[J]. Computational Materials Science, 2016, 117: 221-232.
[8] F. Brückner, D. Lepski, E. Beyer. Modeling the influence of process parameters and additional heat sources on residual stresses in laser cladding[J]. Journal of thermal spray technology, 2007, 16(3): 355-373.
[9] S. Paul, R. Singh, W. Yan and I. Samajdar: Critical Deposition Height for Sustainable Restoration via Laser Additive Manufacturing. Scientific Reports, Vol.8 (2018), p. 14726
[10] S. Denis: Considering Stress-phase Transformation Interactions in the Calculation of Heat Treatment Residual Stresses: Mechanics of Solids with Phase Changes. Journal de Physique IV, Vol. 06(C1) (1997), p. 293–317
[11] X. Wang, L. Hu, Q. Xu, D. Chen and S. Sun: Influence of Martensitic Transformation on Welding Residual Stress in Plates and Pipes. Science and Technology of Welding and Joining, Vol. 22(6) (2017), p. 505-511
[12] J. Goldak, A. Chakravarti and M. Bibby: A New Finite Element Model for Welding Heat Sources. Metallurgical Transactions B, Vol. 15(2) (1984), p. 299-305
[13] L. Costa, R. Vilar, T. Reti and A.M. Deus: Rapid Tooling by Laser Powder Deposition: Process Simulation using Finite Element Analysis. Acta Materialia, Vol. 53(14) (2005), p. 3987-3999
[14] X. Li, K. Chen and D. Huang: Influence of Reinforced Particle on Strength of Particle-reinforced Aluminum Matrix Composites. Aluminium Fabrication, Vol. 2 (2006), p. 9-13 (in Chinese)
[15] N. Li, L. Gao, Z. Yi and F. Zhai: Temperature Field Simulation of TC4/SiC Bilayer in Laser Forming. Solid State Phenomena, Vol.281 (2018), p. 946-951

Processing and Evaluation of Ceramic Filler Reinforced Polymer Matrix Composites

Azzam D. Hassan[a], Usama J. Naeem[b], Imad O. Bachi[c]

a. Materials Engineering Department, College of Engineering, University of Basrah.
b. Mechanical Engineering Department, College of Engineering, University of Basrah.
c. Materials Engineering Department, College of Engineering, University of Basrah.

Emad.bajee@uobasrah.edu.iq

Abstract. Tensile strength, Young modulus, impact strength, creep resistance, recovery measures, and thermal conductivity are used to evaluate the polymer matrix composite containing ceramic particles as degradation of spongy iron by oxidation to a ferric oxide which is used in steelmaking and low-density polyethylene. The results reveal that the mechanical and thermal properties are affected by the ratio of filler particles (360 μm). Six specimens were prepared and tested in time-dependent conditions at a constant temperature. As a result, increasing the filler ratio added to low-density polyethylene does not accelerate creep with nonlinear viscoelastic behavior. The composite with 0.7 percent ferric oxide particles was found to have the best thermal and mechanical qualities.

1. Introduction

The phrase "filler" can refer to a variety of things, such as fibers or particles. Minerals, metals, ceramics, bio-based materials, gases, liquids, and even other polymers are among the materials that can be employed. Fundamentals that affect filler qualities are known as filler properties. New fibers made of ultra-high molecular weight polyethylene were also introduced, with improved properties. Metals, ceramics, and polymers are all used in these composites. While fillers have a minor effect on mechanical properties, some miners increase stiffness while decreasing strength and elongation. Fillers are usually combined with liquid or molten polymers or applied to adhesive polymers to balance their coefficient of thermal expansion to that of the materials to be joined. The amount of filler, the surface behavior of the filler, and the state of preparation all influence the properties of the filled material. Local variations in mobility and special organization of macromolecular chains can be caused by small particles of a filler inserted into a polymer matrix. Adesina [1] discussed how various wastes can be used to create cementitious composites with improved characteristics. Thermal conductivity can be reduced by using recycled rubber and plastics, according to research. Almtori and et al [2] Because of the increased mechanical qualities, a polyethylene liner with a crosslink hard mulch waste tire was employed; the results suggest that the 85 percent ratio is the best. physical properties of hybrid composites reinforced with fillers are discussed in Matykiewicz [3] study. The epoxy matrix primarily improves the adhesion. Abdul Azam and et al [4] The influence of shell filler on the characteristics of linear polyethylene composites was examined. The characteristics of the composite have also improved as a result of the research. As fillers in composites, the dielectric characteristics of several species were examined. Riquelme and et al [5] The dielectric parameter for pure components and composites increase when polyvinylidene fluoride was used as a filler in ceramics. Durowaye and et al

[6] the branch polyethylene matrix composites with coconut and bone waste particles were made. At 15 wt. percent filler addition, properties of water absorption (0.3 percent) improved, suggesting decreased pores/voids, mechanical properties. Sikora and et al [7] were investigated the properties of branching polyethylene matrix using aluminosilicate (halloysite) clay nanotubes. The addition to composite improved heat deflection temperature and ultimate tensile strength, according to research. Kiran and et al [8] metal oxide fibers were used to evaluate of polymer matrix composites. Fillers additives improve the properties of polymer matrix composites. Almtori [9] Polyethylene and polyisobutylene scrap from tire tubes were examined. It was demonstrated that the best thermal insulation was achieved using 40 percent scrap tire tubes and 60 percent polymer, with a thermal conductivity , which was lower than the other. Mohsen [10] study a polyester with asbestos fibers was developed, and the effect of the additive on the thermal conductivity of the specimens was investigated. The results show a decrease in the property of composite materials as the weight percentage of asbestos fiber increases, as well as an improvement in the value of mechanical properties.

Hafed [11] looked into the physical properties, of cement mortar and sawdust composite in based on building components and cost. The mechanical value of the resistors increased proportionally to the amount of sawdust present. This improves mechanical properties while decreasing density and thermal conductivity, resulting in concrete with high thermal insulation complexes. Abdullah and et al [12] studied the epoxy composites reinforced with glass fibers and sandwiched , with 1, 2, and 3 layers of glass fibers and a volume fraction of 25%. In comparison to the other two composites, the property of composites increases as the number of layers increases. Plesa and et al [13] looked at the properties of particularly cross-linked polyethylene and epoxy, with various fillers. It was discovered that the use of micro/nanocomposites in electrical improved properties. Hemanth and et al [14] Glass, carbon, silicon carbide, and alumina fibers were used to reinforce a copolyester elastomer and polyoxymethylene filled polytetrafluroethylene composite. According to mechanical properties studies, short glass fiber improves composite strength while fillers reduce composite tensile and bending qualities.

Araújo and et al [15], Polyvinylidene fluoride composites with filler were used to test the models' suitability the composite dielectric conduct. Predict an increase in property as the filler rises. To obtain desirable properties, a mixture of polypropylene and low density polyethylene matrix composites was fabricated with kaolin ratio . Mustafa [16]. Mechanical properties increased, while strain at break decreased. In Hanna and et al [17] investigation, the ceramic particles to an unsaturated polyester resin. Mechanical properties and water absorption percent all improved. Almtori and et al [18] investigated the fiber glass polyester composite. The results showed that as the percentage of fiber glass in composites increased while the thermal conductivity of the composites decreased.

Recently, there has been an increasing interest in the possibility of altering polyolefins with natural fillers such as cereals, plant stems, nut shells, cellulose, or wood floor.

Daver and et al [19] investigated the effects of composition, compatibilizers, and carbon-nanotubes on the material's Deformation of viscoelastic. Various nanocomposites compositions were put to the test in terms of creep and recovery under continual load. Burger's four-element mechanistic model was applied to represent the composites' creep phase, while a Weibull distribution function was applied to describe the recovery phase of the composites. The goal of Riara and et al [20] study was to make LDPE/cellulose composites which have been compression molded, analyze the influence of concentration of cellulose on LDPE with cellulose reinforcement creep performance, and make a comparison to clean LDPE. The samples' long-term creep behavior (up to 106 s) was predicted using time temperature superposition. By moving The William-Landel-Ferry (WLF) model provided a better description of the shift factors based on short-term data that was used to forecast the long-term behavior of the polymer composites by displaying the curves along the logarithmic time axis. Both the neat resin and the composite material have viscoelastic behavior created under both dry and a wet circumstance was studied using flexural creep-recovery studies, Kontaxis and et al [21]. The impact of applied stress, immersion time, time creeps, and the total quantity of saline absorbed by composites flexural behavior of viscoelastic was investigated. Because of the cumulative effect of absorbed saline from the fibers deboning and easier relaxing of the macromolecules, as the immersion time extends, the creep strain increases. To get unusually wide indentation creep time from a piece of commercial polymer, an instrumented Indentation testing equipment with exceptionally The temperature drift was kept to a minimum (0.2 pm/second), Hou and Jennett [22]. The simplest measure of maximal dimensional change of polymeric material components subjected to long-term stress was the value of infinite modulus. It is better estimated using a new analytic method. The creep and stress relaxation characteristics of m-LLDPE, PP, and their mixes have been studied by Nouri [23] and to see if adding m-LLDPE to the PP affects the relaxation and retardation timings. Marklund and et al works [24] were focused on the macroscale material property modeling, hence microarchitecture and morphology aren't relevant to the results in the study. The material may be characterized with only a few specimens, and the nonlinearity functions that are stress-dependent can be described using Polynomial functions that are easy to understand.

Creep rupture experiments at greater fractions of the maximum load before failure are frequently used to determine the time required to generate failure in Batra's thesis [25]. For each test under a percentage of the rupture load at a specific temperature, creep data is obtained first in terms of extension/strain vs rupture time. These data from a large number of tests at various stresses and temperatures is then used to plot applied stress vs time to failure on a log-log scale, and the data trend can then be used to extrapolate time to failure for general design purposes under lower fractions of rupture loads. Lorandia and et al [26] study, The features of viscoelastic of the composites were examined using mechanical dynamic analysis and creep experiments utilizing Findley and Burger's methods at a different temperatures selected of (30, 60, and 90 Degree) and a loads of (2, 5 and 10 MN/m2). Weibull parameters derived from creep curves were also used to discuss viscoelastic qualities. The viscoelastic parameters derived from both models corroborated that the temperature and stress both influenced the deformation.

The goal of this study is to examine the thermal and mechanical properties of ferric oxide particles as a deterision added to low density polyethylene matrix composites, as well as the effect of ferric oxide particles on low density polyethylene matrix creep recovery behavior under static loading.

2. Experimental work

Experiments on correctly made specimens are highly beneficial for assessing the mechanical and thermal properties of composite materials. The data collected from these tests is put to good use in assessing the various properties of composites utilized in the design process.

3. Rheology & casting

After mixing, the composite was heated to 170°C and spun at 32 rpm on a Haake Rheochrder machine (see Figure 1) before being cast under pressure at 175°C for 9 minutes using a hydraulic tensile machine (see Figure 2) 10 tons. The stock weighed 60 grams and came in 30 * 30 cm2 square sheets

Figure 1. Haake Rheochrder machine

Table 1. The specimen's composite constituents.

Specimen number	polymer (gm)	Filler (gm)	Ratio %
1	60.00	0.00	0.00
2	59.70	0.30	0.50
3	59.58	0.42	0.70
4	59.40	0.60	1.00
5	59.10	0.90	1.50
6	58.80	1.20	2.00
7	57.00	3.00	5.00

Seven composite sheets with various filler ratios were prepared to evaluate the mechanical and thermal properties of the composite (0, 0.5, 0.7, 1.0, 1.5, 2.0, and 5.0) percent, see Table 1. The specimens were prepared using a ZWICK Germany puncher model ZCP 020 ZWICK Germany preparation tool. Tensile strength and Young's modulus were measured at room temperature using an INSTRON model 4411 machine with digital data acquisition capabilities, in accordance with ISO 527-1 and ASTM D638m.
This test was carried out in the marine college lab in Basra, Iraq, as illustrated in Figure 2.

Figure 2. The tensile machine of specimens

The tensile parameters were calculated using general equations.

$$\text{Stress}(\sigma) = \text{load/area} \quad \text{.............(1)}$$

$$\text{Strain}(\epsilon) = \Delta L/L \quad \text{.............(2)}$$

$$\text{Elastic of modulus } E = \sigma/\epsilon \quad \text{.......(3)}$$

The hardness of the composite specimens was determined using an ASTM D 2240-compliant Shore D hardness instrument model ZWICK. This test was carried out in the polymer testing lab of the Basra-Iraq petrochemical company. In addition, the ASTM D256 standard was used to measure the impact strength of the materials. Using a notching machine, each specimen is given an AV-notch with a 45° angle and a depth of 2.0 mm (TMI notch cutter model 43-15-1).
For each composite, three specimens are prepared and examined using the IZOD test process and can be calculated as:

$$S = \frac{E_i}{t(w-c)} \quad \text{.............(4)}$$

Where,
S= impact strength (KJ/m²)
E_i=impact energy (kJ)
t = thickness (mm)
w = width (mm)
c = notch width (mm)

The IZOD test was conducted at Basra-University Iraq's College of Science, Chemical Department Lab. The Lee's Disc apparatus, which was built in the faculty of engineering's laboratory of engineering materials with BS 4745, was used to perform the thermal conductivity test (see Figure 3).

Figure 3. The Thermal conductivity tester

The thermal conductivity (K) was estimated by using Equation below:

$$K = K'\frac{d(T_2 - T_0)(T_1' - T_2')A'}{d'(T_2' - T_0')(T_1 - T_2)A} \quad ....(5)$$

Where:
K'=0.23 w/m.k(Thermal conductivity of poly methyl metha acrylate standard specimen)
d'=6 mm (thickness of standard specimen)
A'=0.49 mm²(Area of standard specimen)
T'_1=100 °C(Input temperature)
T'_2=60 °C (Output temperature)
T_o=21 °C Room Temperature for the standard specimen.

In the experimental study, six large composite samples were subjected to a creep test. This test was conducted in the Maritime Academy's strength of materials laboratory, using a creep machine SM 1006 manufactured by TQ-company in England. Creep testing was carried out at room temperature with a static load. The specimens were loaded in the experimental ranges (0.5, 0.7, 1.4kg). Appropriately chosen loads for the specimen based on the mechanical behaviors of the specimen. Figure 4 depicts the dimensions of the specimen tested.

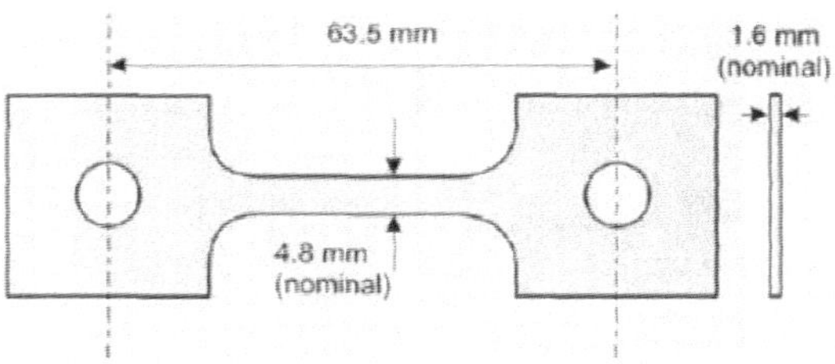

Figure 4. Dimensions of creep specimen

4. Result and Discussions

Filers have been discovered to have an effect on the mechanical and thermal characteristics of composite materials, and an appropriate ratio has been identified to improve mechanical and thermal properties.

Based on the load-elongation curve at yield point, breakpoint, and elastic modulus, differing percentages of filer cause differences in the material's tensile behavior.

As demonstrated in Figure 5, the stress at the yield increases before the filler ratio (0.7 percent) and subsequently drops as the filler ratio grows.

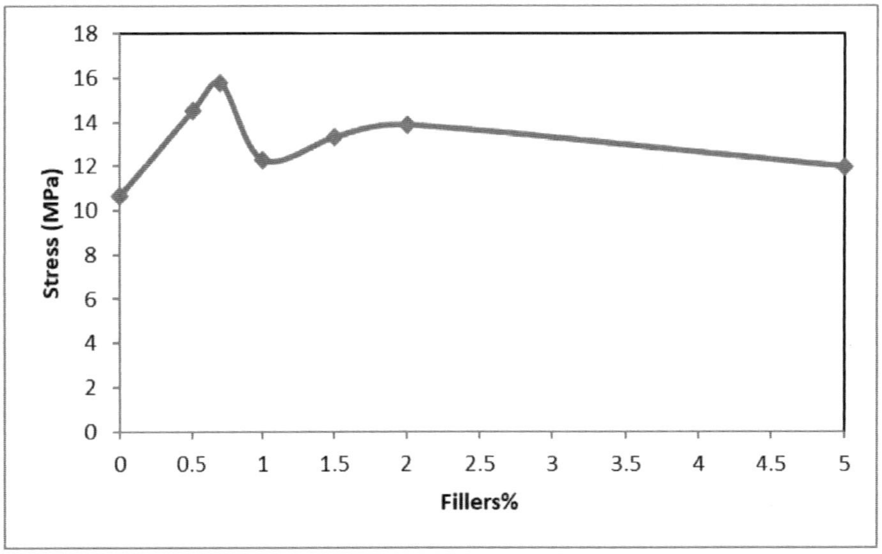

Figure 5. The tensile strength at yield point for the specimens

The behavior of the curve is due to a change in the repeating units of polyethylene, which slows chain motion and restricts molecule movement. The filler ratio has an impact on the yield stress value, as shown in Figure 5. The specimen with the (0.7 percent) ratio obviously has a high yield stress value. Figure 6 illustrates the relationship between the strain and the filler fraction. The (2 percent) ratio has the maximum strain at the yield point, whereas the (0.5 percent) ratio has the least. This statistic can also be used to deduce the specimens' nonlinear behavior, demonstrating that raising the filler ratio does not always imply increasing the strain.

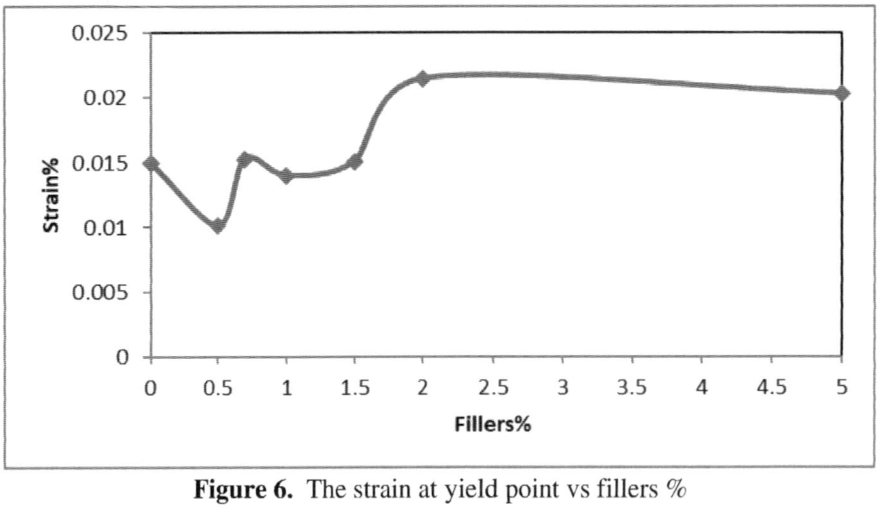

Figure 6. The strain at yield point vs fillers %

Figure 7 shows how the elastic modulus increases as the filler percentage increases, which is higher than that of the virgin specimen instead of 2 and 5percent, and then the elastic modulus decreases as the filler content increases.

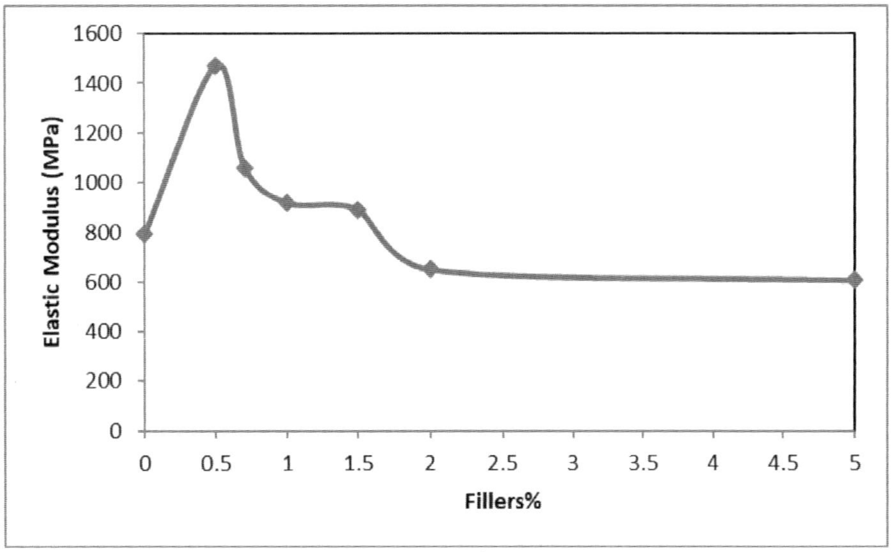

Figure 7. The modulus of elasticity versus fillers content

The (0.5 percent) amount, in contrast to the other filler material ratios, produces a very high elastic modulus. Finally, the graph that depicts the stress-strain relationship curve for all of the specimens in Figure 8 to fill out the tensile parameters. The fracture activity in relation to the ratio of the filler in the specimens is shown in this diagram, and it's clear that the fillers percentage influences the transition from brittle to ductile behavior.

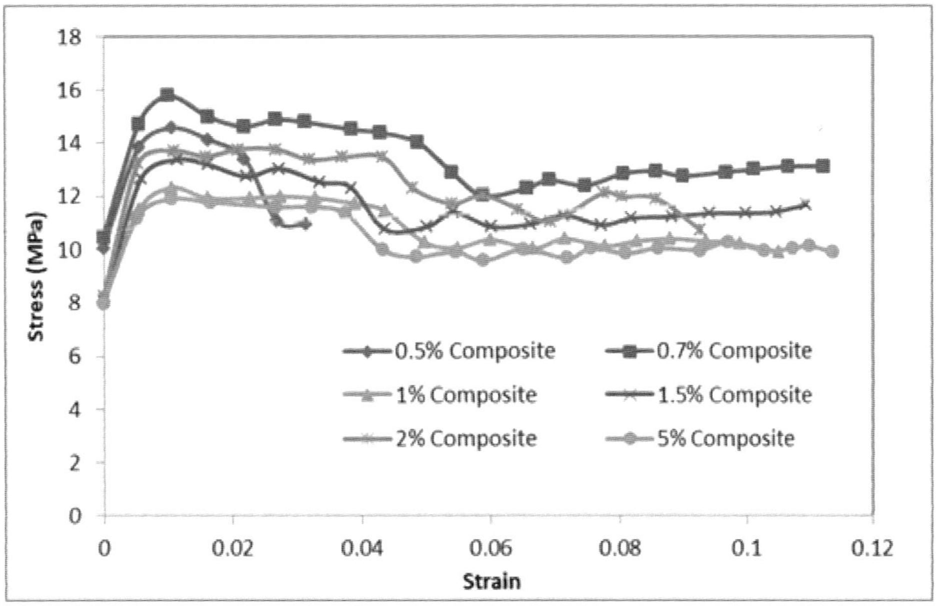

Figure 8. The Stress-strain behavior for the specimens

The branches in polyethylene limit the movement of the free end chain due to the influence of crystallization degree.

Figure 9 shows that the hardness of the specimens does not follow a uniform pattern, but they all have a greater shore D hardness than virgin polyethylene.

Figure 9 illustrates that, in contrast to virgin specimens, all specimens have a high hardness value due to the movement constraint of the free end chains.

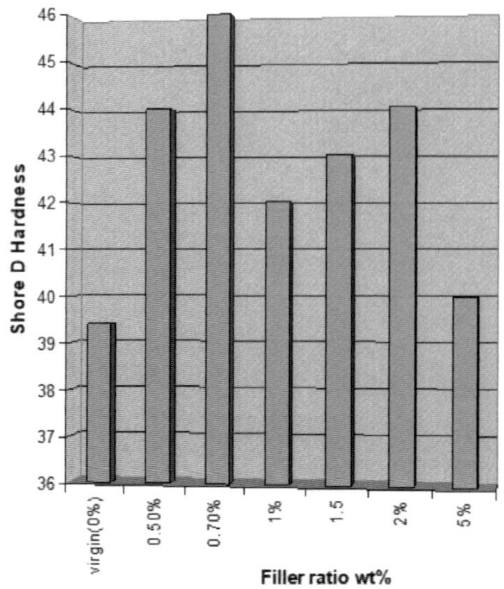

Figure 9. The relation between hardness vs. fillers%

When illustrated in Figure 10, the impact strength of the virgin sheet is greater than all specimens, and it diminishes as the fillers ratio increases due to reduced specimen flexibility or ductility, resulting in more brittle composite specimens.

This increase in impact strength is due to the polymeric chains becoming stiffer, limiting their mobility.

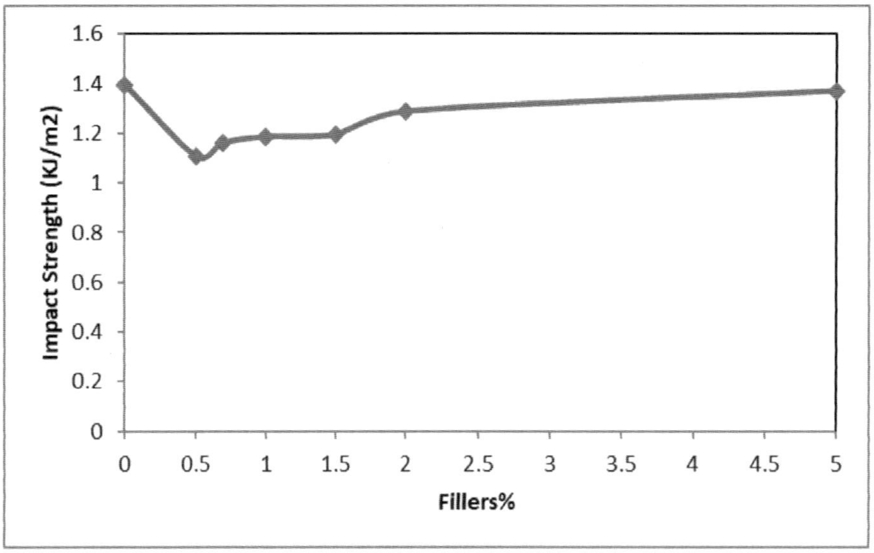

Figure 10. The relation of the fillers % vs. Impact strength

Figure 11 depicts the values of creep and recovery experiments tests. The values were shown have the same axis of extension of the composite specimens under the applied loads of 0.5kg. The 2 percent of the specimens displayed rapid creep activity, as can be seen.

After 180 seconds, the shift in extension becomes extremely small, and at 240 seconds, the imposed weight is has been removed from the hanger.
For a brief period of time (250 seconds), the recovery was swift but thereafter remained practically continuous. The specimen's minimal recovery occurred when the extension reached (2.13mm) after 300 seconds and the applied force was withdrawn; also, the recovery in the (0.7percent) specimen is quite little in comparison to the (2percent) specimen.
Furthermore, as demonstrated in figure 11, the creep in the (1.5%) specimen is lower than the creep in the 1and 5percent specimens. Figure 12 illustrates a comparison of (0.7 percent) specimens under various loads (0.5 and 1.4 kg). Figure 12 demonstrated that in the same specimen, the amount of extension recovery increases as the applied force increases, which is proven to be in good accord with Armstrong's findings [27].

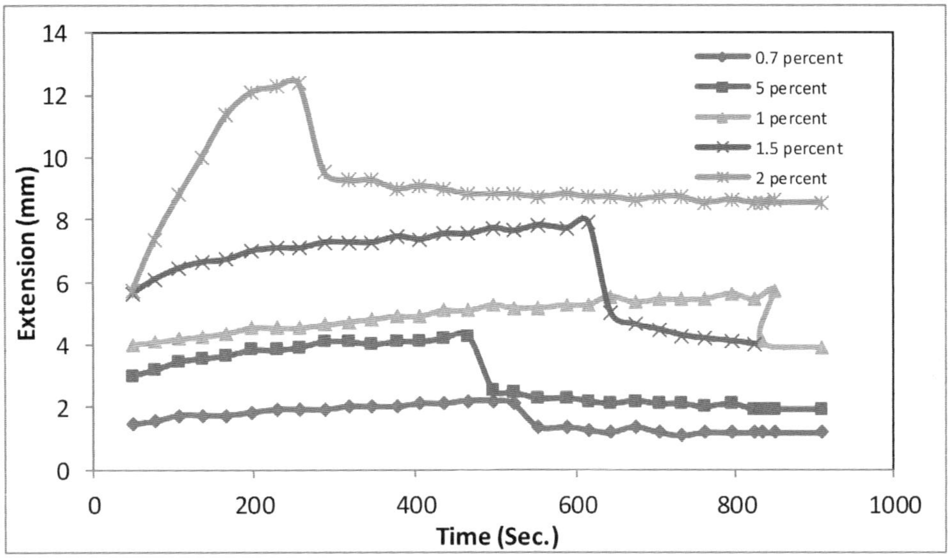

Figure 11. creep recovery behavior, load of 0.5 kg

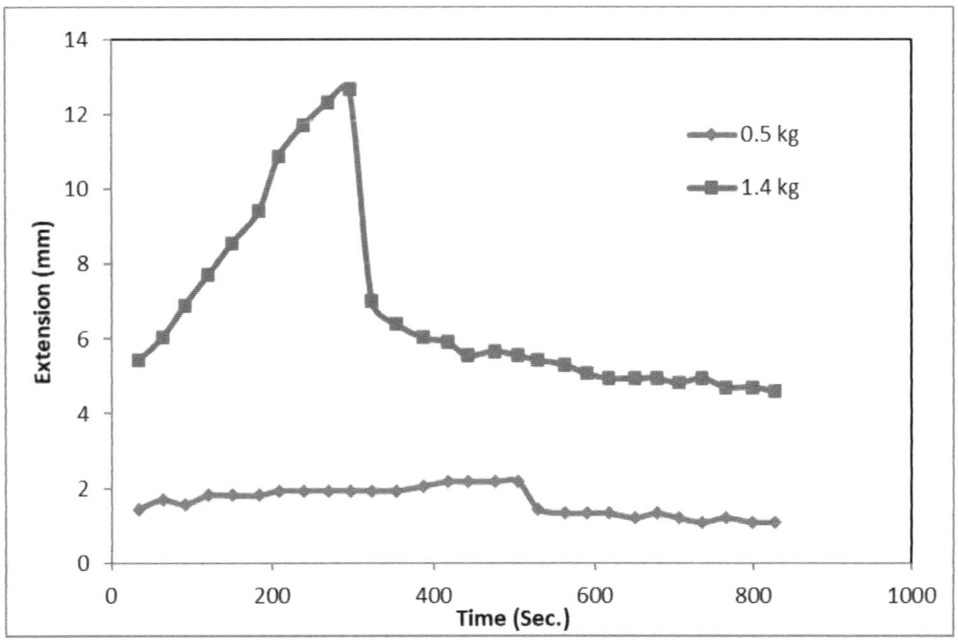

Figure 12. the 0.7% specimen creep recovery behavior.

Figure 13 shows creep behavior for (0.5 percent) specimens when compared to (0.5,1, and 1.5 percent) specimens when the load is constant of (0.7- kg). It concluded that increasing the filler in polyethylene does not speed up creep in composite specimens, implying that creep in composite specimens is viscoelastic in a nonlinear way.

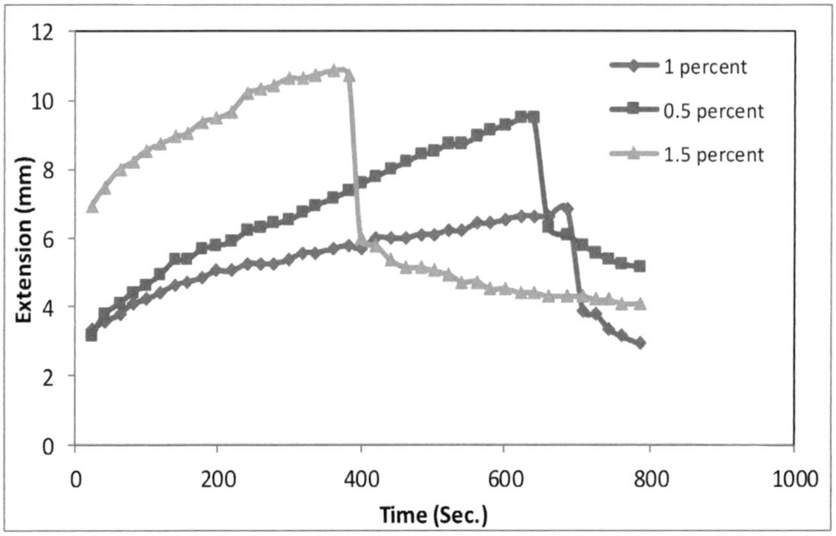

Figure 13. creep recovery behavior, load of 0.7 kg.

Figure 14 shows the log(strain) vs. the log(time) for the applied load (0.5-kg). The best correction equations for the relationship between the logarithm of the experimental transient strain and the logarithm of the transient creep time were used to construct the straight lines shown in Figure 14 at ambient temperature which represented as:

$$\varepsilon=\varepsilon_o+B\sigma^m t^k \text{ ....(6)}$$

Where

Ɛ, Strain of transient-creep

ε_o, Strain of Elastic

σ , Applied-stress (MN/m^2)

t, Time of creep (sec.)

m,k, Constants of stress and time

Equation (6) can be used to explain transitory creep in polymers.

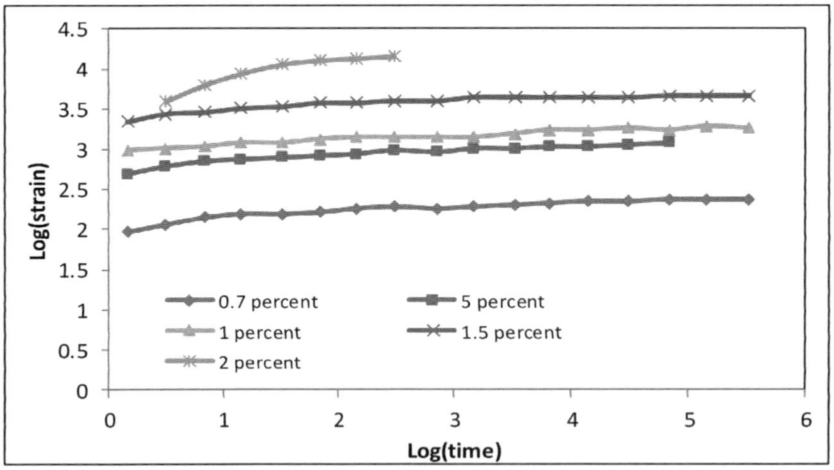

Figure 14. logarithm chart for the time and the strain

Table 2 shows the creep recovery (P.C.R) and the time constant measured using a 0.5kg applied load. The time constant for each specimen is the slope of the straight lines in Figure 14. To measure percentage creep recovery, the PCR equation given as:

P.C.R= ((Creep recovery)/(max. extension – elastic extension))*100% 7

P.C.R= ((2.13-1.1-0.3)/(2.13-0.3))*100% = 39.89 %

Table 2. composite properties.

Ratio%	Time constant ,seconds	PCR%
0.7	0.020	39.90
1	0.015	28.99
1.5	0.013	47.30
2	0.086	28.70
5	0.021	50.20

From Table 2 , can be concluded that for the (2 percent), the time constant of the specimen is the highest, and the creep recovery is the lowest.
The thermal conductivity of composites is a measurement of how well they transport heat.
As illustrated in Figure 15, the influence of fillers on composite thermal conductivity was discovered, with thermal conductivity reducing as the weight % of ceramic filler material rose due to an increase in the polymer matrix's insulating characteristic.

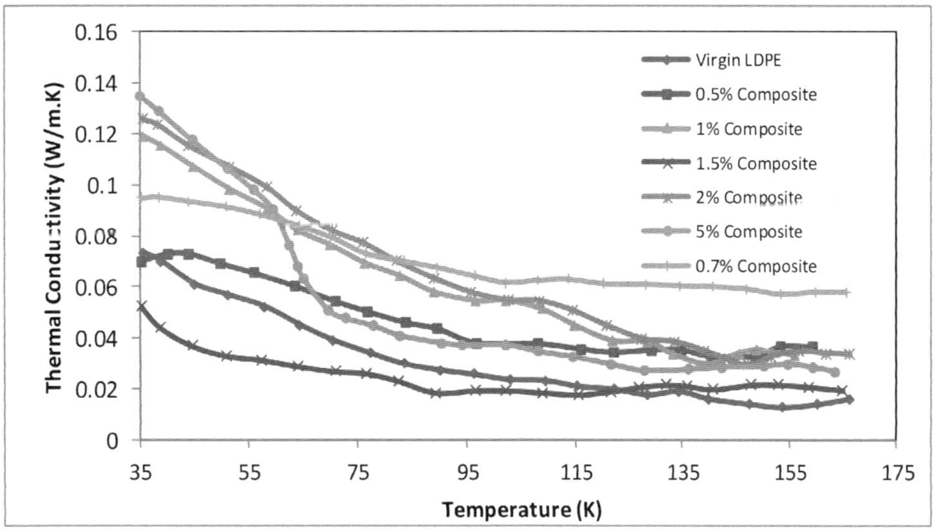

Figure 15. depicts the relationship between thermal conductivity and temperature.

Composites are greater thermal insulators than virgin specimens, which explains why heat conductivity decreases as absorption increases due to the particle size of ceramic fillers. As a result of the insulation at the hot face and variances in the heat flow of the composites, heat transfer absorption in the initial layers occurred.
Thermal conductivity was significantly affected by the 0.7 percent filler material.

Finally, in Figure 15, the virgin and composite specimens are compared. The experimental results may be proved to be superior than the virgin specimen because the fillers reduce thermal conductivity and absorb vibration, which is used to transport heat.

5. Conclusions

It's worth noting the following conclusions:

1. The thermal conductivity for the filler materials decreases as the filler ratio rises and 0.7 percent filler shows a significantly affected.
2. Increasing the filler ratio to the materials increases the value of hardness compared with the virgin sample.
3. The tensile test parameters for the composites were improved and give a higher value from the virgin sample.
4. The (2 percent) specimen has a quick creep rate and the greatest elastic extension.
5. The (0.7 percent) specimen has seen the least amount of creep.
6. The sum of extension recovery for the composite increases as the applied load increases.
7. Creep is not sped up by increasing the filler ratio. Nonlinear viscoelastic materials cause creep in composite specimens.
8. The overall constant time and minimum recovery of creep is found in the sample 2percent, while the maximum creep recovery is found in the 5percent specimen.

6. References

[1] Adeyemi Adesina, "Overview of the influence of waste materials on the thermal conductivity of cementitious composites", Journal of Cleaner Engineering and Technology (2021), https://doi.org/10.1016/j.clet.2021.100046.

[2] Safaa A.S. Almtori, Imad O. Bachi Al-Fahad, Atheed Habeeb Taha Al-temimi, A.K. Jassim, "Characterization of polymer-based composite using neuro-fuzzy model", Journal of Materials Today (2021), https://doi.org/10.1016/j.matpr.2020.12.238

[3] Danuta Matykiewicz, "Hybrid Epoxy Composites with Both Powder and Fiber Filler: A Review of Mechanical and Thermomechanical Properties", Journal of Materials (Basel) (2020), https://doi.org/10.3390/ma13081802

[4] Farah Atiqah Abdul Azam, Nishata Royan Rajendran Royan, Nor Yuliana Yuhana, Nabilah Afiqah Mohd Radzuan , Sahrim Ahmad and Abu Bakar Sulong," Fabrication of Porous Recycled HDPE Biocomposites Foam: Effect of Rice Husk Filler Contents and Surface Treatments on the Mechanical Properties", Journal of polymers (2020), https://doi.org/10.3390/polym12020475

[5] S.A.Riquelme, KoduriRamam, A.F.Jaramillo," Ceramics fillers enhancing effects on the dielectric properties of poly(vinylidene fluoride) matrix composites prepared by the torque rheometer method", Journal of Results in Physics 15, (2019).

[6] Stephen Durowaye, Babatunde Bolasodun, Wasiu Ayoola, Aishat Olarinde andEniola Apena, Kufa Journal of Engineering 10 (3), 134-150 (2019).

[7] Janusz W. Sikora, Ivan Gajdoš and Andrzej Puszka," Polyethylene-Matrix Composites with Halloysite Nanotubes with Enhanced Physical/Thermal Properties", Journal of polymers (2019), https://doi.org/10.3390/polym11050787

[8] M.D.Kiran, H.K.Govindaraju, T.Jayaraju, NithinKumar," Review-Effect of Fillers on Mechanical Properties of Polymer Matrix Composites", Journal of Materials today (2018), https://doi.org/10.1016/j.matpr.2018.06.611

[9] Safaa A. S. Almtori, "Experimental and Theoretical Investigation of the Physical Properties of new Composite Materials", Journal of Materials Science and Engineering (2018).

[10] Zainab H. Mohsein, " Study the Characteristics Thermal and Mechanical for Unsaturated Polyester Resin Forced by Asbestos Fiber", Engineering and Technology Journal 36 (A3), (2018), http://dx.doi.org/10.30684/etj.36.3A.6

[11] Sanaa A. Hafed, " Study of thermal insulation and some mechanical properties for hybrid composites (cement – wood sawdust)", International Journal of Computation and Applied Sciences IJOCAAS 3 (2), (2017).
[12] Hind W. Abdullah, Olfat A.Mahmood, Alaa J. Kadham Algidsawi," The Effect of Thickness and Numbers of Layers On Thermal and Hardness Properties of Epoxy Reinforcement Fiber Glass Composite", Journal of Babylon University 25 (3), (2017).
[13] Ilona Plesa, Petru V. No¸tingher, Sandra Schlögl, " Properties of Polymer Composites Used in High-Voltage Applications", Christof Sumereder and Michael Muhr, (2016), https://doi.org/10.3390/polym8050173
[14] R. Hemanth, M. Sekar, and B. Suresha, "Effects of fibers and fillers on mechanical properties of thermoplastic composites", Indian Journal of Advances in Chemical Science 2, 28-35 (2014).
[15] M. Carvalho Araújo, Carlos Miguel Costa, Senentxu Lanceros-Méndez, "Evaluation of dielectric models for ceramic/polymer composites: Effect of filler size and concentration", Journal of Non-Crystalline Solids 387, 6–15 (2014), https://doi.org/10.1016/j.jnoncrysol.2013.12.005.
[16] Samir Nassaf Mustafa , Diyala journal of Engineering Sciences 5 (2), 162-178 (2012).
[17] Waleed Asim Hanna, Faiza E. Gharib, Ismail I Marhoon, Journal of Minerals and Materials Characterization and Engineering 10(12), 1167-1178 (2011).
[18] Safaa A. S. Almtori , Abdul khaliq A. Hasan and Maitham Y. Atshan , Journal of Basrah Researches (Sciences) 36 (2), (2010).
[19] Fugen Daver, Mladenko Kajtaz , Milan Brandt, and Robert A. Shanks, "Creep and Recovery Behaviour of Polyolefin-Rubber Nanocomposites Developed for Additive Manufacturing", Polymers (2016), https://doi.org/10.3390/polym8120437.
[20] Martin M. Riara, Abdallah S.Merenga, and Charles M. Migwi," Creep and Recovery Behavior of Compression Molded Low Density Polyethylene/Cellulose Composites", Journal of Polymers (2013), https://doi.org/10.1155/2013/209529.
[21] L. C. Kontaxis, A. Georgali, D. V. Portan, and G. C. Papanicolaou, "Flexural creep behavior of epoxy/cotton composite materials before and after saline absorption for orthopedics applications", AIP Conference Proceedings 1932 (1), (2018), https://doi.org/10.1063/1.5024194.
[22] Hou, X. & Jennett, N. M, "Comprehensive investigation of the viscoelastic properties of PMMA by nanoindentation", Polymer testing (2018), https://doi.org/10.1016/j.polymertesting.2018.07.022
[23] M. Razavi-Nouri," Creep And Stress Relaxation Behavior Of Polypropylene, Metallocene-Prepared Polyethylene And Their Blends", Iranian Journal of Chemical Engineering 9 (1), (2012).
[24] Erik Marklund, Johannes Eitzenberger, Janis Varna," Nonlinear viscoelastic viscoplastic material model including stiffness degradation for hemp/lignin composites", Composites Science and Technology (2008), https://doi.org/10.1016/j.compscitech.2008.03.011.
[25] Saurabh Batra, " Creep rupture and life prediction of polymer composites", Thesis, College of Engineering and Mineral Resources at West Virginia University (2009), https://doi.org/10.33915/etd.2035.
[26] Natália Pagnoncelli Lorandia, Maria Odila Hilário Cioffi, Carlos Shiguec, Heitor Luiz Ornaghi," On the creep behavior of carbon/epoxy non-crimp fabric composites", Materials Research (2018) , https://doi.org/10.1590/1980-5373-MR-2017-0768.
[27] William D. Armsrtong, Vipin Kumar," A discrete complex compliance spectra model of the nonlinear viscoelastic creep and recovery of microcellular polymers", Journal of Polymer Science (2000), https://doi.org/10.1002/(SICI)1099-0488(20000301)38:5<691::AID-POLB6>3.0.CO;2-E.

Particle Generation from Aluminum Alloy Surfaces Irradiated by 355nm Pulsed Laser and Multi-shots Laser Pretreatment Method at Low Fluences

Jihua Zhang[1], Yupu He[1], Naihao Guo[2], Hongxu Chu[3], Jingxuan Wang[4,*]

[1]Engineering Training Center, Chengdu Aeronautic Polytechnic, Chengdu, China.
[2]Beijing Institute of Space Mechanics & Electricity, China Academy of Space Technology, Beijing, China.
[3]School of Aeronautical Maintenance Engineering, Chengdu Aeronautic Polytechnic Chengdu, China.
[4]Reserach Center of Laser Fusion, China Academy of Engineering Physics, Mianyang, China.

Corresponding author's e-mail: 373759703@qq.com

Abstract. The studies focus on particle generation from aluminium alloy surfaces irradiated by a Q switched Nd:YAG laser at 355 nm wavelength at low fluencies. The particles number increase with laser fluence but reduces with pulse number. The average initial threshold laser fluence for particles generation is 26.64 mJ/cm2. Larger particle with a diameter more than 5μm are found for fluencies above 76.54 mJ/cm2. After 40 pulsed laser shots at 72 mJ/cm2, it can effectively reduce the generation of metal particles. Our results demonstrated that 7075 aluminum alloy has the best laser resistance and the method of multi-shots laser pretreatment can reduced the generated particles by 76.92%. The SEM results shows that laser pretreatment process lead the passivation effct which smoothed the microstructure on the initial surface. The XPS results shows that zinc aluminate was generated during laser pretreatment process which has good thermodynamic properties. This study has important value for reducing metal particle contamination due to laser irradiation on metal surface in high power laser facilities and improving the service life of optical components.

1. Introduction

Aluminum alloy (AA) have been widely used in high power laser facilities as optical transmission vacuum pipes, optical component holders or diaphragms[1]. Unfortunately, if the problems of scattering light or ghost images are improperly handled in the pratical working of high power laser facilities, the scattering laser will inevitably beam on the aluminum alloy surface[2-3]. Then the metal particles sputter out from the surface and the detached particles may adhere to the optical component surface. This contaminants strongly absorb laser energy to cause optical surface damage[4-5]. This damage on optical components is fatal to high power laser facilities[6-7]. The study of metal particle generation has great research significance and application value for the long-term, stable and efficient operation of laser facilities.

The mechanism of laser interaction with metals has attracted growing interest in recent years. Most studies on the laser surface treatment of metals employed different kinds of lasers and the laser energy is usually limited to high energies to improve specific performance[8-9]. Wojciech Pakieła, Tomasz Tanski, Mirosława Pawlyta, et al had reported the structure and mechanical properties of AlMg5Si2Mn alloy after surface alloying using fiber laser[10]. Excimer laser surface treatment of aluminum alloy was reported by T.M. Yue, L.J. Yan, C.P. Chan, et al about the improvement of AA7075 corrosion resistance[11]. Our preliminarily studies on pulsed laser interaction with AA6061 at low fluences have shown that the particle generation behavior of the aluminum alloys was very different from irradiated at high fluences.

In this paper, we report the result of particle generation from aluminum alloy surfaces irradiated by pulsed laser at low fluences to simulate the scattering laser in high power laser facilities. In order to reduce the number of generated particles from aluminum alloy surface, a multi-shots laser pretreatment method was proposed.

2. Experimental

The experimental samples were prepared using AA5083/6061/7075 specimens, measuring 100mm*100mm*5mm. When preparing the sample required for the experiment, the specimens were cleaned in absolute ethyl alcohol for 5 minutes using an ultrasonic cleaner. After the alcohol was fully volatized, the specimens were obtained.

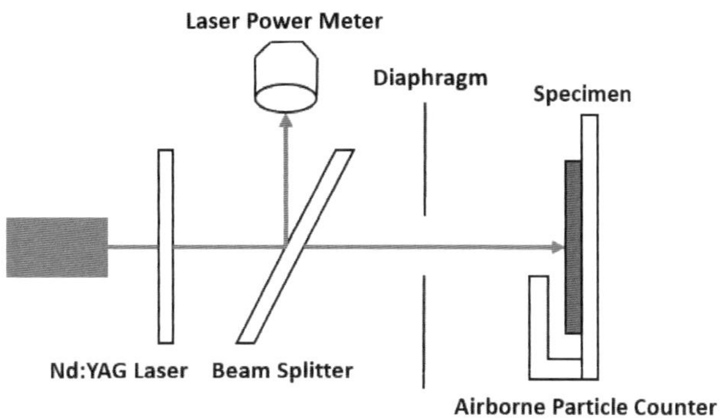

Figure 1. Schematic diagrams of experimental setups.

Figure 1 shows the schematic diagrams of experimental setups for scattering laser induced metal particle contamination experiment on aluminum alloy surface in the atmosphere. A Q switched Nd:YAG pulsed laser was operated at 355nm wavelength. The laser energy per pulse and pulse duration were 5 mJ ~340 mJ and 10 ns, respectively. The spot area was 75 mm2 (ellipse, the long and short axis was 6.3x3.8 mm @1/e2) and the laser fluence used in the experiment is calculated as 5.74 mJ/cm2~454.20 mJ/cm2 to simulate the scattering laser in high power laser facilities. Depending on the purpose of the experiments, an airborne particle counter (flow rate of 40 L/min) was placed on the underside of the substrate.

In the first set of experiments, the effect of particle generation was verified. The cross-control experiments were carried out on the diameter and quantity of the generated particles, using different laser fluence and laser pulses with AA5083/6061/7075 specimens. In the next set of experiments, the laser pretreatment method was verified and the experimental conditions and setups were the same as the first experiments. The generated particles form the specimens were examined by an airborne particle counter and analyzed by data platform software. All the experiments were performed in clean room under ISO class 5 atmosphere condition.

3. Results and Discussions

3.1 Variation of the generated particles with laser process parameters

In order to study the effect of different laser fluencies on the number and diameter of generated particles, the experiment was conducted on AA6061 specimen. The results were summarized in Figure 2.

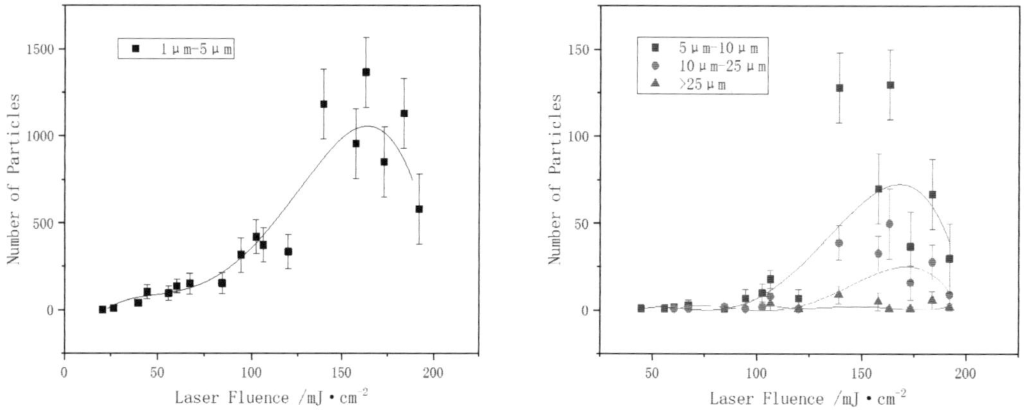

Figure 2. The effect of different laser fluence on the number and diameter of generated particles in AA6061 specimen focused on particles diameter (a) 1μm-5μm (b) larger than 5μm

Figure 2a shows that the number of generated particles is significantly positively correlated with laser fluence. In the operation of high-power laser facilities, particles larger than 1μm can cause obvious laser damage to optical components. So this study focused on particles of this size. Particles of 1μm-5μm accounted for about 90% of the total generated particles. When the laser fluence is 26.64 mJ/cm2, AA6061 specimen began to generate metal particle contamination. Larger particle with a diameter more than 5μm are found for fluencies above 76.54 mJ/cm2.The larger the particle diameter was, the larger the particle generating threshold. Peak number of particles was got at 163.08 mJ/cm2.

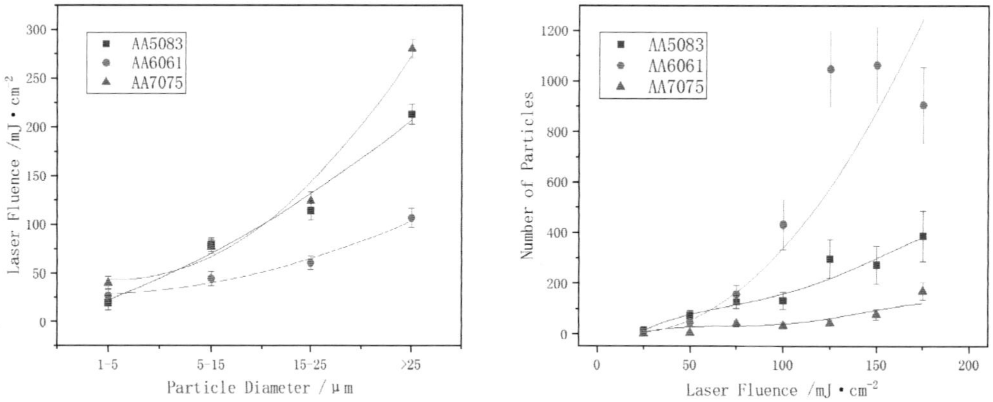

Figure 3. The comparison of (a) generating threshold and (b) particle numbers of AA5083/6061/7075 specimen

As shown in Fig. 3, the overall law of irradiated particle generation characteristics of three materials is consistent. The average initial generating threshold of the three specimens was 28.58 mJ/cm2(Figure 3a). The AA6061 specimen had the lowest generating threshold. For the particles large than 5μm, the generating threshold of 6061 aluminum alloy was only 50.93% -57.23% of the other specimen. For the particles of all sizes, AA7075 specimen had the highest generating thresholds. It can be seen from Figure 3b that the AA7075 was always the specimen with the least number of generated particles at each different laser fluencies.

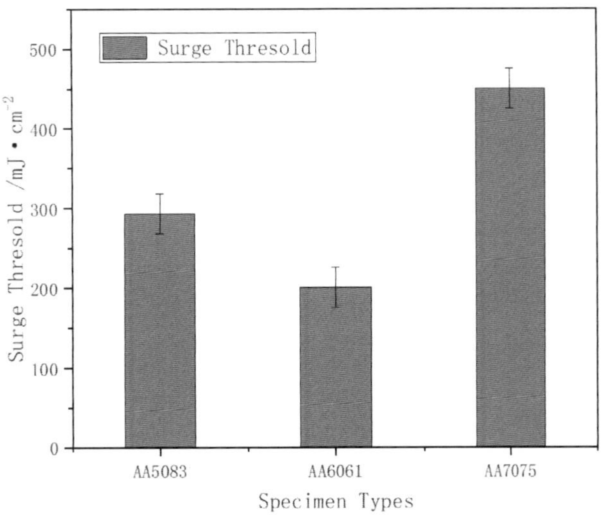

Figure 4. The surge thresholds of AA5083/6061/7075 specimen

During the experiment, it was found that when the laser fluence increased to some certain values, the number of generated particles would sharply increase by one order of magnitude and this growth can be defined as the surge. The surge thresholds of three specimens are shown in Figure 4. The AA7075 specimen had the highest surge threshold about 450 mJ/cm2 and the AA6061 specimen had the lowest surge threshold about 200 mJ/cm2.

The above experiments show that AA7075 specimen had the highest generating thresholds with all particle sizes, the least number of generated particles and the highest surge threshold in three aluminum alloys. These results show that AA7075 specimen has obvious effect to reduce the generated particles irradiated at low fluences.

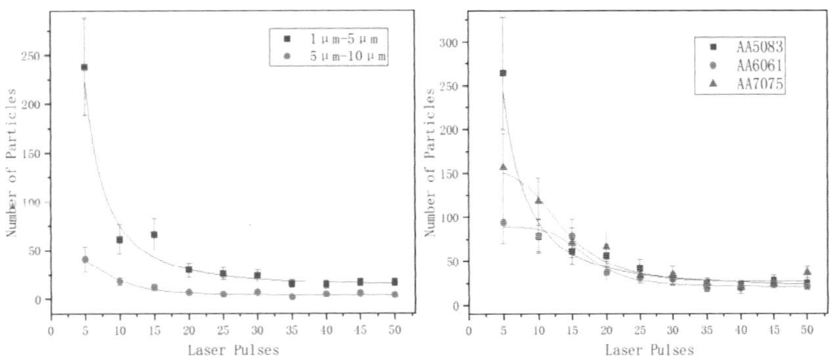

Figure 5. The particle generation of (a) AA6061 specimen and (b) all specimens after multi-shots pulsed laser at 72 mJ/cm2

Previous studies have shown that the scattering laser fluence is always below 100 mJ/cm2 in high-power laser facilities. Therefore, 72 mJ/cm2 was chosen as the preferred laser fluence in the next experiment. As shown in Figure 5a, the diameter of generated particles was concentrated within 5μm in 5 shots. When the laser pulses were issued within 20 shots, the number of generated particles was sharply reduced and was one order of magnitude less than before. This reduction rate was slowed down after 20 shots. After 40 shots, particle number remained substantially stable. The generated particle number was reduced to 3.5% at this time. Figure 5b shows that the generated particle number of three specimens after multi-shots processing was basically the same.

This result is consistent with the practical operation of the high power laser facilities. Characterizatic methods shows that after years of operation of facilities, there are fewer and fewer metal particles on the surface of the optical components. This is not only related to the gradually improving stray light control methods, but also to this multi-shots processing.

3.2 The method of multi-shots laser pretreatment

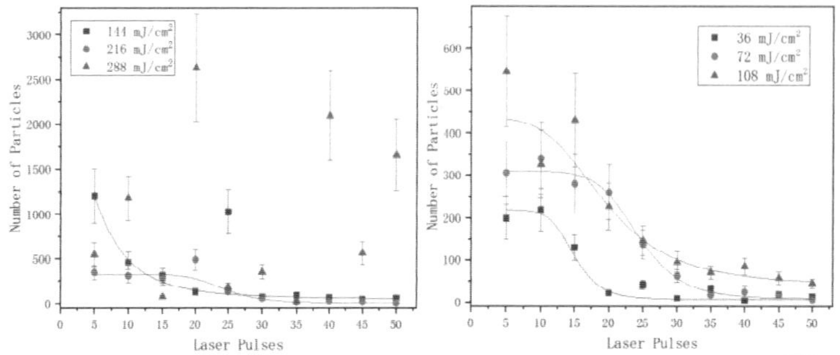

Figure 6. (a) The first shot in laser pretreatment at different fluences (b) The subsequent 40 shots at different fluences of AA6061 specimen

For the AA6061 specimen with a surge threshold of 200 mJ/cm2, single-pulse laser at 144 mJ/cm2, 216 mJ/cm2, and 288 mJ/cm2 was used to irradiate the aluminum alloy surface. And then 40 shots laser were processed at 72 mJ/cm2 (Figure 6a). The laser pretreatment with the first shot at 216 mJ/cm2 which is about 10% higher than the surge threshold, had the least generated particles. The generated particle number gradually decreased with the increase of laser pulses, and eventually stabilized at 40. Under the condition that the laser fluence of the first shot was unchanged, the higher the laser fluence of subsequent 40 shots, the more metal particles were generated (Figure 6b). In this study, 72 mJ/cm2 was chosen as the preferred laser fluence for the subsequent 40 shots to simulate the scattering laser in high-power laser facilities.

After irradiating the aluminum alloy surface with a laser fluence 10% higher than the surge threshold, and then 40 shots laser treatment at preferred 72 mJ/cm2, the number of generated particles is far less than before this processing. This is the laser pretreatment method.

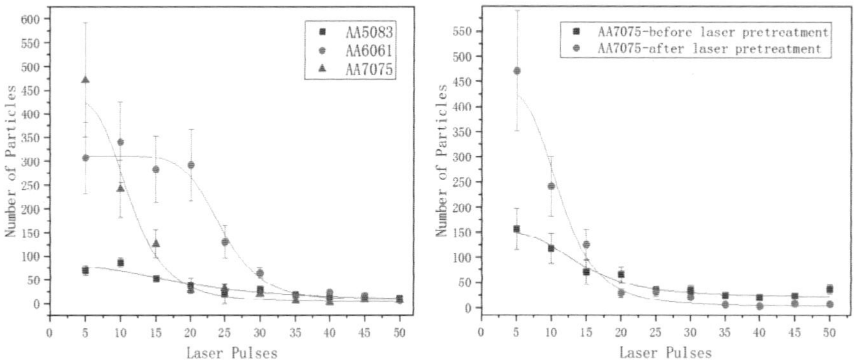

Figure 7. (a) The comparison of laser pretreatment of AA5085/6061/7075 specimens (b) Particle generation of AA7075 before and after laser pretreatment

Figure 7a shows that the AA7075 specimen produced the least number of metal particles about only 10. The particle diameter was less than 5μm. The laser resistance of AA7075 specimen was improved the most by laser pretreatment and it produced the least metal particles. Compared with 40 shots without laser pretreatment at 72 mJ/cm2, this method reduced the generated particles by 76.92% of the AA7075 specimen (Figure 7b).The results shows that this laser pretreatment method can effectively reduce the generation of metal particles.

3.3 Discussions of SEM and XPS Results

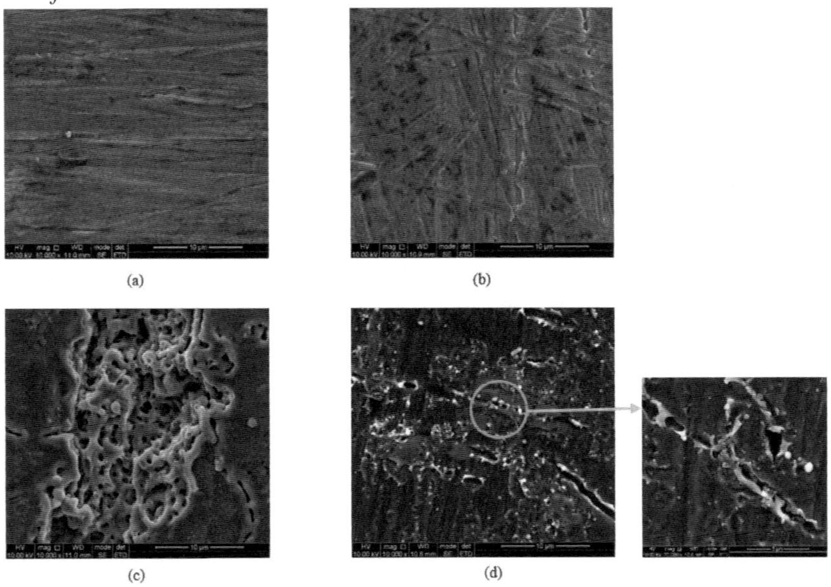

Figure 8. The SEM results of AA7075 specimen at each processing stage

As shown in Figure 8a, the initial aluminum alloy surface was smooth and there were evenly distributed gullies on the surface. Figure 8b shows that when the laser fluence reached the generating threshold, the gullies were more cluttered. Micrometer-sized gaps can be observed. This was caused by detached particles from the surface due to the laser thermal effect. When the laser fluence reached the surge threshold, honeycomb ablative pit and loose structures appeared on the surface (Figure 8c). These structures detached form surface during laser processing. This was what caused the sharply rising in generated particle numbers. Figure 8d showed that after the laser pretreatment, the ablative pits were repaired and bulbous structures were formed due to the plastic extrusion bulge. A large number of dimple and micropores appeared on the surface. The loose structure had been detached from the surface of the aluminum alloy.

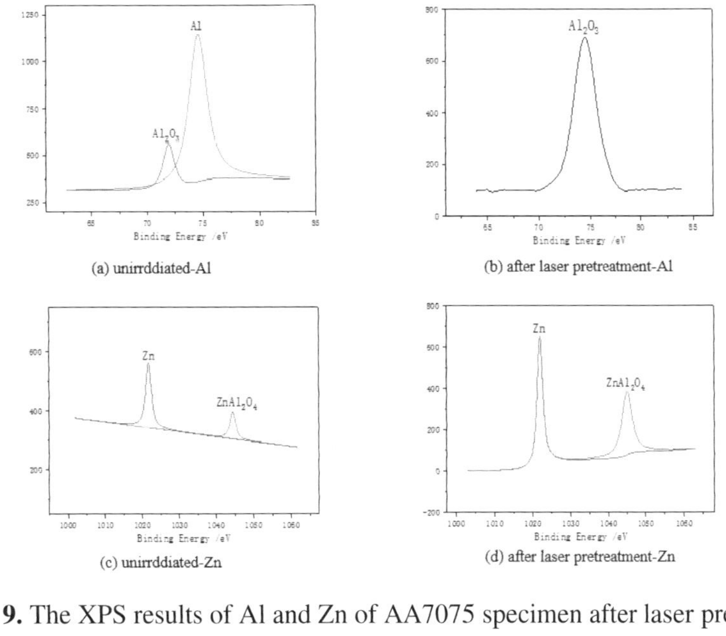

Figure 9. The XPS results of Al and Zn of AA7075 specimen after laser pretreatment

As shown in Fig. 9a and 9b, the aluminum on the sample surface had completely converted into aluminium oxides. Based on the SEM results, it can be found that low fluence pulsed laser caused passivation on aluminium alloys. The aluminium oxide film separated the substrate from the environment and prevented the alloy particles shedding during laser irradiation. 7075 aluminum alloy is based on Al-Zn-Mg-Cu system, in which the zinc content is usually 5%, far more than other alloy composition. Fig. 9c and 9d show that large amount of zinc aluminate was generated on the sample surface after laser pretreatment. The spinel structure of zinc aluminate has excellent mechanical and thermal properties, which effectively reduced the generated particles. This results shows that the laser pretreatment method has obvious effects to improve the laser resistance of aluminum alloy.

4. Conclusion

This work demonstrates that a Nd:YAG laser ($\tau = 10$ ns, $\lambda = 355$ nm) was used to simulate the metal particle generation form5085/6061/7075 aluminum alloys surface irradiated by scattering laser in high power laser facilities. Under the laser irradiation at low fluences, 7075 aluminum alloy has the least generated particles. In addition, the results of SEM and XPS confirmed that the multi-shots laser pretreatment method can further reduce the generated particle from aluminum alloy surface in pratical operation in high-power laser facilities.

5. References

[1] Liu B L and Xie F X 2019 Research on position control system of mechanical structure of high power laser device. *Laser Journal.* **40**(7): 152-155.

[2] Patterson B A and Wells M 2003 Analysis of Scattered Light for VISTA. SPIE, 4842: 116-127.

[3] Roni S, Steve B, Michael M, etc. 2003 Stray radiation in airborne FLIR system. SPIE, 4820: 676-687.

[4] Manenkov A A. 2008 Fundamental mechanisms of laser-induced damage in optical materials: Understanding after 40-year research. SPIE, 7132: 713202.

[5] Genin F Y, Michlitsch K, Furr J, et al. 1997 Laser-induced damage of fused silica at 355 and 1064 nm initiated at aluminum contamination particles on the surface. SPIE, 2966: 126-138.

[6] Sommer S C, Stowers I F and Van Doren D E. 2003 Clean construction protocol for the national ignition facility beampath and utilities. *Journal of the IEST*, **46**(1): 85-97.

[7] Pitts J H. 1988 Structural damage and analysis of the NOVA final focusing lenses. *Damage in Laser Materials*, **1**(9): 775-781.

[8] Chi Y M, Gu G C, Yu H J and Chen C Z 2018 Laser surface alloying on aluminum and its alloys: A review. Optics and Lasers in Engineering, 100: 23-37.

[9] Butt M Z, Ammal Javed, M. Waqas Khaliq, et al. 2017 Impact of 1064 nm–10 ns pulsed laser on the surface morphology, structure, and hardness of Pd80Ni20 alloy. *Int J Adv Manuf Technol*, 90:1857–1869.

[10] Wojciech Pakieła, Tomasz Tanski, Mirosława Pawlyta, et al. 2018 The structure and mechanical properties of AlMg5Si2Mn alloy after surface alloying by the use of fiber laser. J. *Applied Physics A*, 124:263.

[11] Yue T M, Yan L J, Chan C P, et al. 2004 Excimer laser surface treatment of aluminum alloy AA7075 to improve corrosion resistance. J. *Surface and Coatings Technology*, 179:158–164.

Magnetic anisotropy of a selective laser melted Fe-Si-B alloy

DANIL ERUTIN[1, a *], EVGENII BORISOV[2,b], ANATOLY POPOVICH[3,c], VADIM SUFIIAROV[4,d]

Russia, 195251, St.Petersburg, Polytechnicheskaya, 29

[a]erutin@inbox.ru, [b]evgenii.borisov@icloud.com , [c]popovich_aa@spbstu.ru, [d]vadim.spbstu@ya.ru

Abstract. Selective laser melted 1CP alloy sample was obtained. Sample is partially amorphous, otherwise it consist of α–Fe(Si) and ordered Fe3Si. In the direction parallel to the axis of the sample the coercivity is slightly lower than in the perpendicular direction and the saturation magnetization is greater. Magnetic anisotropy in sample is not weakened by held annealing: heating of 10 C°/min to 440C°, holding for 30 minutes, followed air cooling. After heat treatment coercivity in the direction parallel to the sample axis decreased from 50 Oe to 38 Oe (34%).

1. Introduction.

Selective laser melting (SLM) is an additive manufacturing technology of layer-by-layer melting of powder layer of material with thickness of 20-60 μm with the laser beam energy [1-4]. Nowadays, there is one very interesting application of the SLM-technology: production of soft-magnetic bulk metallic glasses.

Metallic glasses (amorphous metal) are the solid metals without a long-ranged atomic ordering [5]. Such material condition can be obtained with high-rate cooling from the liquid via fixation of atomic positions in initial state. Amorphous metal have no crystalline lattice and it exists without the typical crystal defects which are the source of properties anisotropy. Ferromagnetic amorphous metal exhibit better magnetic properties than ferromagnetics with crystalline lattice: lower coercivity, higher magnetic permeability, saturation magnetization and electrical resistivity. Better level of soft magnetic properties of amorphous ferromagnetics makes them perfect for using as a magnetic core of electromagnetic device because of a small values of magnetic field energy losses [6].

The problem of obtaining partially amorphous samples from Fe-based soft magnet alloys using selective laser melting has being investigated by many researchers [7, 8, 9, 11–18]The efficiency of amorphous ferromagnetics is largely due to the almost complete absence of magnetic anisotropy in them. Therefore, investigating the mechanisms that influence the magnetic anisotropy of the resulting products is an important step on the way to achieving the maximum level of magnetic characteristics.

2. Experimental details.

The samples were manufactured in the SLM280HL selective laser melting system.

The structure of the sample was examined with a Tescan Mira 3M scanning electron microscope.

Etching of the sample was proceed in a solution based on isopropyl alcohol with addition of 10-% of nitric acid.

The phase composition was analysed using a Bruker D8 Advance X-ray diffractometer with CuKα irradiation.

The magnetic hysteresis curves of the specimens were measured by Lake Shore 7410 vibration sample magnetometer at room temperature and under applied different magnetic fields from -18000 to +18000 Oe.
Used 1CP powder material consists of spherical and rounded particles with particle size between 10 μm and 80 μm, flow rate of 13,00 s/50 g, apparent density of 4,21 g/cm^3 and skeletal density of 7,41 g/cm^3.

3. Experimental results.

An array of columnar elements (with a height of 20 mm) of Fe-based ferromagnetic alloy 1CP powder (iron, boron, silicon, phosphorus, carbon, cobalt, barium) was manufactured via SLM. The following process parameters were used: laser power P = 90 W, scanning speed V = 1200 mm/s, powder layer thickness t = 25 μm. The appearance of the array is shown in Fig. 1, (a), the appearance of the array element is shown in Fig. 1, (b).

a)

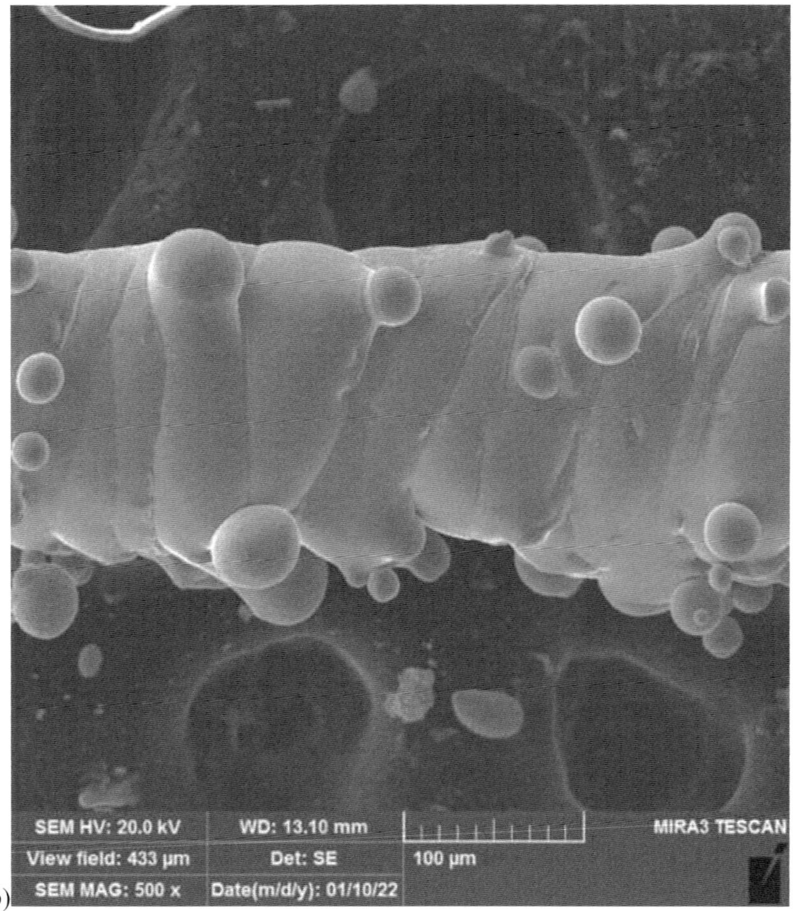

b)

Figure 1. a - array appearance, b - array element appearance

The microstructure of the array element is shown in Figure 2.

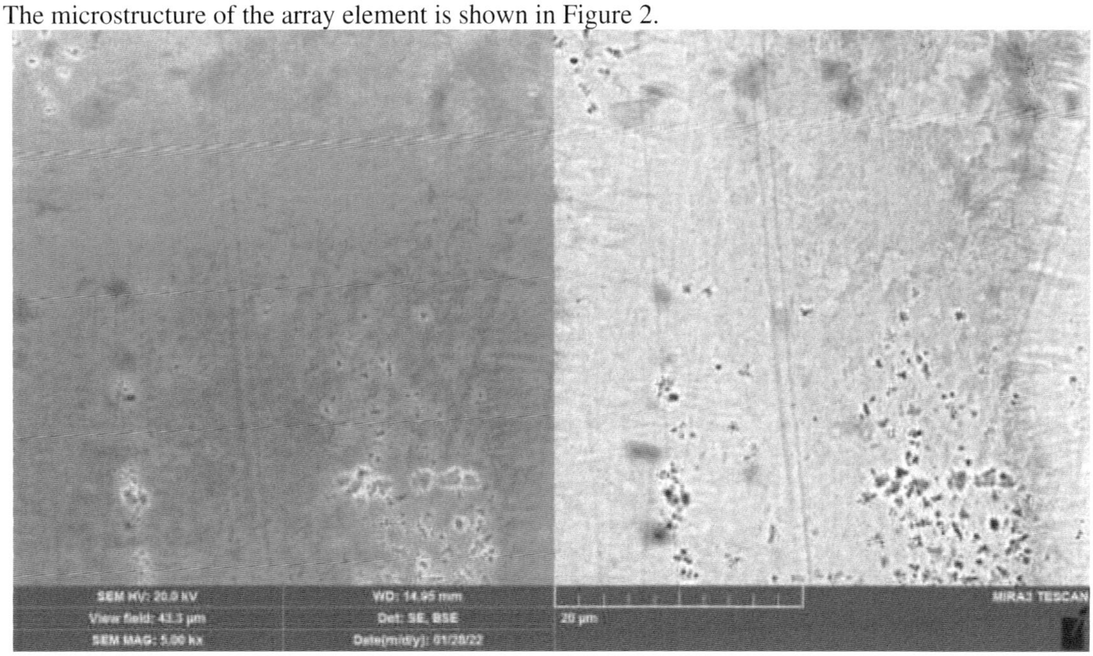

Figure 2. Microstructure of an array element

The SEM images shown in Figure 2 show that the element structure is represented by threecomponents. The dark cavities remaining after etching are distributed both in the matrix and at the fusion boundary. Bright highlights are only located along the fusion boundary.
The X-ray diffraction pattern of the sample is shown in Fig. 3.

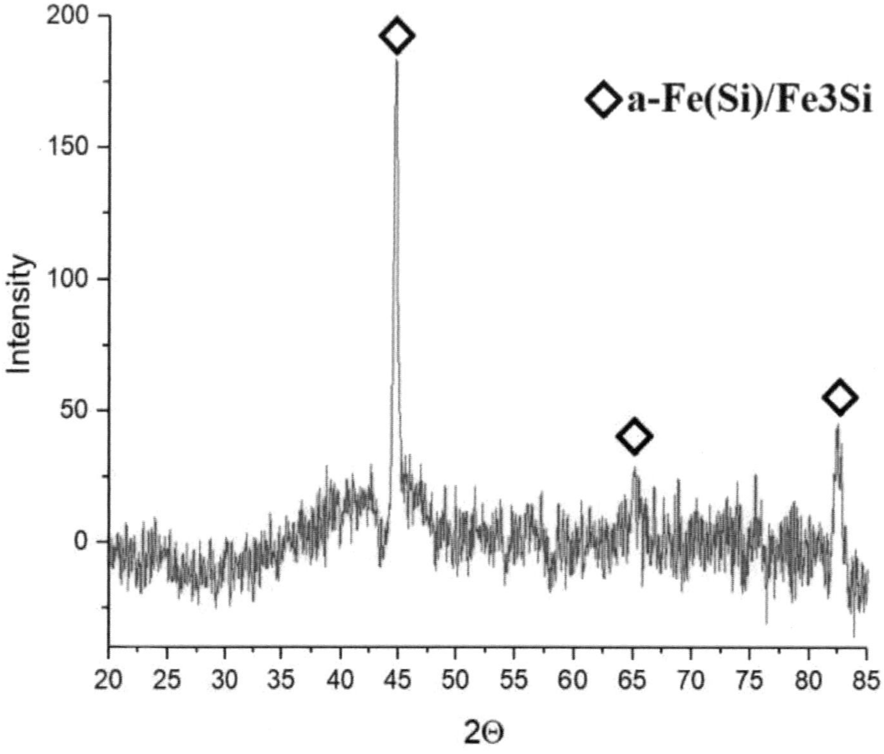

Figure 3. X–Ray diffraction pattern of the sample

X-ray data indicate the presence of α–Fe(Si) solid solution in the sample. A small halo indicates the amorphous phase in the sample. The images of the sample microstructure in Fig. 2 show that there are etched cavities, which presence indicates the presence of another phase. This phase can be identified as an ordered Fe3Si. The coherence of the boundary between this phase and α–Fe(Si) explains the absence of explicit Fe3Si peaks on the diffraction pattern.
Magnetization curves of the sample taken when the sample axis is parallel to the magnetic field strength vector and when the sample axis is perpendicular to the magnetic field strength vector are shown in Fig. 4. (a, b).

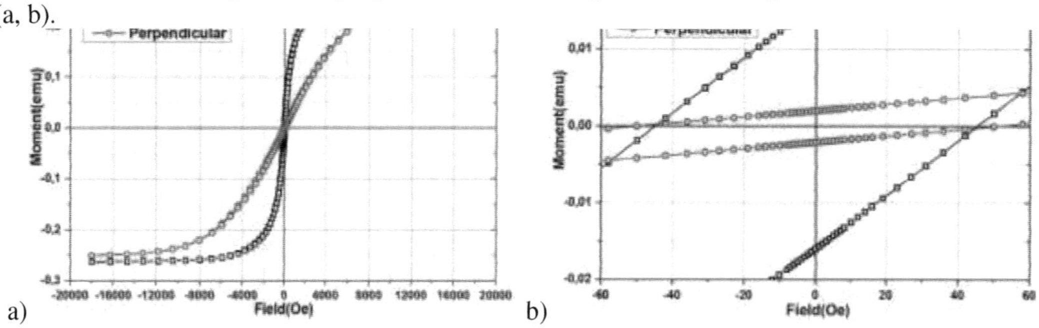

Figure 4. Magnetic hysteresis curves of the sample taken in parallel and perpendicular directions:a - general view of the magnetization curves; b – coercivity area

The magnetization curves of the sample reveal the anisotropy of its magnetic properties: in the direction parallel to the axis of the sample the coercivity is slightly lower than in the perpendicular direction and the saturation magnetization is greater.

The sample was then heat treated according to the regime recommended for amorphous tape made of the researched alloy [10]: heating at a rate of 10 C°/min to 440 C°, holding for 30 minutes, followed by cooling outside the furnace.

The magnetic hysteresis curves of the sample after the heat treatment are shown in Fig. 5. (a, b)

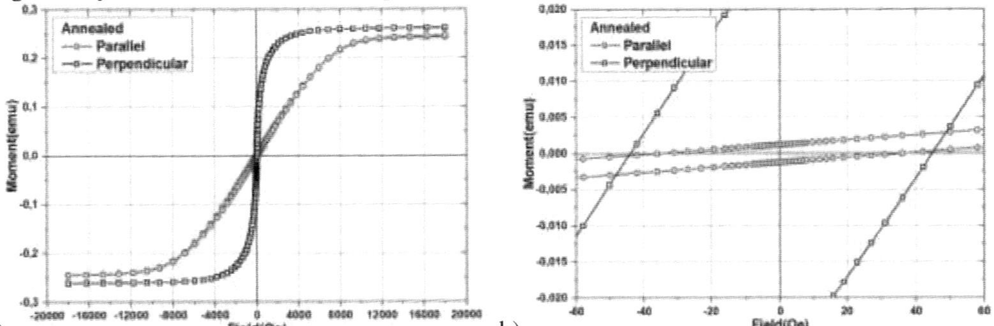

a) b)

Figure 5. Magnetic hysteresis curves of the sample after heat treatment, taken in parallel and perpendicular directions: a - general view of the hysteresis curves; b – coercivity area

The annealing did not affect both the anisotropy in the sample and the magnetic properties of the sample in the direction perpendicular to its axis. However, after annealing the coercivity in the direction parallel to the sample axis decreased from 50 Oe to 38 Oe.

4. Conclusion.

The anisotropy of the magnetic properties of the sample obtained is due primarily to the presence of crystalline phases in its structure: α–Fe(Si) and ordered Fe3Si. The annealing according to the used regime does not influence the magnetic anisotropy in the sample; however,it has a positive influence on the magnetic properties of the sample in the direction parallel to its axis. The coercivity in this direction decreased from 50 Oe to 38 Oe (by 34%). Such decrease of coercivity is caused by the presence of an amorphous phase in the sample, the magnetic properties of which are strongly dependent on the level of internal stresses. In the direction perpendicular to the sample axis, this was not observed, which is most likely due to the peculiarities of the small amount of amorphous phase present in the sample in terms of its volume.

This research allows to use obtained results to produce samples and products with tailored magnetic properties.

5. References.

[1] Alleg, S. Effect of the Laser Scan Rate on the Microstructure, Magnetic Properties, and Microhardness of Selective Laser-Melted FeSiB / S. Alleg, R. Drablia, N. Fenineche // Journal of Superconductivity and Novel Magnetism. – 2018. – № 31. – P.3565–3567.

[2] Popovich, A. & Sufiiarov, Vadim & Polozov, Igor & Grigoriev, A.. (2019). Selective Laser Melting of the Intermetallic Titanium Alloy. Russian Journal of Non-Ferrous Metals. 60. 186-193. 10.3103/S1067821219020081.

[3] Sufiiarov, Vadim & Borisov, Evgenii & Polozov, Igor. (2020). SELECTIVE LASER MELTING OF COPPER ALLOY. Materials Physics and Mechanics. 43. 65-71.10.18720/MPM.4312020_8.

[4] Grigoriev, Alexey & Polozov, Igor & Sufiiarov, Vadim & Popovich, Anatoly. (2017). In-situ synthesis of Ti2AlNb-based intermetallic alloy by selective laser melting. Journal of Alloys and Compounds. 704. 434-442. 10.1016/j.jallcom.2017.02.086.

[5] Suzuki, K. Amorphous metals / K. Suzuki, H. Hujimori, K. Hashimoto; Edited by Ts. Per. from Japan. - Moscow: Metallurgy, 1987. - 328 c.

[6] Selective laser melting vitrification of amorphous soft magnetic alloys with help of double-scanning-induced compositional homogeneity / Y. G. Nam [at al.] // Materials Letters. – 2020. – № 261. – P. 1–4.

[7] Fabrication of Fe-based bulk metallic glass by selective laser melting: A parameter study / H.Y. Jung [at al.] // Materials and Design. – 2015. – №86. – P.703–708.

[8] New approach to amorphization of alloys with low glass forming ability via selective laser melting/ Ł. Zrodowski [at al.] // Journal of Alloys and Compounds. – 2019. – № 771. – P. 769–776.

[9] Selective laser melting vitrification of amorphous soft magnetic alloys with help of double-scanning-induced compositional homogeneity / Y. G. Nam [et al.] // Materials Letters. – 2020. – № 261. – P. 1–4.

[10] Rapidly hardened tape of soft-magnetic amorphous alloys and magnetically soft composite material (nanocrystalline alloy). Technical specifications. TS 14-123-149-2009.

[11] Żrodowski, Ł.; Wysocki, B.; Wróblewski, R.; Kurzydłowski, K.J.; Święszkowski, W. The Novel Scanning Strategy For Fabrication Metallic Glasses By Selective Laser Melting. In Proceedings of the Fraunhofer Direct Digital Manufacturing Conference 2016, Berlin, Germany, 16–17 March 2016.

[12] Zou, Yongming; Wu, Yaosha; Li, Kefeng; Tan, Chaolin; Qiu, Zhaoguo; Zeng, Dechang. Selective laser melting of crack-free Fe-based bulk metallic glass via chessboard scanning strategy. Mater. Lett. 2020, 272, 127824.

[13] Ouyang, Di; Wei, Xing; Li, N.; Li, Yicheng; Liu, Lin. Structural evolutions in 3D-printed Fe-based metallic glass fabricated by selective laser melting. Addit. Manuf. 2018, 23, 246–252.

[14] Ozden, M.; Morley, N. Laser Additive Manufacturing of Fe-Based Magnetic Amorphous Alloys. Magnetochemistry 2021, 7, 20.

[15] Jiang, Qi; Zhang, Peilei; Jie, Tan; Zhishui, Yu; Tian, Yingtao; Ma, Songyun; Wu, Di. Influence of the microstructure on mechanical properties of SLM additive manufacturing Fe-based bulk metallic glasses. J. Alloys Compd. 2021, 894, 162525.

[16] Liang, Shun-Xing; Wang, Xueqing; Zhang, Wenchang; Liu, Yu-Jing; Wang, Weimin; Zhang, LaiChang. Selective laser melting manufactured porous Fe-based metallic glass matrix composite with remarkable catalytic activity and reusability. Appl. Mater. Today 2020, 19, 100543.

[17] Shen, Yiyu; Li, Yingqi; Chen, Chen; Tsai, H.L. 3D printing of large, complex metallic glass structures. Mater. Des. 2017, 117, 213-222.

[18] Bordeenithikasem, P.; Hofmann, D.; Firdosy, S.; Ury, N.; Vogli, E.; East, D. Controlling microstructure of FeCrMoBC amorphous metal matrix composites via laser directed energy deposition. J. Alloys Compd. 2020, 857, 157537.

Acknowledgments
The research was supported by a grant from the Russian Science Foundation № 21–73– 10008, https://rscf.ru/project/21-73-10008.

Using Ammonia as Future Energy: Modelling of Reaction Mechanism for Ammonia/Hydrogen Blends

Hua Xiao[1,*], Aiguo Chen[1], Minghui Zhang[2], Yanze Guo[1,3], Wenxuan Ying[1,3]

[1] School of Naval Architecture and Ocean Engineering, Guangzhou Maritime University, Guangzhou, China
[2] Department of Basic Courses, Guangzhou Maritime University, Guangzhou, China
[3] School of Materials and Energy, Guangdong University of Technology, Guangzhou, China

Corresponding author's e-mail: xiaohua@gzmtu.edu.cn

Abstract. To utilize ammonia-based fuels, it is fundamental to understand chemical mechanisms of combustion process, in which reaction characteristics of such a chemical are described in detail. Detailed chemical-kinetics mechanism of ammonia was modelled by an automatic reaction mechanism generation program to investigate characteristics of premixed combustion for ammonia/hydrogen fuel mixture. To develop an accurate model for practical combustion applications, validation of the reaction mechanism was carried out in terms of laminar flame speed under different conditions. Results suggested that the established mechanism model has satisfying performance under different ammonia/hydrogen ratio conditions. Moreover, comparison with other mechanism models demonstrated that the developed model can be used to describe flame propagation of ammonia/hydrogen fuels. Then characteristics of laminar flame speed were predicted under various ammonia concentration and equivalence ratio conditions. Sensitivity analyses showed that ammonia mole fraction has a prominent impact on kinetics of flame speed for ammonia/hydrogen blends. Flame structure analyses showed that hydrogen can enhance ammonia flames with higher light radical concentrations whilst deteriorate NOx emission in exhaust gases.

1. Introduction

As environmental pollution and climate change problem has been drawing attention all over the world, to develop low-carbon energy source is regarded as the future choice for energy system [1]. Just like hydrogen, ammonia is a carbon-free chemical with less issues in transportation and storage, which are the chief barriers to hydrogen's wide utilization [2]. Especially with the ever-increasing concerns about climate change problems and energy demand [3, 4], ammonia has been receiving great attention recently. As ammonia has been widely used in different industrial processes, there is already enough experience and existing infrastructures for the storage, production and handling. Moreover, ammonia is able to be produced from various renewable energy such as solar, wind power, etc., which makes it potentially a kind of zero carbon green fuel.

Many groups have studied the potential of ammonia utilization as a fuel in different combustion facilities [5-7]. Though generally good performance has been demonstrated in these combustion studies, considerable difficulties still need to be settled for ammonia, e.g., low flame speed, narrow range of flammability, high ignition resistance, etc. In this situation, it is necessary to enhance ammonia combustion with promoters in order to achieve effectively utilize ammonia in energy systems. Previous studies have explored using different kind of fuels as promoters for ammonia combustion applications, such as diesel [8], hydrogen [7], DME [9], etc. The results in these studies have generally demonstrated the potential of ammonia blends in practical combustion applications. Thus, significant carbon emissions reduction can be realized by co-firing ammonia with other fuels as ammonia is zero-carbon fuel. Among the fuel blends studied, ammonia mixed with hydrogen has been considered to have a good potential to enhance ammonia combustion.

To develop ammonia/hydrogen as fuel in practical combustion application, it is important to understand fundamental combustion of ammonia/hydrogen mixtures. There are already numerous studies which investigated the chemical reaction kinetics of ammonia/hydrogen combustion experimentally and numerically [10-13]. For instance, He et al. [10] investigated the ignition kinetics of $NH_3/H_2/O_2$ mixtures at different temperature (950-1150 K) and pressure (20 to 60 bar) conditions. By comparing different reaction mechanism models, it is found that none of the reaction mechanism in literature can calculate the auto-ignition delay time satisfactorily. Rocha et al. [11] investigated the 10 reaction mechanisms for NH_3/H_2/air combustion in terms of different properties such as lame speed and auto-ignition, and concluded that the reaction mechanism of NH_3/H_2 oxidation is not accurate enough. Xiao et al. [12] studied the modelling of NH_3/H_2 fuel under practical gas turbine operational conditions, in which improved mechanism was proposed to better predict the combustion properties of such fuel blends. Though previous mechanisms shows good performance for ammonia/hydrogen combustion under certain laboratory conditions, studies show that performance of still needs more test and improvement, especially for more practical conditions [14-16].

As ammonia co-fired with promoter fuels is usually adopted for practical combustion application, the complexity of detailed mechanisms production for ammonia-based fuels becomes much larger compared with single fuel. Computational chemistry, which is developed on the basis of quantum chemistry, electronic structure theory and semi-empirical models has been developing fast recently as computational power of computer increases. This leads to stronger ability of solving the needed kinetic and thermodynamic constants for more complex molecular systems. Therefore, automated methods including automatic detailed mechanism generator can be used to develop large detailed mechanisms based on the recurring and systematic use of chemical principles. The Reaction Mechanism Generator (RMG) is a software based on general kinetic principles of reaction families and libraries [17], which can automatically investigate possible elementary reactions and small species for a certain reacting mixture using a flux-based algorithm. RMG can generate detailed reaction kinetic mechanisms for various species including hydrogen, oxygen, carbon, nitrogen, etc., especially it was enhanced with nitrogen chemistry recently [18]. With advantages of more efficient and systematic whilst less error-prone, RMG can take over the time-consuming details of exploring each elementary reaction and species by hand. Diverse kinetic model development studies have proven RMG an efficient tool in mechanism development [19, 20].

Therefore, in this work a kinetic reaction mechanism was generated by RMG to model the combustion chemistry of ammonia/ hydrogen mixtures. Flame propagation has been analyzed to better understand the premixed ammonia/hydrogen combustion characteristics. The impact of ammonia substitution on ammonia/hydrogen combustion has been numerically studied with the proposed detailed reaction mechanism model, which can promote the implementation of ammonia as a primary fuel.

2. Methodology

2.1 Chemical mechanism generation

In previous studies, detailed mechanism models of ammonia combustion were generally created through keeping track of all elementary reactions and intermediate species and incorporating related sub-chemistry by hand, which is usually an error-prone process. However, when it comes to ammonia mixed with other fuel mixtures, complexity of the detailed chemistry modelling can become considerately larger, especially in the situation that the chemical mechanisms need to be used in various kinds of conditions such as different pressures, temperatures, equivalence ratios, etc. Thus, production of ammonia-based reaction mechanisms needs to be carefully and systematically studied to develop a model that includes thousands of elementary reactions and species. The use of an automatic reaction mechanism generation code can handle the challenges associated with producing mechanisms by hand relatively easily.

In the present study, the software package RMG-py v2.1.0 developed by the MIT group was employed to model combustion processes through automatic detailed mechanism generation. RMG-py is an object-oriented program of Python version. The algorithm creates the detailed mechanism models of elementary reactions by using a typical understanding of how molecules react. In the procedure of ammonia/hydrogen combustion reaction mechanism generation, a homogeneous isobaric reactor model was employed to discover all possible 'edge' species and elementary reactions. Reaction models were produced based on an algorithm excluding species from the mechanism model using reaction fluxes. As species reacted in the core, the program relied on two data structures of graphs and trees. Graphs can characterize chemical structures while trees were used to illustrate thermodynamic and kinetic data. Kinetics were estimated using its reaction family templates or determined directly through parameters in its kinetic library. Thermochemistry of species was determined using Benson group additivity method or just determined through RMG's thermal chemistry library. If RMG found a matching reaction and species in libraries, the kinetic and thermochemistry data were used. Otherwise, the program applied its own database of known reaction and templates rate rules to estimate rate coefficients and group additivity methods for thermochemical properties. Furthermore, RMG was also capable of estimating transport coefficients and it computed pressure-dependent rate coefficients automatically, thus identifying chemically-activated reaction paths. The general process of kinetic model generation is illustrated in Fig. 1.

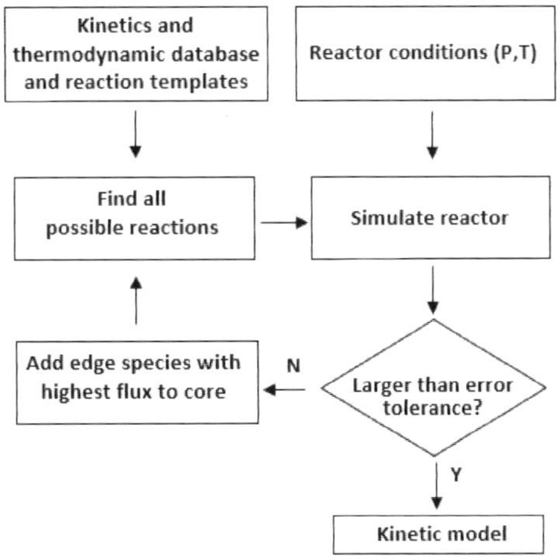

Figure 1. Model generation process by RMG.

A mechanism model was generated according to the above-mentioned algorithm and database. In the procedure, to generate detailed mechanism for ammonia/hydrogen combustion, mechanisms such as the Burke et al. [21] and Lopez et al.[22] kinetic libraries were used as the initial reaction libraries. Thermodynamic libraries of RMG including most-updated data for radicals of the H_2/O_2 chemistry were used in building the kinetic model. Stoichiometric conditions for a NH_3/H_2 with ratio of 50/50 were used, which was the ratio utilized by others whilst employing a tangential swirl burner [23]. The selected tolerance for the mechanism generation was 0.02, as ammonia chemistry in the model was found to be independent on smaller error tolerances. Finally, a RMG-generated mechanism was obtained consisting of 41 reactive species and 341 reactions, which is available in the Supporting Information.

2.2 Laminar flame modelling

In modern combustion applications e.g. internal combustion engines and gas turbines premixed flame propagation is utilized as the combustion mode. Unstretched laminar flame speed is important characteristic for premixed combustion, as it can help understand the combustion of different fuels and assess stable limits of the mixture. Therefore, modelling of freely propagating flames is important to help gain a deep understanding of combustion process for flame experiments. To investigate laminar flame, a steady, isobaric, quasi-one-dimensional flame model is usually used for flame laminar freely propagation studies [24, 25].

In the present work, premixed flames were numerically simulated using the open-sourced computational code of Cantera [26]. The 1-D laminar flames were modelled for ammonia/hydrogen fuel blends at different inlet temperature and pressure conditions. As described in previous literature [27], this model can be used to perform analyses on parameters such as flame speed, radical concentrations, reaction rates, sensitivity analyses, etc. Flame speed is defined as the speed of flame front relative to the unburnt mixture. Temperature is calculated from solving the energy equations as no heat losses are assumed. Calculation of flame speed is based on the heat transport and temperature profile prediction is an integral part in the model. Since the composition of the mixture will have an impact on the flame speed, calculations of various ammonia/hydrogen fuel blends are conducted so as to verify the accuracy of chemical reaction mechanisms. Experimental data of Li [28] are used in the work for the assessment of laminar flame speed by the numerical simulations, in which a variety of ammonia/hydrogen fuel compositions and equivalence ratio conditions are studied.

3. Results and Discussion

3.1 Comparison between modelling and experimental results

To verify the performance of established chemical reaction mechanism, unstretched flame speed calculations were performed for premixed combustion of ammonia/hydrogen blends. With ammonia mole fractions various from 40% to 61.5% in the fuel blends and equivalence ratio various from 0.6 to 1.4, simulation results were obtained to be validated with experimental data from literature [28]. Moreover, several popular reaction mechanisms in previous ammonia chemistry studies were also tested to illustrate the accuracy of the mechanism proposed in this study.

Generally, almost of the mechanisms tested have showed the same trend of unstretched flame speed as the experimental results. The value of unstretched flame speed peaks around an equivalence ratio of 1.1. In Fig. 2, NH_3 fraction of 61.5% in fuel blend was calculated for the unstretched flame speed study. As shown in the figure, the newly developed mechanism has predicted slightly higher values of unstretched flame speed than that of experimental results under fuel lean conditions, whilst under fuel rich conditions slower flame speeds were predicted. The best agreements between calculations were achieved by the present established mechanism and Tian's mechanism, while the GRI 3.0 mechanism and the Konnov mechanism show relatively poorer performance. Fig. 3 and Fig. 4 show simulation results of unstretched flame speed for NH_3 fraction of 50% and 40% in fuel blends respectively. In Fig. 3, the established mechanism in the present study and the Konnov mechanism have shown better predictions despite of some deviations from experimental data, especially under fuel rich conditions. For 40% of initial NH_3 fraction in fuel blends, almost all the four mechanisms have predicted lower flame speed than experimental data as shown Fig. 4. In the fuel lean region, the established reaction mechanism shows the best prediction accuracy while the best prediction performance is obtained by the Konnov mechanism under high equivalence ratio conditions.

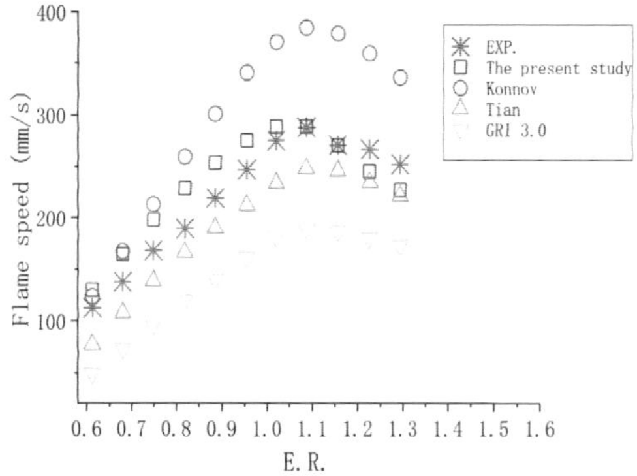

Figure 2. Unstretched flame speed of 61.5% NH_3 in fuel blends. Experiments as in [28].

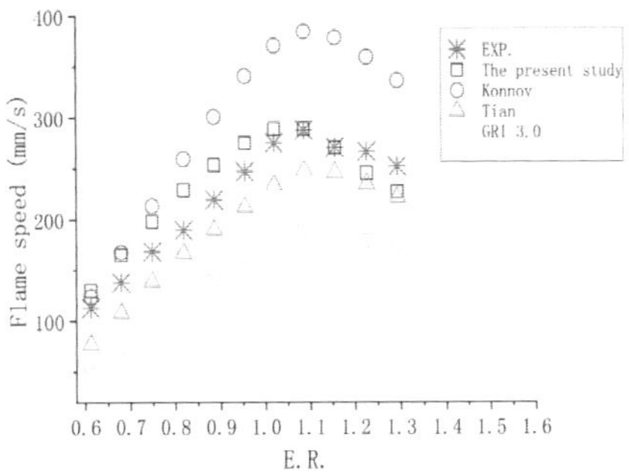

Figure 3. Unstretched flame speed of 50% NH_3 in fuel blends. Experiments as in [28].

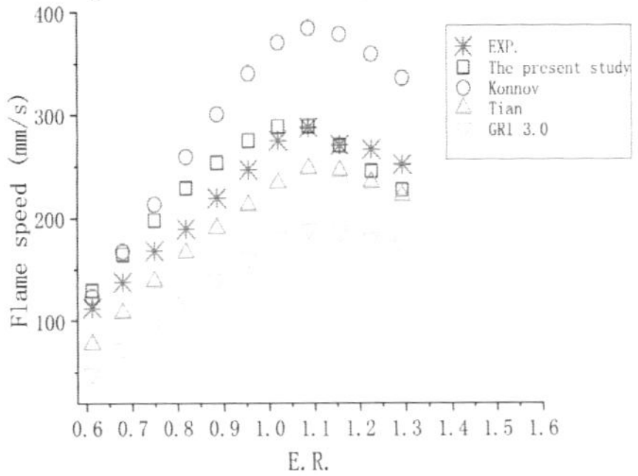

Figure 4. Unstretched flame speed of 40% NH_3 in fuel blends. Experiments as in [28].

The results indicate that the developed reaction mechanism performs generally good accuracy in terms of unstretched flame speed prediction for NH_3/H_2 blends within the range of fuel blend compositions tested, whilst the prediction of the GRI 3.0 mechanism is relatively poor. This indicates that the established chemical reaction mechanism is capable of producing good prediction on unstretched flame speed for different ammonia/hydrogen fuel blends. Therefore, it will be utilized in the next sections to test the characteristics study of ammonia/hydrogen combustion.

3.2 Characteristics of laminar flame speed

To gain a comprehensive understanding on ammonia/hydrogen flame, a numerical simulation prediction was performed for NH_3/H_2 flame propagation characteristics using the reaction mechanism established by RMG in this study. Vary of unstretched flame speed is illustrated in Fig. 5, with equivalence ratios ranging from 0.6 to 1.4 at various ammonia/hydrogen ratios. For the NH_3/H_2 blends studied, unstretched laminar flame speed increases within the range of low equivalence ratio conditions while declines under low equivalence ratio conditions. The value of unstretched flame speed peaks at around the equivalence ratios of 1.1~1.2. It can be observed that flame speed rises when ammonia mole fraction in fuel blend decreases for a certain equivalence ratio and the phenomenon becomes more remarkable at the stoichiometric conditions. This is because passing stoichiometry, unreacted ammonia recombines with NO species in the mixture, leading to promoting of unstretched flame speed compared with low equivalence ratio conditions. It also can be supported by previous relevant studies in which the impact of equivalence ratio was analyzed on unstretched flame speed of ammonia [27]. This suggests that in practical combustion applications slightly fuel rich stoichiometry conditions are more favorable than lean conditions for better flame propagation performance, whilst the combustion efficiency and ammonia emission will be the most prominent problems under such conditions, which is out of the scope of this paper.

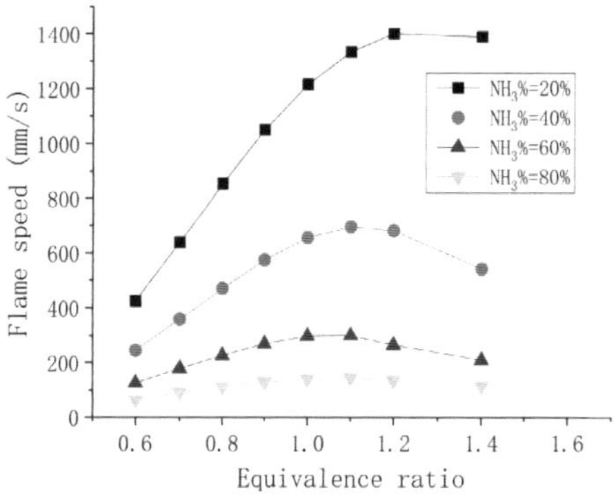

Figure 5. Unstretched flame speed of ammonia/hydrogen fuels as a function of equivalence ratio.

To understand the impact of ammonia mole fraction, unstretched flame speed is plotted against ammonia mole fraction at equivalence ratios ranging from 0.6 to 1.4, as shown in Fig. 6. It shows that from pure hydrogen (NH_3%=0) to pure ammonia, unstretched flame speed decreases when NH_3 fraction in fuel blends increases. Moreover, under high equivalence ratio conditions the impact of ammonia mole fraction shows more impact on the unstretched flame speed than that of fuel lean conditions. It can also be observed that the inhibiting effect of ammonia on the flame propagation is higher under relatively higher ammonia concentration conditions.

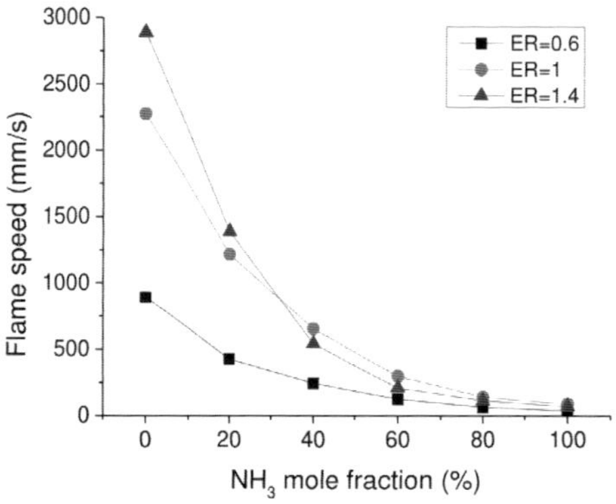

Figure 6. Unstretched flame speed of ammonia/hydrogen fuels as a function of ammonia mole fraction.

From the above results of different ammonia/hydrogen blends, it can be concluded that unstretched flame speed tends to significantly influenced by ammonia substitution in fuel blends. Reason of such results can be mainly attributed to ammonia's relatively lower reactivity compared with hydrogen. Therefore, in order to have a deep understanding on the difference between ammonia/hydrogen blends combustion, sensitivity analyses were performed. As shown in Fig. 7 and Fig. 8, the most-sensitive elementary reactions in unstretched flame speed simulation of 60% and 40% ammonia in fuel blends were compared under stoichiometric conditions. It can be seen that the elementary reaction $H+O_2 \Leftrightarrow O+OH$ shows the largest impact on unstretched flame speed modeling for the two fuel blends. The difference is that NH_2 radical has played a more prominent role in the calculation of 60% ammonia fuel blend than that of 40% ammonia, i.e. the reaction $NH_2+O \Leftrightarrow H+HNO$ is the second most sensitive elementary reaction for unstretched flame speed calculation in the 60%NH_3 case whilst the impact is much smaller for 40%NH_3 fuel blends. In the case of 60%NH_3, high ammonia mole fraction case shows higher effect in promoting flame propagation, as in ammonia oxidation procedure NH_2 produced from NH_3 is quite an important intermediate.

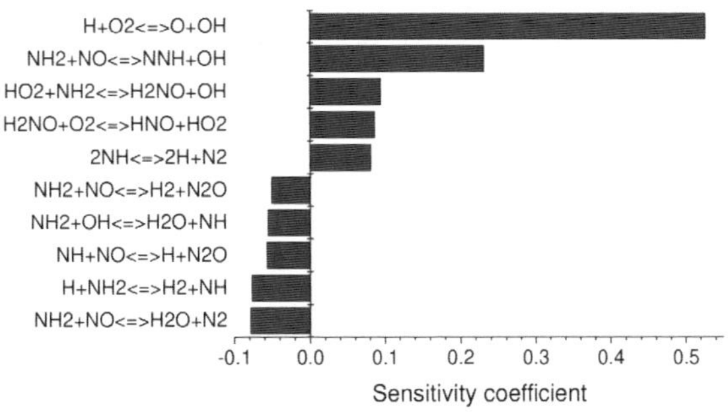

Figure 7. Sensitivity analysis of unstretched flame speed (60%NH_3).

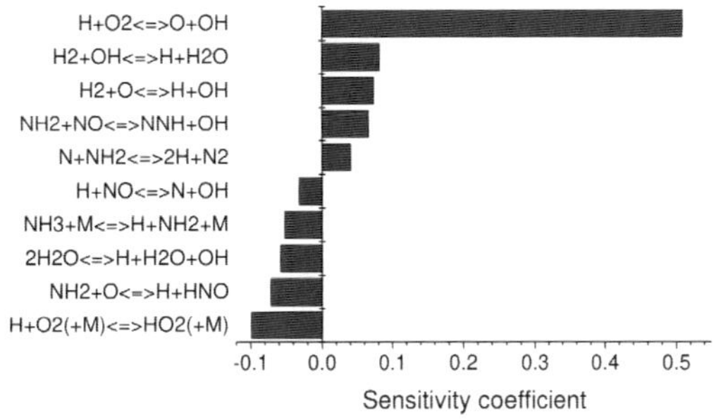

Figure 8. Sensitivity analysis of unstretched flame speed (40%NH_3).

3.3 Laminar flame structure analysis

To gain a better understanding on the ammonia/hydrogen combustion characteristics, especially the impact of hydrogen addition in ammonia, flame structure analyses can be quite helpful in understanding the combustion process of ammonia/hydrogen flames. As identified in the previous section, predictions of laminar flame speed using the mechanism developed by RMG can provide satisfactory agreement with experimental data, therefore the developed mechanism was utilized to explore laminar flame structure of ammonia mixed with hydrogen.

As shown in Fig.9 and Fig.10, pure ammonia and ammonia/hydrogen premixed flame structures were predicted respectively at ambient temperature and pressure and stoichiometric conditions. As shown in these figures, fuels, oxidizer, products and some important intermediate species mole fractions were described corresponding to height above the burner. Comparing fig. 9 with fig. 10, a relatively much stronger flame is observed for NH_3/H_2 flame than NH_3 as adding hydrogen into ammonia obviously shows increasing concentration of OH, H and O radicals, which are the most important radicals impacting flame speed and the main reaction zone. As expected, when more reactive hydrogen is added to ammonia, more accumulation can be found for OH radicals at the flame front leading to higher reaction rates of the flame, thus flame speed increases compared with a pure ammonia flame.

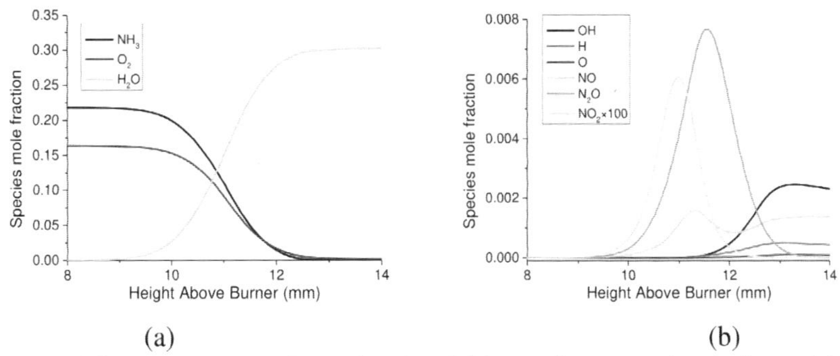

(a) (b)

Figure 9. Predicted flame structure of premixed stoichiometric ammonia-air flame (a)profiles of NH_3, O_2 and H_2O (b) profiles of OH, H, O, NO, etc.

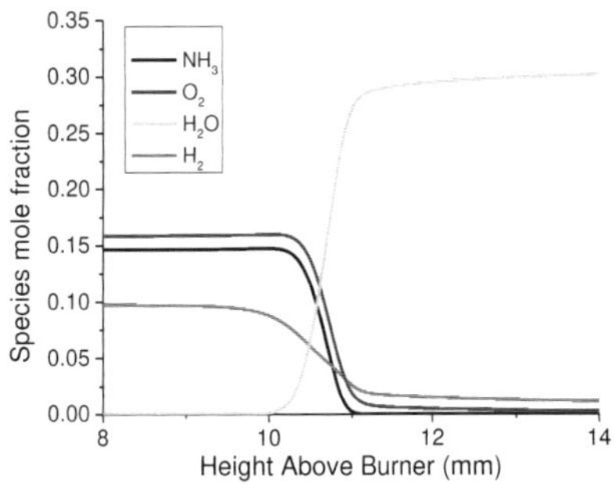

(a)

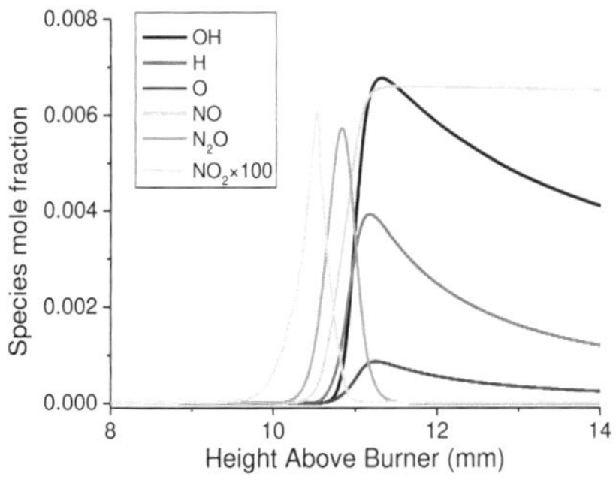

(b)

Figure 10. Predicted flame structure of premixed ammonia/hydrogen-air flame at stoichiometric condition (60%NH_3/40%H_2) (a) profiles of NH_3, O_2, H_2O and H_2 (b) profiles of OH, H, O, NO, etc.

It is also indicated that NO is the primary pollutant in nitrogen-bond emissions as NO_2 and N_2O concentration is two orders of magnitude smaller than NO. A considerable increase of NO emissions at the exhaust in ammonia/hydrogen flames compared with pure ammonia flames indicates a deteriorating emission performance when adding hydrogen, although H_2 helps to enhance flame propagation. The reason of NO increase can be attributed to the effect of thermal NOx formation mechanism. With hydrogen added to ammonia, increasing reactivity of the fuel mixture leads to higher temperature thus contribution of thermal NOx tends to grow with an increase of hydrogen concentration. Also, it is known that some other reactions based on H, O and OH radicals as promoters of nitrogen oxides can also contribute to the increase of this emission. Further information can be found somewhere else [29].

4. Conclusions

In this work, a detailed chemical reaction mechanism model was generated by using RMG program for premixed combustion of ammonia/hydrogen. The established chemical reaction mechanism was proved to be capable of simulating fundamental properties of ammonia/hydrogen fuels combustion for different ammonia mole fractions and various equivalence ratios from fuel lean to fuel rich conditions. It was demonstrated that RMG code can produce reasonable predictive reaction mechanism models for ammonia/hydrogen flame propagation. Calculations of characteristic ammonia/hydrogen flame speeds showed that ammonia mole fraction has a significant inhibiting impact on unstretched flame speed. Flame structure analyses showed light radicals changes with the variation of ammonia mole fractions. Simultaneously, propagation of ammonia/hydrogen flame is inhibited with ammonia mole fraction increasing, suggesting the need of stable combustion operation techniques using these blends. In future, the mechanism will be tested and further optimized under a great variety of conditions.

5. References

[1] Foxon TJ. A coevolutionary framework for analysing a transition to a sustainable low carbon economy. Ecological Economics. 2011;70:2258-67.
[2] Lee JH, Lee SI, Kwon OC. Effects of ammonia substitution on hydrogen/air flame propagation and emissions. International Journal of Hydrogen Energy. 2010;35:11332-41.
[3] Zamfirescu C, Dincer I. Ammonia as a green fuel and hydrogen source for vehicular applications. Fuel Processing Technology. 2009;90:729-37.
[4] Wang W, Herreros JM, Tsolakis A, York APE. Ammonia as hydrogen carrier for transportation; investigation of the ammonia exhaust gas fuel reforming. International Journal of Hydrogen Energy. 2013;38:9907-17.
[5] Grannell SM, Assanis DN, Bohac SV, Gillespie DE. The Fuel Mix Limits and Efficiency of a Stoichiometric, Ammonia, and Gasoline Dual Fueled Spark Ignition Engine. Journal of Engineering for Gas Turbines and Power. 2008;130:042802.

[6] Frigo S, Gentili R. Analysis of the behaviour of a 4-stroke Si engine fuelled with ammonia and hydrogen. International Journal of Hydrogen Energy. 2013;38:1607-15.

[7] Westlye FR, Ivarsson A, Schramm J. Experimental investigation of nitrogen based emissions from an ammonia fueled SI-engine. Fuel. 2013;111:239-47.

[8] Reiter AJ, Kong S-C. Combustion and emissions characteristics of compression-ignition engine using dual ammonia-diesel fuel. Fuel. 2011;90:87-97.

[9] Gross CW, Kong S-C. Performance characteristics of a compression-ignition engine using direct-injection ammonia–DME mixtures. Fuel. 2013;103:1069-79.

[10] [Valera-Medina A, Gutesa M, Xiao H, Pugh D, Giles A, Goktepe B, et al. Premixed ammonia/hydrogen swirl combustionunder rich fuel conditions for gas turbines operation. International Journal of Hydrogen Energy. 2019;44:8615-26.

[11] He X, Shu B, Nascimento D, Moshammer K, Costa M, Fernandes RX. Auto-ignition kinetics of ammonia and ammonia/hydrogen mixtures at intermediate temperatures and high pressures. Combustion and Flame. 2019;206:189-200.

[12] da Rocha RC, Costa M, Bai X-S. Chemical kinetic modelling of ammonia/hydrogen/air ignition, premixed flame propagation and NO emission. Fuel. 2019;246:24-33.

[13] Xiao H, Valera-Medina A, Bowen PJ. Modeling Combustion of Ammonia/Hydrogen Fuel Blends under Gas Turbine Conditions. Energy & Fuels. 2017;31:8631-42.

[14] Li J, Huang H, Kobayashi N, He Z, Nagai Y. Study on using hydrogen and ammonia as fuels: Combustion characteristics and NOxformation. International Journal of Energy Research. 2014;38:1214-23.

[15] Lee S, Kwon OC. Effects of ammonia substitution on extinction limits and structure of counterflow nonpremixed hydrogen/air flames. International Journal of Hydrogen Energy. 2011;36:10117-28.

[16] Nozari H, Karabeyoğlu A. Numerical study of combustion characteristics of ammonia as a renewable fuel and establishment of reduced reaction mechanisms. Fuel. 2015;159:223-33.

[17] Xiao H, Howard M, Valera-Medina A, Dooley S, Bowen PJ. Study on Reduced Chemical Mechanisms of Ammonia/Methane Combustion under Gas Turbine Conditions. Energy & Fuels. 2016;30:8701-10.

[18] Xiao H, Valera-Medina A. Chemical Kinetic Mechanism Study on Premixed Combustion of Ammonia/Hydrogen Fuels for Gas Turbine Use. ASME. J. Eng. Gas Turbines Power. 2017;():. doi:10.1115/1.4035911. 2017.

[19] Gao CW, Allen JW, Green WH, West RH. Reaction Mechanism Generator: Automatic construction of chemical kinetic mechanisms. Computer Physics Communications. 2016;203:212-25.

[20] Dana AG, Buesser B, Merchant SS, Green WH. Automated Reaction Mechanism Generation Including Nitrogen as a Heteroatom. International Journal of Chemical Kinetics. 2018;50:243-58.

[21] Gao CW, Vandeputte AG, Yee NW, Green WH, Bonomi RE, Magoon GR, et al. JP-10 combustion studied with shock tube experiments and modeled with automatic reaction mechanism generation. Combustion and Flame. 2015;162:3115-29.

[22] Merchant SS, Zanoelo EF, Speth RL, Harper MR, Van Geem KM, Green WH. Combustion and pyrolysis of iso-butanol: Experimental and chemical kinetic modeling study. Combustion and Flame. 2013;160:1907-29.

[23] Burke MP, Chaos M, Ju Y, Dryer FL, Klippenstein SJ. Comprehensive H_2/O_2 kinetic model for high-pressure combustion. International Journal of Chemical Kinetics. 2012;44:444-74.

[24] Lopez JG, Rasmussen CL, Alzueta MU, Gao Y, Marshall P, Glarborg P. Experimental and kinetic modeling study of C 2 H 4 oxidation at high pressure. Proceedings of the Combustion Institute. 2009;32:367-75.

[25] Kee RJ, Grcar JF, Smooke MD, Miller J, Meeks E. PREMIX: a Fortran program for modeling steady laminar one-dimensional premixed flames. Sandia National Laboratories Report. 1985.

[26] Li J, Zhao Z, Kazakov A, Dryer FL. An updated comprehensive kinetic model of hydrogen combustion. International journal of chemical kinetics. 2004;36:566-75.

[27] David Goodwin HM, Raymond Speth. Cantera: An object- oriented software toolkit for chemical kinetics, thermodynamics, and transport processes. http://www.cantera.org, 2016. Version 2.2.1.
[28] Hayakawa A, Goto T, Mimoto R, Arakawa Y, Kudo T, Kobayashi H. Laminar burning velocity and Markstein length of ammonia/air premixed flames at various pressures. Fuel. 2015;159:98-106.
[29] Valera-Medina A, Xiao H, Owen-Jones M, David WIF, Bowen PJ. Ammonia for power. Progress in Energy and Combustion Science. 2018;69:63-102.

Acknowledgments
Dr. H. Xiao is grateful for Prof. William H. Green and his group at MIT, including Dr. Alon Grinberg Dana, for the hospitality and for the helpful instructions regarding the usage of the Reaction Mechanism Generator software. The authors gratefully acknowledge the support from the China Postdoctoral Science Foundation funded project (No. 2019M653037).

Ship Matrix-Deck and Analysis of Its Mechanical Properties

Zhuoyuan Chen[1,*], Aizhong Qin[1], Peiyang Li[1], Zhihua Gao[1], Jianbo Zhao[1], Xiaoyong Liu[1]

[1]China State Shipbuilding Corportion 713th Institute, Zhengzhou 450000, China

*E-mail: 573122978@qq.com

Abstract. Ship deck is generally made of steel materials. With the development of ship technology and the demand of ship lightweight design, a lightweight and modular ship deck is needed. In this paper, a ship deck is designed based on the form of matrix; The deck is mainly made of engineering plastics and filled with load-bearing materials. The overall strength of the material is analyzed by FEM simulation. According to the simulation results, the compressive strength of this grid composite structure is basically equivalent to that of C30 concrete, but the weight can be reduced by 30% compared with C30 concrete The matrix-deck has the characteristics of high strength and low density, and is suitable for occasions with high requirements for material strength and material quality.

1. Introduction

Polyphenylene oxide(PPO) is a high-strength engineering plastic developed in the last century, which has good comprehensive properties, but the manufacturing process is complex[1]. The strength and thermal stability of pure PPO will deteriorate over time. In order to solve this problem, PPO and high impact polystyrene (HIPS) are usually mixed to obtain modfied polyphenylene oxide(MPPO). Stack[2] analyzed the thermal stability of PPO materials and MPPO; The results shows that the thermal decomposition temperature of MPPO is significantly higher than PPO; The reason is that HIPS molecules interrupt the molecular transfer in the high temperature decomposition of PPO, so as to improve the overall thermal stability of the material. HIPS is a highly elastic material, and its elastic deformation has strong nonlinearity.

Bucknall[3] proposed a method to quantitatively study the toughening of HIPS according to the characteristics of HIPS. By comparing the transverse and longitudinal strain components of HIPS under high tensile stress, the contribution mechanism of each component leading to HIPS deformation is determined; The results show that the deformation of HIPS under high tensile stress is mainly composed of elastic deformation and crazing.

Foam concrete(FC), which is made of blowing agent in ordinary concrete, has the characteristics of low density. It is widely used as a non-load bearing material in engineering. Because of the existence of pores of FC, modeling of FC is very difficult. Nambiar[4] analyzed the influence of porosity, permeability and pore size on the strength of FC. A series of experiments were carried out to study the effect of the above parameters on the strength and durability of the FC. The results show that the porosity, permeability and pore size affect the strength and density of the FC, while the shape of the pores has no effect on the strength of the FC. And fitting the function relationship of each parameter on the strength of FC. Zuhua Zhang[5] is used to study the thermal stability and energy saving of FC. The potential scale effect of FC in application is analyzed. Due to the difficulty in controlling manufacturing process and performance degradation process of the FC, the foamed concrete can not be widely used at present. Zuhua Zhang had proposed a method to control the quality and degradation process of FC, which could effectively reduce the uncertainty of FC characteristics in manufacturing process. His research findings can make the degradation process modelable, and the degradation process predictable, which greatly improve the usability of FC in engineering. WANG Hui[6] proposed a new type of FC. The new FC can be used for filling materials in large-span soft rock tunnels. Because of its low density, it can effectively reduce the settlement of the tunnel crown and greatly improve the stability and reliability of the tunnel. The research of WANG Hui shows the feasibility of foamed concrete as filling material in practical engineering, which has good reliability and durability.

Nambiar[7] in subsequent studies found that the type and radius of aggregate had a significant effect on the strength of FC. The smaller the radius is, the higher the strength of FC. The higher quality of the aggregate, the greater strength/density ratio. In order to further reduce the density of FC, Patchara Onprom[8] verified the feasibility of using coal ash instead of sand as aggregate. Using different proportion of coal ash instead of sand to make FC test blocks, The test results show that adding coal ash in FC can significantly reduce the density of FC and increase its water absorption. But the compressive strength of FC will decrease correspondingly. Shaohai Wang[9] directly used the graphite schist with radius of 0.1mm as the aggregate to manufacture FC. The results show that the strength/density ratio is medium and have obvious weight reduction effect. By adding animal angle particles in FC, Dong L[10][11] significantly reduced the manufacturing cost of FC, and the FC had good strength. The above three studies show that the composition of the aggregate has great influence on the performance of FC. Selecting the material with small density, small particle size and high strength as the aggregate of FC can significantly improve the comprehensive performance of FC. According to the high strength of MPPO and the low density of foamed concrete, a composite structure of light weight plate is proposed. The plate is designed for ship's deck, which has the characteristics of high reliability and durability; The weight of the ship can be effectively reduced by replacing traditional steel deck with this plate.

2. Design of composite structure

The overall structure of the plate is shown in Fig. 1. Its surface and subface are asymmetric design. Fig. 1(a) is the structural diagram of the subface; The main load-bearing structure of the plate is a grid MPPO frame; There are several rectangular cavities between the frame; The cavity is filled with FC. Because the cost of MPPO material is much higher than that of FC, this design can reduce the cost and density of the plate greatly under the condition that the strength of the plate does not decrease.

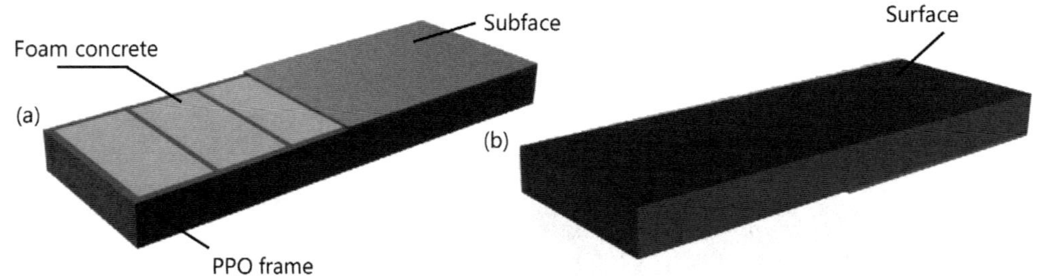

Fig. 1. Schematic diagram of the composite plate's structure.

The surface of the plate is a MPPO plate integrated with the frame, as shown in Fig. 1(b). After the MPPO frame is filled with FC, the polyurethane wear-resistant coating is sprayed on the surface and subface. After the coating is solidified, a complete plate is completed. In order to show the internal structure of the plate in Fig. 1, only half of the polyurethane coating is drawn on the subface.

The structural design with asymmetric subface and surface has the following advantages. First, because the compressive strength of FC is only 1-2Mpa, if the polyurethane wear-resistant coating is sprayed directly on the surface of FC, when the coating is subjected to pressure, the pressure of the polyurethane coating can not conduct well due to the small thickness and flexibility of the polyurethane coating. Then, FC will be subjected to a large concentrated stress, resulting in the destruction of FC. In order to solve this problem, the asymmetry design of the subface and the surface is put forward. A MPPO plate is added to the surface which is directly under pressure, so that the pressure can be transferred unimpeded to the frame, and the stress concentration of the FC is eliminated. As a non-bearing face, the subface does not need to add MPPO plate, which reduces the production cost and density.

3. Force analysis of the plate

3.1. Mechanical property modeling of FC

In order to establish the mechanical model of FC, the pores' distribution of FC is first described. According to[7][11], it is generally believed that the pores of the FC conform to normal distribution. The mathematical expression of its distribution is as follows[7]:

$$R(x)=\frac{1}{\sqrt{2\pi}\sigma}e^{-\frac{(x-\mu)^2}{2\sigma^2}},(0<x<+\infty) \quad (1)$$

Where $R(x)$ represents the probability of occurrence of pores with radius of x; x is the radius of the pore; μ and σ are the expectation and variance of normal distribution.

The bubble content of FC should also be expressed. The porosity can be obtained by the ratio of the density of non gas containing hardened slurry to that of gas containing hardened slurry[11]:

$$\omega=1-\frac{\rho}{\rho_0} \quad (2)$$

Where ω is porosity; ρ is the density of gas hardened slurry; ρ_0 is the density of non gas hardening slurry. By converting the normal distribution expression into the standard probability density formula:

$$R(x<X\le x+\Delta x)=\phi\left(\frac{x+\Delta x-\mu}{\sigma}\right)-\phi\left(\frac{x-\mu}{\sigma}\right) \quad (3)$$

When $x=0$:

$$R(0<X\le \Delta x)=\phi\left(\frac{\Delta x-\mu}{\sigma}\right)-\phi\left(\frac{\mu}{\sigma}\right) \quad (4)$$

Assunming the number of pores per unit volume is m, the following expression can be obtained[7]:

$$\sum_{i=1}^{\infty}\left\{\left[\phi\left(\frac{i\Delta x-\mu}{\sigma}\right)-\phi\left(\frac{(i-1)\Delta x-\mu}{\sigma}\right)\right]\cdot m\cdot\frac{\pi x^3}{6}\right\}=\omega \quad (5)$$

Assuming that porosity ω and normally distributed expectations μ and variance σ has following relationship:

$$\omega=f(\mu,\sigma) \quad (6)$$

Simultaneous formula(6), formula(5) and formula(2) can obtain:

$$\sum_{i=1}^{\infty}\left\{\left[\phi\left(\frac{i\Delta x-f^{-1}(\omega,\sigma)}{\sigma}\right)-\phi\left(\frac{(i-1)\Delta x-f^{-1}(\omega,\sigma)}{\sigma}\right)\right]\cdot m\cdot\frac{\pi}{6}\cdot\left[\left(i-\frac{1}{2}\right)\Delta x\right]^3\right\}=\omega \quad (7)$$

The pore distribution of FC can be obtained by this formula. The average pore's size and variance of FC can be calculated by this distribution. The pore's size of FC can be described more accurately. According to the formula of compressive strength of porous materials, the compressive strength of FC is as follows:

$$T=T_0(1-\omega)^{\beta} \quad (8)$$

T is the compressive strength of FC. T_0 is the compressive strength of concrete without foaming agent; ω Is porosity; β Is the compressive index, which is obtained by test.

3.2. Mechanical property modeling of MPPO

MPPO is a high polymer composite, and its mechanical properties are obviously different from ordinary molecular materials. The deformation of polymer materials is mainly creep and viscoelastic deformation. Creep is a phenomenon that the deformation of materials develops gradually with the passage of time under the action of a certain temperature and external force. The mechanical properties of polymer are closely related to its historical loads. In order to analyze the influence of historical load on MPPO deformation, Boltzmann superposition principle needs to be introduced. The principle is as follows: the deformation of polymer is a function of all historical loads; Each historical load has an independent contribution to the deformation, which satisfies the linear superposition principle. The total strain can be expressed as:

$$\varepsilon(t)=\sum_{i=1}^{n}\sigma_i J(t-\tau_i) \quad (9)$$

Where in $t=\tau_1,\tau_2\ldots\tau_n$ applying discontinuous stress $\sigma_1,\sigma_2\ldots\sigma_n$. If the stress changes continuously, the stress increment can be expressed by integral as[2]:

$$\varepsilon(t)=\int_{-\infty}^{t} J(t-u)\frac{\partial\sigma(u)}{u}du \tag{10}$$

Considering the lower integral limit, it is necessary to consider the whole history of polymers; However, in many cases, the tested polymer is not in the "zero state", so it is necessary to estimate its previous initial state, calculate its "zero state" response and "zero input" respectively, and then obtain the total strain of the polymer after superposition. Consider the scenario that the polymer is not in the "zero state". The total strain can be expressed as[2]:

$$\varepsilon(t)=\int_{-\infty}^{0} G(t-u)\frac{\partial\sigma(u)}{u}du+\int_{0}^{t} G(t-\tau)\frac{\partial\sigma(\tau)}{\tau}d\tau \tag{11}$$

The viscoelastic deformation of polymers is generally described by the constitutive model of materials. The commonly used viscoelastic models of polymers include Maxwell model, Kelvin Vogit model and Zener model. The three models are suitable for different viscoelastic scenarios; The specific selection needs to be determined according to the parameters of the material itself. No matter which model is adopted, the constitutive relationship of linear viscoelastic behavior is:

$$a_0+a_1\frac{d\sigma(t)}{dt}+a_2\frac{d^2\sigma(t)}{dt^2}+\cdots=b_0+b_1\frac{d\varepsilon(t)}{dt}+b_2\frac{d^2\varepsilon(t)}{dt^2}+\cdots \tag{12}$$

Through the constitutive model of polymer materials, the elastic strain under specific stress can be calculated; By accumulating the creep obtained in the early stage, the total strain of the polymer when a certain force is applied at a certain time can be determined.

4. FEM simulation results and analysis

According to the mathematical method in the previous chapter, the mechanical properties of MPPO and FC with regular shape can be analyzed; However, the composite structure proposed in this paper is complex, and there is a coupling relationship between MPPO and FC, which needs to be analyzed by FEM.

In this chapter, static FEM simulation is carried out for composite materials, and FEM simulation is carried out on MPPO frame, FC and whole panel respectively. The difference of the results proves the feasibility and rationality of the designed structure. The mesh type of FEM model is regular hexahedron; The cells' type are C3D10H. The boundary condition of PC and MPPO is "friction" and its friction coefficient is 1.5. The input load is uniformly distributed static pressure

4.1. Simulation results and analysis of MPPO frame

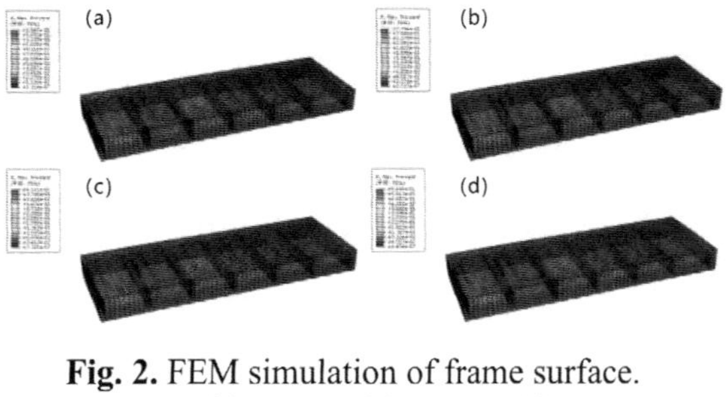

Fig. 2. FEM simulation of frame surface.
(a)10Mpa. (b)20Mpa. (c)30Mpa. (d)40Mpa.

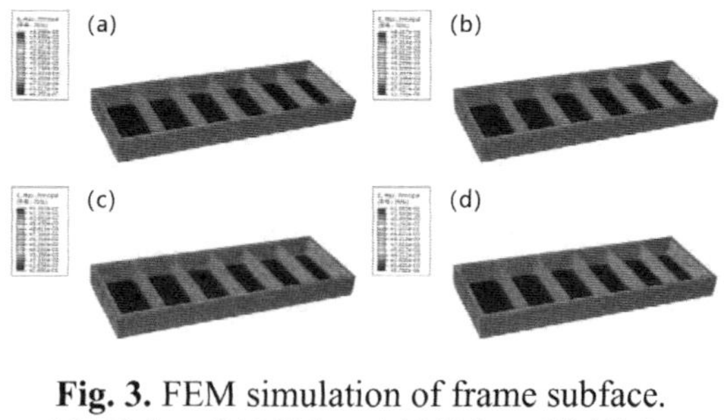

Fig. 3. FEM simulation of frame subface.
(a)10Mpa. (b)20Mpa. (c)30Mpa. (d)40Mp.

Fig. 2 shows the strain results of MPPO frame surface under stress. Apply pressure of 10, 20, 30 and 40Mpa to the frame surface respectively (Fig. 2(a)(b)(c)(d)); It can be seen from the results that the maximum strain of the frame is only 0.56mm under the pressure of 40Mpa; It is proved that the grid structure has sufficient strength, can effectively undertake the pressure from the surface, and has good force conduction. Fig. 3 shows the results when the frame subface directly bears pressure; Applying 10, 20, 30 and 40Mpa pressures to the subface respectively (Fig. 3(a)(b)(c)(d)). It can be seen that the cross section of the frame's wall is very small, so the strain is only 0.16mm under the pressure of 40MPa. In actual use, the subface is only used as the contact face which does not directly bear the pressure, so the subface also has good support performance.

4.2. Simulation results and analysis of the plate under stress on the subface

Fig. 4 is the strain diagram of the plate's subface under stress. Fig. 4(a), (b) and (c) show the strain of the coating when the pressure is 5, 10 and 20Mpa respectively. Fig. 4(d) (e) and (f) show the strain of internal structure when the pressure is 5, 10 and 20Mpa respectively.

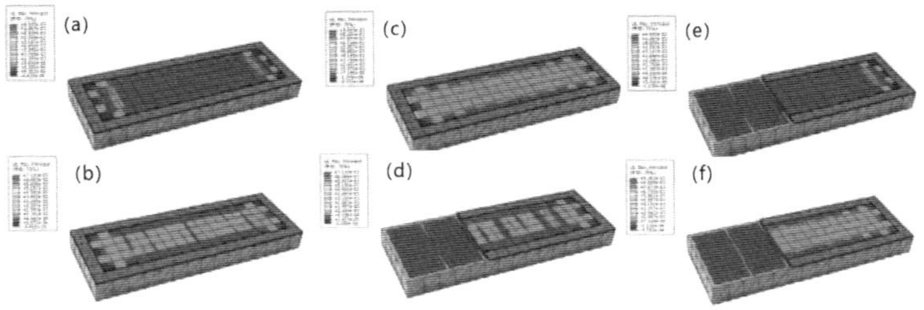

Fig. 4. FEM simulation of plate's subface. (a)5Mpa subface. (b)10Mpa subface. (c)20Mpa subface. (d)5Mpa internal. (e)10Mpa internal. (f)15Mpa internal.

According to the FEM results, when the whole plate is under pressure; The boundary of plate coating will bear large stress, resulting in maximum strain. However, because the coating material is polyurethane which has flexibility. It can be seen that the edge strain is 0.01mm under the pressure of 20Mpa; It is much lower than the maximum strain of polyurethane material. The strain of the internal structure is far less than 0.01mm, and its main support is provided by MPPO; Foamed concrete only plays a role of filling. However, because polyurethane is an organic material, with the cyclic application of stress, its material may age and affect its mechanical properties. Moreover, if the load is not uniformly applied, the FC may be damaged due to the greater local stress. Therefore, this surface is not suitable for direct stress face; It is suitable for the supporting face at the bottom, which will not be subjected to a large concentrated stress, and the FC will not be destroyed easily.

4.3. Simulation results and analysis of the plate under stress on the surface

Fig. 5 and Fig. 6 show the FEM results when the surface of the plate is stressed. Fig. 5(a)(b)(c)(d) shows the surface strain at the pressure of 10, 15, 20 and 30Mpa respectively. Fig. 6(a)(b)(c)(d) shows the internal strain when the pressure is 10, 15, 20 and 30Mpa respectively.

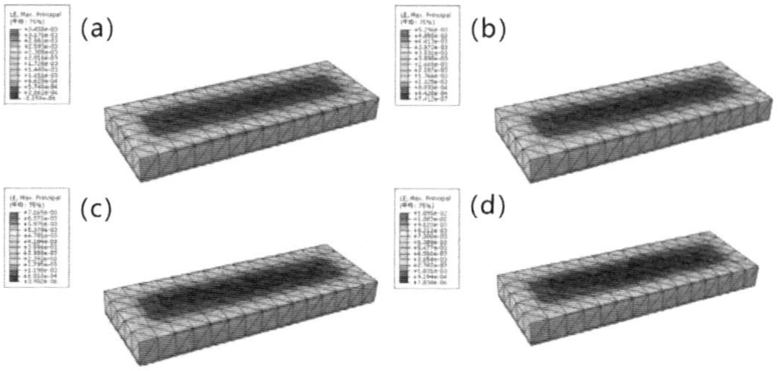

Fig. 5. FEM simulation of plate's surface.(a)10Mpa. (b)15Mpa. (c)20Mpa. (d)30Mpa.

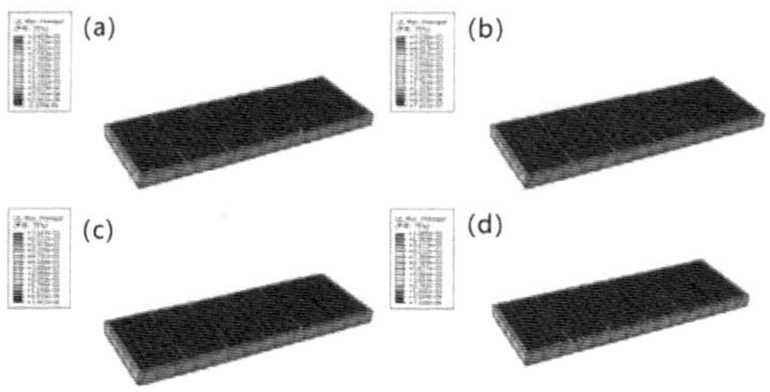

Fig. 6. FEM simulation of internal structure. (a)10Mpa. (b)15Mpa. (c)20Mpa. (d)30Mpa.

It can be seen from the FEM results that the strain of MPPO surface directly under pressure is also concentrated in the edge part, but its strain is significantly reduced. Moreover, it can be seen from Fig. 6 that the internal strain component is evenly distributed. It shows that the internal stress distribution can be significantly improved by adding 5cm MPPO plate on the frame. And the internal stress can be effectively reduced, so that FC will not be destroyed easily.

5. Summary

In this paper, a composite light plate is designed, and its mechanical properties are analyzed by FEM. The results show that the designed asymmetric subface and surface structure can significantly improve the stress distribution in the plate. In order to lighten the weight of the plate, the FC with light weight is filled. Because of the low strength and brittleness of FC, the stress in the plate is particularly important. Good structural design can effectively improve the internal stress distribution and reduce the mutation of internal stress, so as to protect the internal FC from failure. The simulation results in Chapter "FEM simulation results and analysis" can also strongly explain the above conclusion. In this paper, the feasibility of its structure is verified by FEM simulation, which provides the possibility for the lightweight design of non-metallic plates. The reliability and durability of the material need to be tested through experiments. This composite structure broadens the idea of composite plate design and provides a broad prospect for lightweight plate design in the future.

6. References

[1] Bucknall, C. B. , and D. Clayton . "Rubber-toughening of plastics - Part 2 Creep mechanisms in HIPS/PPO blends." Journal of Materials ence 7.2(1972):202-210.0
[2] Stack, S. , O. O'Donoghue , and C. Birkinshaw . "The thermal stability and thermal degradation of blends of syndiotactic polystyrene and polyphenylene ether." Polymer Degradation & Stability 79.1(2003):29-36.
[3] Rowell, R. L. . "Polymer applications of renewable resource materials." Journal of Colloid & Interface Science 97(1983).
[4] Nambiar, Ekk , and K. Ramamurthy . "Air - void characterisation of FC." Cement & Concrete Research 37.2(2007):221-230.

[5] Zhang, Z. , et al. "Geopolymer Foam concrete: An emerging material for sustainable construction." Construction & Building Materials 56.APR.(2014):113-127.
[6] Hui, et al. "Development of a new type of Foam concrete and its application on stability analysis of large-span soft rock tunnel." Journal of Central South University (2012).
[7] Nambiar, Ekk , and K. Ramamurthy . "Influence of filler type on the properties of FC." Cement & Concrete Composites 28.5(2006):475-480.
[8] Patchara, O. , C. Krit , and C. Raungrut . "Influence of Bottom Ash Replacements as Fine Aggregate on the Property of Cellular Concrete with Various Foam Contents." Advances in Materials Science & Engineering 2015(2015):1-11.
[9] Wang, and S. Hai . "Preparation of Foam concrete from Graphite Tailing." Advanced Materials Research 356-360(2011):1994-1997.
[10] Dong, L. , et al. "Preparation of Novel Foaming Agent and its Application in FC." Advanced Materials Research 785-786(2013):305-307.
[11] Nambiar, Ekk , and K. Ramamurthy . "Models relating mixture composition to the density and strength of Foam concrete using response surface methodology." Cement & Concrete Composites 28.9(2006):752-760.

Design and Research on the Structure of the Radar Antenna Array Unfolding and Retracting System

Guocan Tao, Jingshi Fang, Biao Li*, Haitao Shi

The 38th Research Institute of China Electronics Technology Corporation, Hefei, China

*ahutbiao@163.com

Abstract: In order to realize the efficient and rapid erection and flexible unfolding of super large active array meter wave radar with an area of more than 100 square meters, this study takes a meter wave radar antenna array unfolding and retracting system as the research object, makes a detailed structural design of the unfolding and retracting system, and establishes a three-dimensional model of the unfolding and retracting system on this basis to to Simulate the stiffness and strength of it. The changes of stiffness and strength of the array under different wind loads and wind angles are compared and analyzed. The results show that the unfolding and retracting system has high reliability under the condition of rapid erection and mobility, and has a good application and popularization prospect in the automatic unfolding and retracting system of super large radar array.

1. Introduction

Meter wave radar has the characteristics of long wavelength and has unique advantages in discovering anti stealth and anti radiation missiles. It has attracted universal attention all over the world. Developed countries have put the development of meter wave radar technology in one of the most important positions in radar early warning and detection system, especially in the case of vast territory and complex border lines like our country. The demand for highly mobile vehicle mounted large aperture meter wave radar is more urgent[1-4].

For low-frequency meter band radar, if we want to obtain higher detection accuracy and longer detection distance, that is, to achieve the long-range early warning ability of stealth aircraft, we must have a larger antenna aperture. The main function of the super large array unfolding and retracting system is to provide the installation interface for the antenna array, expand / retract the antenna array to the specified size position, complete the conversion of working state and transportation state, and meet the requirements of high mobility transportation. At present, the commonly used unfolding and retracting systems are less than 100 square meters[5-8]. Therefore, in order to realize the efficient and rapid erection and flexible unfolding of super large active array meter wave radar with an area of more than 100 square meters, there is an urgent need for the rapid development of new unfolding and retracting system design and engineering implementation technology[5-10].

In this study, a super large array unfolding and retracting system of meter wave high maneuvering radar is proposed. Based on the design technology of multi-level cross link linkage mechanism, the system realizes the unfolding and retracting, locking, side block folding and overturning, locking technology and so on. The lightweight and high maneuverability of large meter wave radar are realized by multi-level cross link motion synchronization control and multi link mechanism stiffness and strength design technology. This research is of great military significance for keeping up with the latest international radar development technology, benchmarking the highest radar development level, developing a new generation of radar system, and improving the combat and unfolding capability of China's active phased array radar, especially the meter wave.

2. Overall structure design

The super large array unfolding and retracting system is mainly composed of basic framework, side block turnover mechanism, side block locking mechanism, array unfolding and retracting mechanism, unfolding and retracting locking mechanism and synchronization rod, as shown in Figure 1. The foundation framework provides support, the side block turnover mechanism realizes the 90°turnover of the unfolding and retracting mechanism in the unfolding and retracting system, and the side block locking mechanism is the mechanical limit locking before and after the 90° turnover of the unfolding and retracting mechanism. The array unfolding and retracting mechanism realizes the horizontal expansion and retraction of the unfolding and retracting mechanism. The unfolding and retracting locking mechanism is the mechanical limit locking of the unfolding and retracting mechanism, and the synchronization rod ensures the synchronization of the unfolding and retracting mechanism.

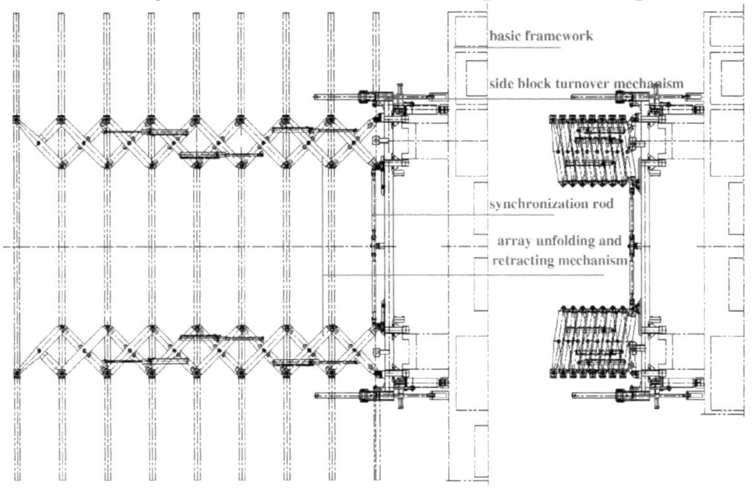

Figure 1. Schematic diagram of the overall structure

3. Technical design of unfolding and retracting system

3.1. Design of array unfolding and retracting mechanism

As shown in Figure 2, in this study, the array expansion and retraction mechanism is modularized. The array expansion and retraction mechanism module mainly includes basic skeleton, upper and lower cross connecting rod linkage mechanism, synchronization rod, unfolding cylinder, etc. Among them, the size and number of cross links (as shown in Figure 3) and the number and distribution of unfolding cylinders can be designed adaptively according to the size of side block antenna.

The array includes a column frame, on which an antenna vibrator and a reflection net are installed, and the column frame is installed on the expansion and retraction mechanism of the upper and lower array planes respectively. The upper and lower cross link linkage mechanism groups are multi-level cross link linkage mechanisms. Each level is connected with the column skeleton through the connecting flange. The first level cross link linkage mechanism is connected with the foundation skeleton through the fixed hinge points A1 and A2 and the linear sliding hinge points B1 and B2, in which the linear sliding hinge points B1 and B2 realize the sliding movement through the linear guide rail installed on the foundation skeleton. The upper and lower cross link linkage mechanism groups are driven by two groups of unfolding cylinders respectively, and the linear sliding points B1 and B2 are connected through the synchronization rod Ensure the synchronous movement of B1 and B2, so as to realize the synchronous expansion and retraction of the array unfolding mechanism.

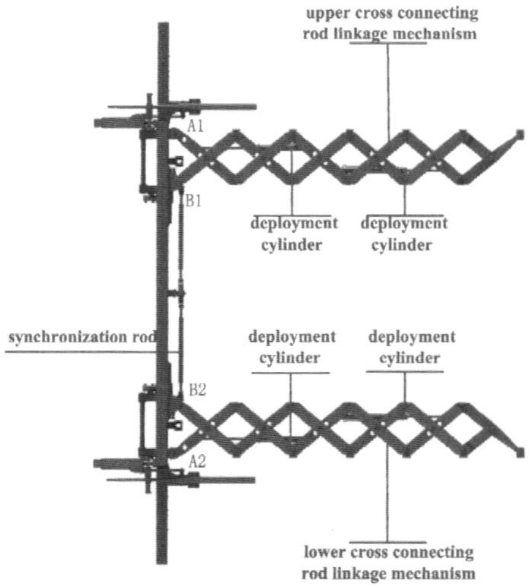

Figure 2. Array unfolding mechanism

Figure 3. Schematic diagram of single-stage cross link mechanism

3.2. Design of unfolding and retracting locking mechanism

The array unfolding mechanism needs to be mechanically locked by the automatic locking mechanism in both the unfolding and retracting states. Considering that the unfolding cylinder is horizontally arranged and the brake lock is set in the driving system to resist the pulsating load caused by wind load, the array unfolding mechanism does not receive continuous retraction force in the working state, and the locking mechanism mainly overcomes the abduction force caused by self weight in the driving state. Under these two working conditions, the force of the locking mechanism is small, which is a simplified mechanism, The system adopts the locking scheme of locking the sliding hinge point through the hydraulic bolt mechanism, as shown in Figure 4.

Figure 4. Schematic diagram of unfolding / retraction locking mechanism

3.3. Design of side turnover mechanism

The design of side turnover mechanism mainly considers the following three points: a) avoiding interference, b) The hinge points of the turnover cylinder shall be arranged to maximize the turnover force arm, c) the turnover angle shall be positioned accurately to ensure that the unfolding accuracy of the array meets the requirements under the working state. The side block turnover mechanism mainly includes: back frame, foundation framework, unfolding mechanism and turnover cylinder.

As shown in Figure 5, based on the above design principles, the turnover cylinder is arranged on the outside of the upper and lower groups of cross link linkage mechanisms of the array unfolding mechanism, and the lower point of the turnover cylinder on the back frame is arranged downward as far as possible on the premise of avoiding interference. In order to increase the stroke, the turnover cylinder is designed in the form of trunnion to facilitate the accurate adjustment of the positioning angle under the working state. The cylinder rod earrings are designed to be adjustable.

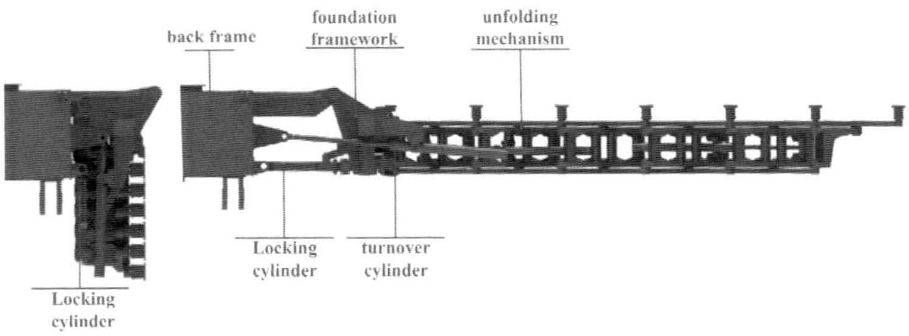

Figure 5. Schematic diagram of side block turnover

3.4. Design of side block turnover locking mechanism

The design of the unfolding locking mechanism is shown in Figure 5. After unfolding, a stable triangular structure is formed through the locking of the latch cylinder. The locking of the side block turnover mechanism in the retracted state is designed as the driving cylinder bolt structure, and the driving cylinder bolt mechanism is designed at the outer side of the array unfolding mechanism, which is convenient for inspection and maintenance.

4. Finite element model simulation analysis

The antenna array is mainly composed of multiple column skeletons and two pairs of retractable cross link mechanisms. The column skeleton and the cross link mechanism are fixed by bolt connection, and each column skeleton has a connection point with the upper and lower cross links. The finite element model of the test array is established by ANSYS, and the simulation analysis is carried out. The deformation of array antenna under wind load and self weight is analyzed.

According to the characteristics of the type, the structural mechanics modeling adopts the corresponding shell element shelli81 and body element solidi85. The elastic modulus, Poisson's ratio, density and other attribute values of each material need to be input to obtain the ANSYS finite element model, as shown in Figure 6. The finite element model includes 556149 elements and 531179 nodes.

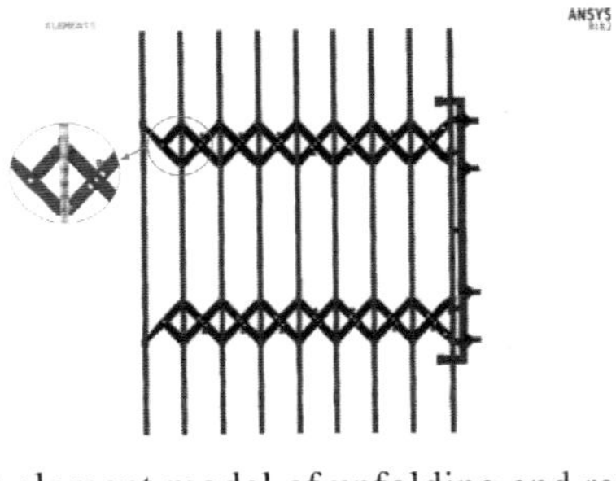

Figure 6. Finite element model of unfolding and retracting system

4.1. Self weight state

Figure 7. Cloud diagram of deformation of telescopic mechanism

In the unfolding and folding system, only the self weight load is applied, and the deformation of the unfolding and folding system is analyzed. As shown in Figure 7, the deformation cloud diagram of the unfolding and folding system is shown. It is easy to know from the figure that the maximum deformation of self weight is 26.2mm, including 10mm in the longitudinal direction and 23.7mm in the vertical direction, which is decomposed into 17.5mm in the normal direction of the array.

4.2. Wind load status

Figure 8 is the deformation cloud diagram of the unfolding and folding system (self weight + 25m / s wind speed, wind direction angle of 90 °). It is easy to know from the figure that the maximum deformation of the outer rod of the array under the self weight + 25m / s wind load is 42.2mm, including 30mm in the longitudinal direction The vertical direction is 27.5mm, and the normal deformation of the array is 37.6mm.

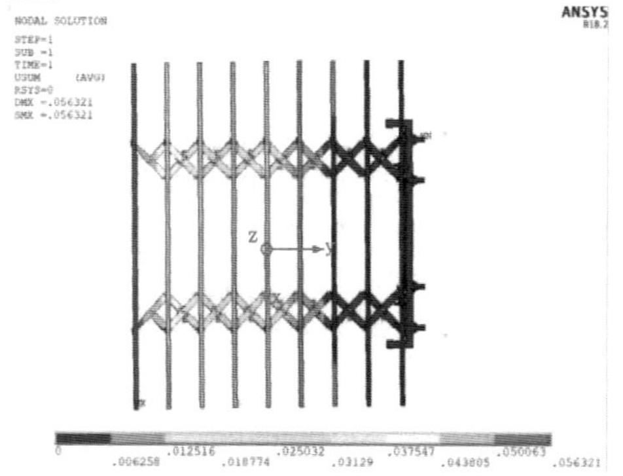

Figure 8. Cloud diagram of deformation of telescopic mechanism

Figure 9 shows the deformation of the antenna array and the deformation of the maximum stress with the wind direction angle under different wind speeds. It can be seen from the figure that when the wind speed is constant, with the increase of the included angle between the wind direction and the array, the maximum displacement of the array in the *X* direction gradually decreases, and the total displacement of the array in the *Y*, *Z* directions and increases gradually; When the angle between the wind direction and the array is certain, the displacement, total displacement and stress in each direction gradually increase with the increase of wind speed, and the displacement, total displacement and stress in each direction gradually increase with the increase of wind speed.

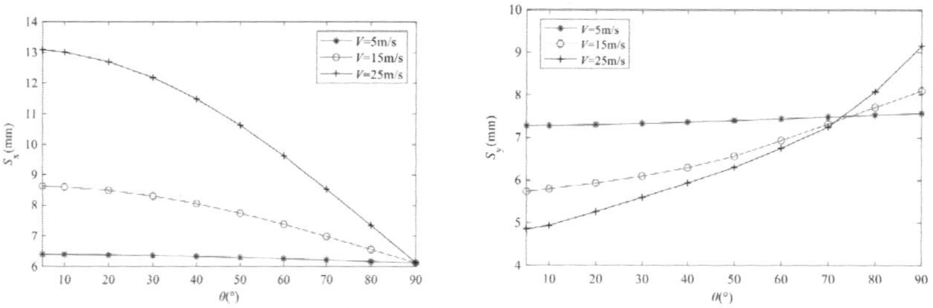

(a) Deformation in X direction (b) deformation in Y direction

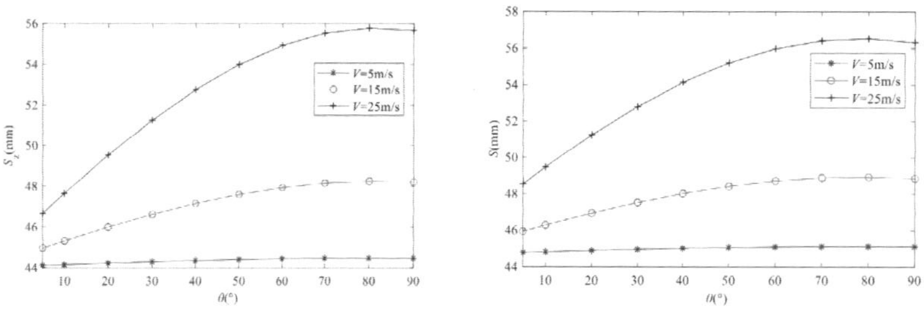

(c) Z-direction deformation (d) total direction deformation.

Figure 9. The deformation of the antenna array and the deformation of the maximum stress with the wind direction angle under different wind speeds.

5. Conclusion

Based on the spatial cross link mechanism, a unfolding and retracting system suitable for super large array is designed in this study. The unfolding and retracting locking, side block overturning and locking of the unfolding and retracting system are designed in detail. On this basis, the finite element simulation model is established, and the changes of stiffness and strength of the unfolding and retracting system under self weight state and different wind speed and wind direction angle are analyzed. The results show that the unfolding and retracting system can better achieve the working target of meter wave radar with a large array of more than 100 square meters, has high mobility and rapid erection ability, and the maximum stress and strain of the system meet the working requirements. This study provides a theoretical basis for the high mobility and precision measurement of super large array meter wave radar, and can be applied in practice, which has high popularization value and reference significance.

6. Reference

[1] Du Yong. Overall structure design of a three coordinate target indication radar [J]. Journal of ordnance and equipment engineering, 2015, 36(007):28-31.
[2] Zhou Chengzhe, Zhuo Hongli, Wang Hao. Antenna design of high maneuvering two-dimensional phase scanning meter wave three coordinate radar [J]. Modern radar, 2021, 43(12):6.
[3] Zhang zengtai, Fang Jingshi. Design of lifting and turnover mechanism of a radar antenna [J]. Radar science and technology, 2009, 007(004):312-315.
[4] Wei G A , Ying-Chuan Q I , Ping-Yong X U , et al. Design of Hydraulic Driving System in Lift Mechanism of Radar Antenna[J]. Journal of Air Force Radar Academy, 2006.
[5] Yan N , Guo Y J , Wang H . Characteristics Research of Hydraulic Lifting System Based AMESim[J]. Journal of Sichuan Ordnance, 2015.
[6] Lou H W , Chun-Jiang D U . Design and Simulation of Radar Antenna Lifting Mechanism[J]. Electro-Mechanical Engineering, 2016.
[7] Huang K , Sun H H , Qiao L . Structural design of an elevated mobile radar antenna lifting rod[J]. Radar & ECM, 2019.
[8] Tang Weimin, Guo Yajun, Wang Lei. Design and analysis of radar back frame turnover mechanism based on aemsim [J]. Journal of Sichuan ordnance industry, 2022(11).
[9] Guo Yajun, Fang Jingshi, Wang Zhihai. Optimization analysis of a radar back frame turnover mechanism [J]. Journal of Ningbo Institute of engineering, 2019(1):6.
[10] Peng guopeng, Huang Haitao, Ye Beifa. Research on automatic unfolding mechanism and mechanical characteristics of a segmented radar protection frame [J]. Machine tools and hydraulics, 2021, 49(16):4.

Numerical Analysis about the Dynamic Properties of a Pressure Relief Valve and Improvement Measures

Haibo Dong[1], Shaoteng Li[1], Chunguang Xu[2], Jun Liu[3]

[1]Beijing Institute of Space Mechanics & Electricity, Beijing 100094, China; Laboratory of Aerospace Entry, Descent and Landing Technology, CASC, Beijing 100094, China
[2]School of Aeronautics and Astronautics, Sun Yat-sen University, Guangzhou, 510006, China
[3]School of Aeronautics and Astronautics, Dalian University of Technology, Dalian, 116024, China

donghb508@163.com

Abstract. In order to investigate the influence of internal flow field on the dynamic characteristics of a pressure reducing valve, based on the new discrete geometric conservation law and high-order coupling algorithm, numerical simulations of the dynamic characteristics have been performed. The three-dimensional unsteady integral forms of ALE equation is solved by finite volume scheme based on the spring analogy method and dynamic grid. The novel techniques of virtual mesh ventilation method is adopted to solve the problem when multiple bodies move from contact to separation and the mesh topology changes. The results show that the outlet pressure of valve oscillates within wide range upstream pressurization rate, the average pressure is close to the theoretical value of static performance design; numerical simulation reappears vibration divergence which is similar to the test phenomenon, flutter phenomenon is caused by the interfere between aerodynamic force and dynamic systems. By modifying the method, the fault can be eliminated.

1. Introduction

A spring-loaded pressure reducing valve consists of a compression spring, which is a key device used to protect pressure vessels and systems[1]. The function of the valve is to limit the maximum pressure that can exist in a system. Under ideal conditions, the valve is designed to reduce high inlet pressure to a lower level of outlet pressure which has to be maintained within a wide operating range. The traditional development process must go through the sample production, test and improvement process, this makes longer period and higher cost. As a consequence, with the development of digital computers, computational fluid dynamics (CFD) has increasingly being used to supplement experiments for product development. CFD is becoming a well-established practice also in analysis and design, since it can give a clear insight into its operation mode, which can't be achieved by experimental tests and measurements. The numerical experiments through CFD also save development time by reducing the number of experiments as one can perform parametric studies without building prototype[2-3].

In recent years, many scholars around the world have carried out a series of numerical simulations on the dynamic characteristics of valve. Beune et al.[4] used fluid-structure interaction method to study the influence of valve dynamics on steady flow performance through a safety valve. Song X. G. et al.[5] investigated transient flow characteristics and dynamic behavior of a spring-loaded pressure safety valve in three dimensions. The literature generated pure structural meshes to ensure the correlative mesh without negative volumes. Daines et al.[6] predicted the valve flow coefficient as a function of valve plug position prior to testing a modified control valve. Srikanth and Bhasker[7] used ANSYS-CFX software to simulate valve element mesh motion. The literature predicted pressure history was significantly affected by the velocity of moving contact in the puffer chamber.

Therefore, as mentioned in the above literatures, CFD numerical simulation can play a very crucial role in the analysis of pressure safety valve. Due to the complexity of valve internal structure and small space scale, it is very difficult to measure data in internal space during the opening and closing process, CFD can predict flow rate and pressure loss caused by internal complex flow; at the same time, due to the greater gas compressibility and the lower sound velocity, it often forms the supersonic flow phenomenon such as shock wave and choking during the valve opening and closing process, severe aerodynamic load changes may affect valve structure and lead to valve malfunction, CFD can also perform the failure analysis[8].

However, many studies are simulated with CFD commercial software, and some problems in the process of study tend to certain simplified processing. In order to reduce geometrical complexity and to limit calculation time, many researches use axially symmetric simplified model to predict the internal flow field. Since most of safety valves are not axisymmetric, symmetrical simplified model can't describe the internal flow characteristics. Currently, dynamic analyses of moving valves are rarely performed and most analyses involving valve operation are limited to either obtaining steady state performance-related flow coefficient curves or simplified analysis of surge equations in pipeline systems. Without dynamic analyses, these researches can't investigate transient events and chatter. Due to large deformations of the computational domain intermediate meshes are necessary to keep the mesh quality appropriate. Using a single deformable grid, such as structural grids, which is not applicable for the large deformation calculation.

During the ground experiments, due to the unreasonable internal structure design, the pressure reducing valve would be in failure state after working for a little while and caused the whole pipeline system couldn't work normally. The specific fault states could be summed up in two kinds: one kind was the valve would be violent vibration with harsh noise; the other kind was the valve would remain opening, the fluid had been kept flowing through pipeline without obvious velocity decrease. Because of the pressure reducing valve size, the diaphragm is only 82mm in diameter, it couldn't measure the internal pressure, speed and other data information inside the value, it had to repeat the failure state only through numerical simulation; at the same time, according to the design requirements, when the inlet pressure was different, the valve outlet pressure must be kept at about 1.7 MPa, the error is less than 5%.

The main objective of modeling the spring-loaded pressure reducing valve in the present work is to use CFD to obtain an estimate of the transient flow characteristics and dynamic behavior, and diagnose the root cause of failure. In order to describe the complex configurations within the valve, we use unstructured hybrid meshes composed of any combination of tetrahedral, pyramids, prisms, or hexahedra. Without using any of the finite element software, the CFD codes are based on the finite volume discretization on unstructured hybrid grid. These literatures[9-10] show the calculative ability of the codes. In this paper, we have developed a three-dimensional computational fluid dynamics model in combination with a dynamics equation. When fluid-domain involves movement of internal objects, transient flow simulations are highly dependent on grid quality and skew angles between mesh points. We have also extended our multi-element unstructured framework to include timing studies for transient valve operation by including valve movement and capturing the flow instabilities associated with it. Our approach accounts for this change in grid topology through a unique methodology of mesh movement to accommodate the valve[10]. In the present work, we also use a novel technique to realize real opening and closing of the valve. Some common commercial softwares, such as Fluent and ANSYS-CFX, can't realize the function. We will introduce the virtual mesh ventilation method in later chapters.

2. Geometrical configurations and numerical method

The pressure reducing valve studied in this paper is simplified from that used in a rocket system. Figures 1 shows the structure schematic diagram of the valve, in the state of compression, the main spring makes diaphragm upward deformation, the diaphragm is only 82mm in diameter, when it contacts the valve stem and they push the valve core moving up, the auxiliary spring and flapper control its movement displacement, the high pressure gas from the inlet enters into the low pressure chamber through the gap which is formed between valve core and flapper, it could generate the choking phenomenon from the gap that plays the current-limiting and decompression effect. Gas into the increases the pressure; the pressure difference on both sides of diaphragm makes main spring occur deformation and valve core move down, valve opening gets smaller in order to reduce inlet flow. The flapper controls the movement displacement of diaphragm and main spring. High pressure gas not only can flow into the unloading chamber, but also through the hole in the middle valve core into the low pressure cavity in order to change aerodynamic load. Although the flux is very small, the effect can't be ignored.

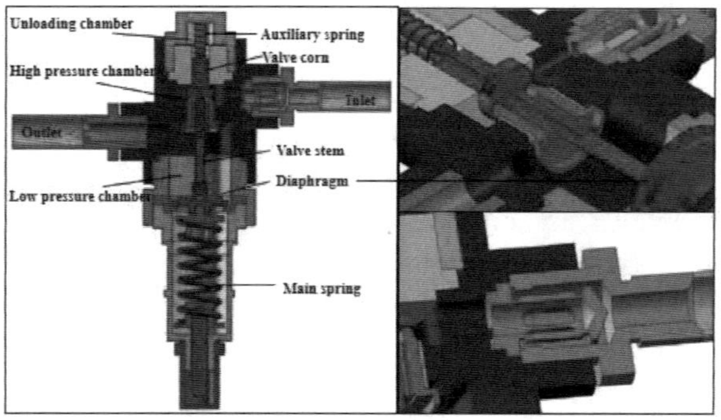

Figure 1. Schematic diagram of pressure reducing valve

Gas viscosity is smaller than the liquid. In the case of supersonic flow, the load changes caused by shock and expansion wave are much greater than the contribution of viscosity in the local space of gas valve, showing the time derivative term in the NS equation is bigger than the influence of viscous term. In order to improve the calculation efficiency, using three-dimensional Euler equations in this paper, the case also verifies this simplified method is feasible. The calculation method can refer to these literatures[11-12], which are not listed in this paper.

The computational grid uses tetrahedral unstructured grid to discrete internal complex spatial structure, in the inlet and outlet, data monitoring points are shown in figure 2. The model used to control mesh deforming is improved spring analogy with boundary improvement and torsional effect improvement[13]. Describing the contact and separation process between valve core and valve seat, the volume V_i starts from 0, it is contrary to the stability condition. In order to improve the efficiency of calculation, it is usually used in commercial software that reserved within the grid filling viscous fluid to simulate the close condition which can't simulate the real physical process, need to improve its methods. Literature[13]has developed a new technology of virtual grid and achieved good effect in the simulation, its basic principle as shown in figure 3. Before generating initial grid, it designs a layer of grid in contact separation position or small scale cracks, according to the moving objects in the grid, sets the virtual baffle as the wall surface like the shadow part in the figure, while the other parts still keep flow through.

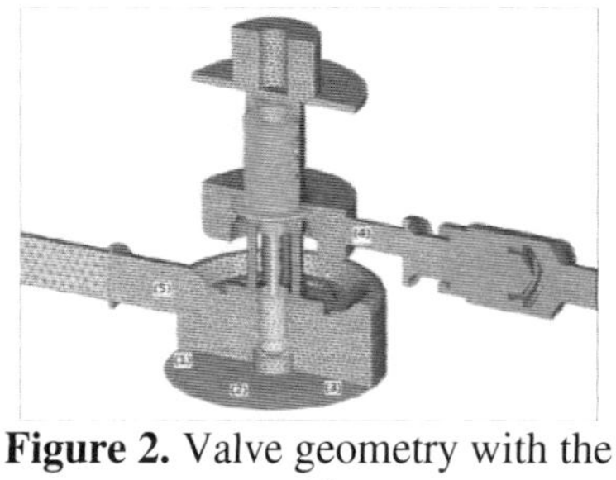

Figure 2. Valve geometry with the grid.

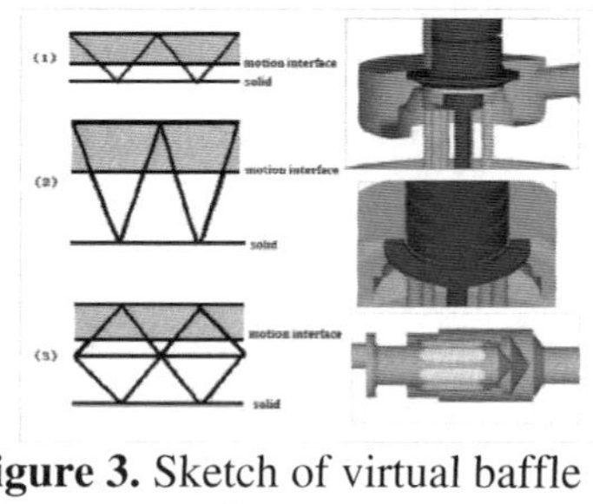

Figure 3. Sketch of virtual baffle faces.

This new technology can simulate the case including infinitesimal crack width of zero, the process from contacting to separating. At the same time, it will not affect the calculation for the small grid scale.

When the position of the moving object is equal to the length of the virtual baffle grid, normal grid deformation is used to describe the width increase. While the tensile distance is too big to deteriorate the grid quality, it usually adopts grid updates to finish reconfiguration.
This paper sets up the virtual baffle in three positions: 1) the gap between diaphragm and flapper, the opening is less than 0.4041mm until fully closed position; 2) the gap between valve core and the ring of valve seat, it makes the high pressure gas pass to the unloading chamber, then the aerodynamic load will be effected, therefore, in the seat setting 0.45mm grid width to simulate the flow through 0.009mm crack width; 3) before entering into the high pressure chamber, high pressure gas pass through a porous filter structure, according to the previous experimental data, it has obvious pressure drop and also needs to set up virtual baffle.

3. Results of simulation

3.1 The flow field characteristics when the high pressure gas entering into the valve

The high pressure gas flow through the pipeline which is connected with the inlet into the high chamber, the computation time starts from the instant of solenoid valve starts working, the working time is range from 10ms to 100ms. The inlet pressure has a linear increase in solenoid valve working period and the rapid pressurization process generates a shock wave in the pipeline. When the shock wave meets the smaller pipe diameter and the filter wall, it will generate reflection shock wave and return to the inlet. Following the increase of inlet pressure, the pressure in low pressure chamber is also raised and drives the diaphragm and the valve core with the auxiliary spring to move downward. Finally, the valve opening decreases gradually. When the inlet pressure gets its maximum value, it will maintain in a certain range with a small vibration both in low and high pressure chamber. Due to the limitation of computational domain, the reflection shock wave quickly reaches the entrance and interferes with the inlet boundary condition which will lead the simulation failure. In order to solve this problem, this paper proposes a one-dimensional flow model with inlet and outlet boundary conditions. Particular way as: in a three-dimensional computational domain, a long enough one-dimensional section line model is connected with inlet and outlet condition, it sets two rows of virtual grid point to deliver the flow flux between the one-dimensional model and three-dimensional model. Total upstream pressure of standard case is 21MPa, initial flow field is the standard atmospheric parameters(0.1MPa). At the entrance of the one-dimensional model, total pressure increases from 0.1MPa linear to 21MPa in a given time. The model has 4000 grid points and the length is 10m at the entrance, it has 1000 grid points and the length is 2.5m at the exit. Shown as in figure 4.

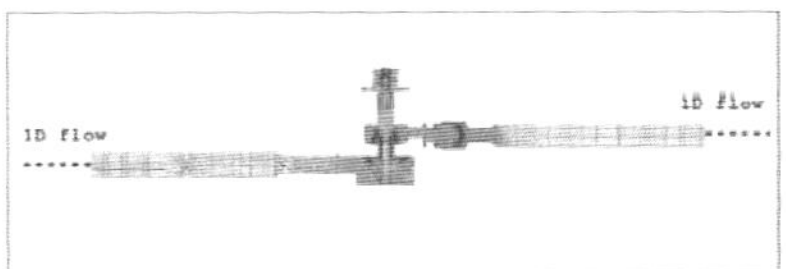

Figure 4. Computational grid for the model.

It is shown as in figure 5 that time-averaged Mach number distribution along the plane of symmetry – it can be seen that the shock wave reflect upstream at the entrance, after high pressure gas into the low pressure chamber, pressure is increased and released from the outlet. When the inlet pressure gets its maximum value, pressure reducing valve goes into the stable working condition; the valve opening is very small, nearly close. The high pressure gas via flange structure enters into the unloading chamber, then into the low pressure chamber by the slender hole.

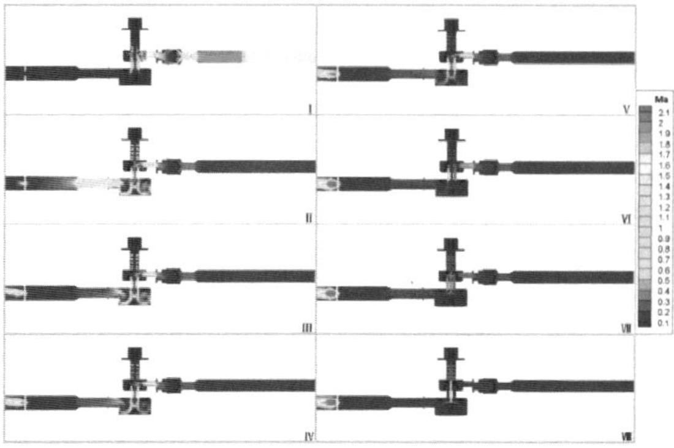

Figure 5. Sequence of Mach number distribution along the symmetry plane of the valve.

Figure 6 shows the pressure contours and it can also see the same phenomenon like figure 5.With inlet pressure to the maximum, the unloading chamber pressure is increased too.

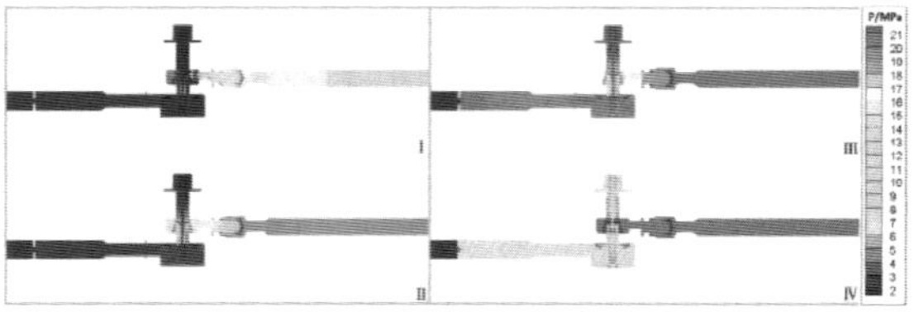

Figure 6. Sequence of pressure distribution along the symmetry plane of the valve.

Figure 7 shows the pressure contours of filter, the upstream flow generates reflection and diffraction wave, it declares virtual baffle can successfully simulate the resistance effect and the high pressure gas flowing process, at this stage, the flow flux enters into the low pressure chamber is quite small and forms pipe choking. It generates pressure gradient on both sides of the filter virtual damper and has significant pressure difference at the flange.

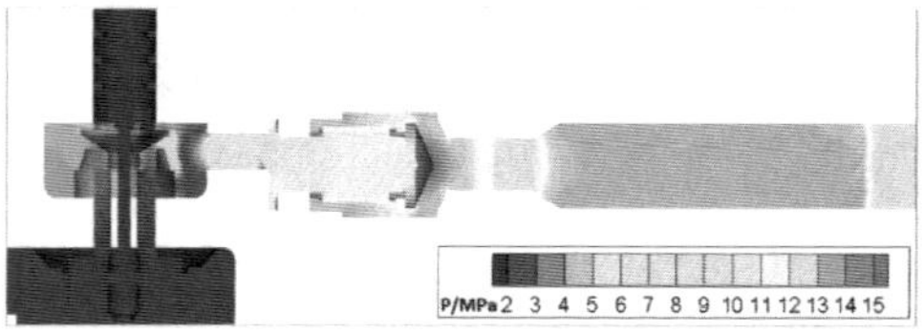

Figure 7. Stress contours within the piping system.

Fig.8 shows the pressure curves of probe points at outlet and diaphragm, the four curves are almost unanimously and it is the subsonic flow in low pressure chamber with mainly uniform pressure distribution, therefore the pressure value of point5 can be used to represent the whole chamber pressure. In accordance with the original calculation model, the main spring preload force is 1050N with the theoretical value pressure of the chamber is 1.7MPa, calculation slightly above average. If the main spring preload force is 678N with the chamber theoretical value is 1.2MPa, the pressure curves which is also fluctuates around the theoretical value are similar to Fig.8. Considering the viscous effects, the calculation is higher than the theoretical value, which is also reasonable.

3.2 The influence of opening pressure for flow field characteristics

In the stable working case, the pressure of upstream piping is close to the high pressure chamber. In order to investigate the reducing pressure effect, according to the working condition, the total upstream pressure take 21MPa, 15MPa, 8MPa and 2MPa to calculate, respectively.
Extracting probe points' data, the pressure curves at different time is shown in Fig.9.

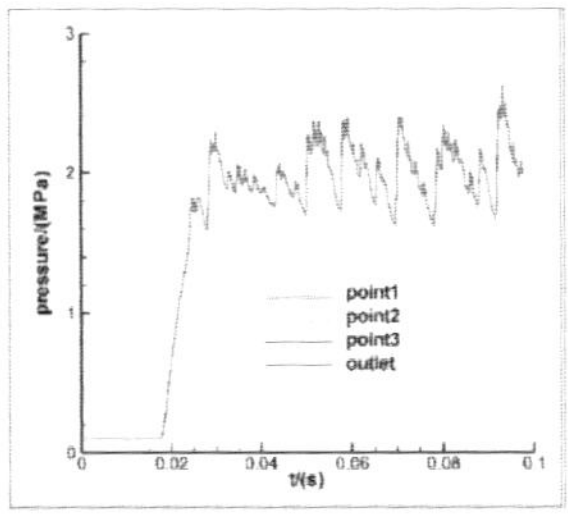

Figure 8. Pressure curves of observation points.

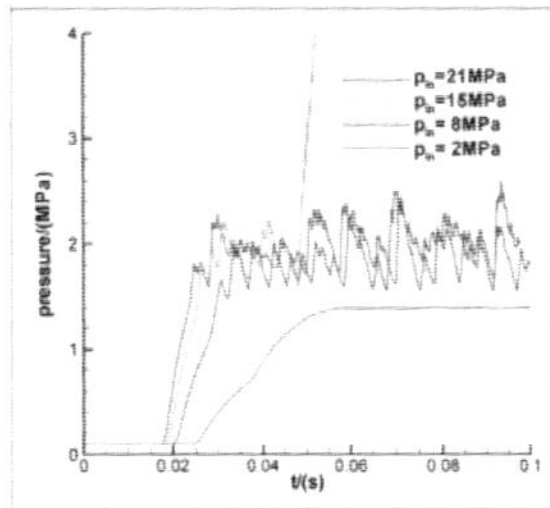

Figure 9. Pressure curves of low-pressure chamber.

Although the upstream pressure changes from 21MPa to 8MPa, the pressure still fluctuates around 1.7MPa in the low pressure chamber. When upstream pressure is 2MPa, it is close to design index, pressure difference is so small that the aerodynamic force of core is not enough to overcome spring preload force, keeping the full open state, there is also a stable solution for flow field. Due to the total pressure loss in the pipeline, pressure fluctuates around 1.5MPa, implements the decompression on flow loss alone.

Fig.10 shows the displacement of valve core with diaphragm at different opening pressure. The original point illustrates the closed case, the displacement below origin represents both of them are on the detached state. Contrarily, the displacement above origin represents they are moving together. When opening pressure is larger, elastic elements keep in vibration form. Different opening pressure lends to the different mean value, the higher opening pressure, the smaller average amplitude of the valve core and the smaller valve opening. The case of 2MPa opening pressure, the pressure reducing valve is not normal operation because the pressure in low pressure chamber is so small that diaphragm is almost stationary.

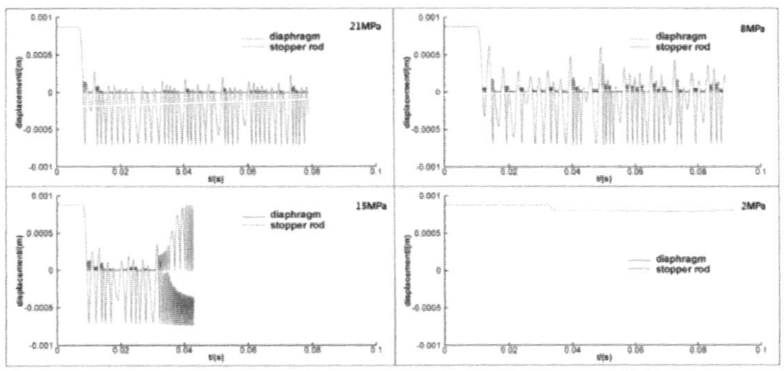

Figure 10. Displacement curves of valve core and diaphragm at different opening pressure.

The upstream 15MPa case, the stress contours are similar to Fig.7, but pressure is more than 7MPa in low pressure chamber, a complete failure case. The elastic elements displacement and pressure curves are shown in Fig.11(a) and (b), respectively. The results represent that both valve core and diaphragm are working at a small vibration case in a period of time, then divergent vibrating, pressure rises sharply. The phenomenon also appears in the case of 15MPa with 30ms, shown in Fig.11(c) and (d). By the two fault cases, speculated that using the original design of the low pressure chamber, outlet orifice is so close to the low pressure chamber that pipeline volume is relatively small, and pressure in the high pressure chamber is high, pressure change rate is sensitive to valve opening. If pressure vibration is same as the structural natural vibration frequency will lead to flutters and valve damage.

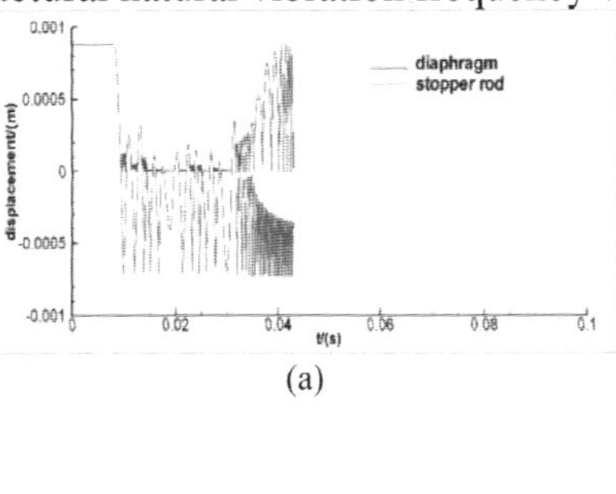

(a)

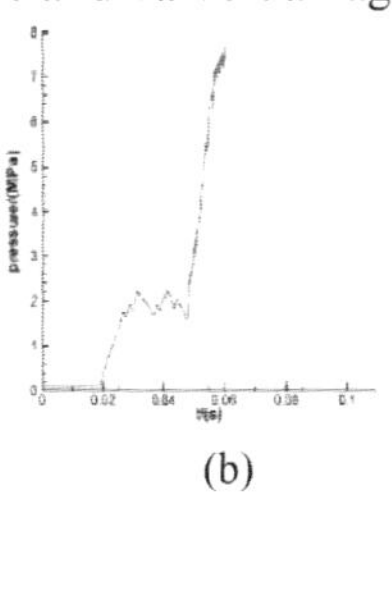

(b)

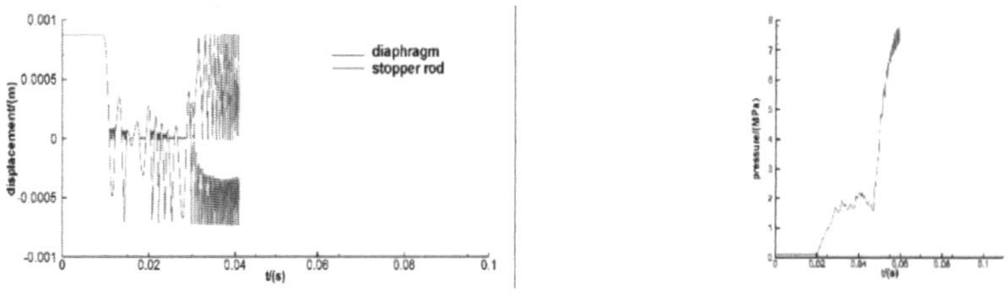

(c) (d)

Figure 11. Plots of the solution showing: (a) displacement curves of valve core and diaphragm and (b) pressure curves of low pressure chamber.

3.3 The influence of opening duration time for flow field characteristics

The elastic elements displacement and pressure curves are shown in Fig.12, respectively. Following the increasing of inlet pressure, the higher pressure in the high pressure chamber, the smaller valve opening, the vibrating curve illustrates reducing process; the longer time, the slower process.

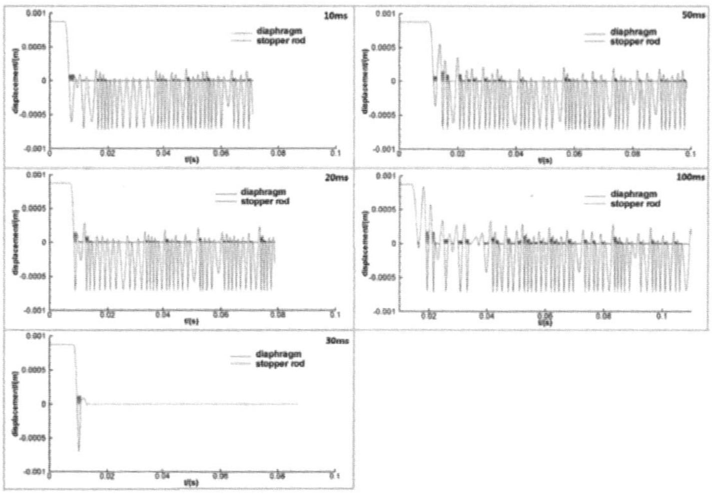

Figure 12. Displacement curves of valve core and diaphragm at different opening time.

There is also an abnormal working state in the 30ms case. The reason for this is that diaphragm collides with flapper and both of them continue moving upward, momentum loss makes them keep contact with same speed until closed. Finally, pressure in low pressure chamber is stable in a certain value, but not the design value. Orifices always exhaust gases to outside, in the mean while; gases in the unloading chamber pass the flange into the low pressure chamber where can keep pressure balance. Fig.13 shows the stable working conditions. The abnormal working state is due to the large air leakage which can lead to valve failure.

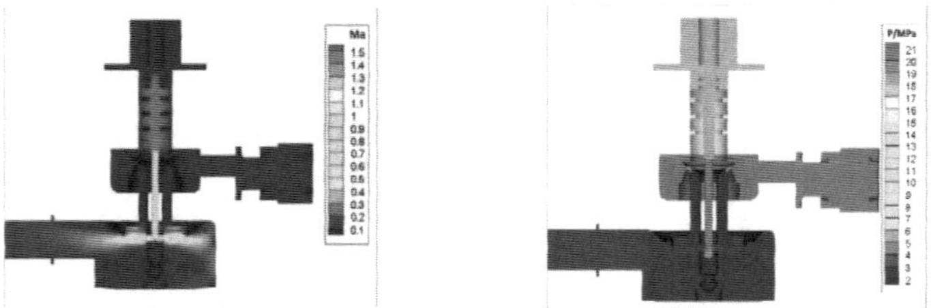

(a) (b)

Figure 13. Plots of the solution showing: (a) mach number and (b) pressure.

The results of the 8MPa case are similar to the 21MPa case. The influence of opening time for valve characteristics is similar to the case 21MPa, but the average valve opening is much bigger.

4. Conclusions

(1) Based on unstructured dynamic grid calculation of the fluid-structure coupling method can be used for high-fidelity dynamics simulation of valve, the numerical example shows that the new discrete geometry conservation law and interface algorithm has higher accuracy;
(2) The virtual baffle technology well solves problem of space topology change inside the valve, the results reasonable show pressure change caused by the filter and flange structure in flow field, separation and contact process between different components;
(3) Under normal working state, the dynamic characteristics show oscillation and the average of the outlet pressure is close to design theoretical value, for different inlet gas pressurization rate, it also has a good adaptability;
(4) The lower upstream pressure, the higher reliability, outlet pressure is lower than the design value, products have the minimum inlet pressure;
(5) The pressure reducing valve will appear flutter phenomenon caused by interfere between aerodynamic force and dynamic systems, the fluid-structure interaction phenomenon is in line with the mechanism through the qualitative analysis, it may take appropriate measures to eliminate the flutter phenomenon;
(6) Under some individual pressurization rate conditions, valve is out of work, the main reason is the discharge effect of flange gap and the width of cracks needs to pay attention to when testing.

5. References

[1] Song X. G., Wang L., Park Y. C., 2010, "Transient Analysis of a Spring-Loaded Pressure Safety Valve Using Computational Fluid Dynamics (CFD)," ASME J Press Vess., 132(5), pp. 054501-054501-5.
[2] Himadri C., Arindam K., Binod K. and Tapas G., 2012, "Analysis of Flow Structure Inside a Spool Type Pressure Regulating Valve," Energy Conversion and Management., 53, pp. 196-204.
[3] Dossena V., Marinoni F., Bassi F., Franchina N., Savini M., 2013, "Numerical and Experimental Investigation on the Performance of Safety Valves Operating with Different Gases Review," International Journal of Pressure Vessels and Piping., 104, pp. 21-29.
[4] Beune A., Kuerten J.G.M., Heumen M.P.C., 2012, "CFD Analysis with Fluid–Structure Interaction of Opening High-Pressure Safety Valves," Computers & Fluids., 64, pp. 108-116.
[5] Song X. G., Cui L., Park Y. C., 2010, "Three-Dimensional CFD Analysis of a Spring-Loaded Pressure Safety Valve from Opening to Re-Closure," Proceedings of the ASME 2010 Pressure Vessels & Piping Division / K-PVP Conference, Bellevue, Washington, 25024, pp. 1-9.
[6] Russell L., Jody L. and Peter R., 2003, "Computational Analysis of Cryogenic Flow through a Control Valve," Proc. 4th ASME_JSME Joint Fluids Engineering Conference, Honolulu, Hawaii, 45120, pp. 221-226.
[7] Srikanth C., Bhasker C., 2009, "Flow Analysis in Valve with Moving Grids through CFD Techniques," Advances in Engineering Software, 40, pp. 193-201.
[8] Vineet A., Ashvin H., Jeremy S. and Peter A., 2006, "Simulations of Instabilities in Complex Valve and Feed Systems," Proc. 42nd AIAA/ASME/SAE/ASEE Joint Propulsion Conference & Exhibit, Sacramento, 4758, pp. 1-10.
[9] Liu Jun, Bai Xiao-zheng, Guo Zheng, et al. A new method for transferring flow information among meshes[J].Computational Fluid Dynamics Journal, 2007, 15(4): 509-514.
[10] Guo Z., Liu J. and Qu Z., 2004, "New Computational Method to Determine Trim Angle of Attack for Reentry Vehicles," JOURNAL OF SPACECRAFT AND ROCKETS, 41(6), pp. 1063-1065.
[11] Dong H, Zhang F, Xu C, et al, 2018. "An improved uncoupled finite volume solver for simulating unsteady shock-induced combustion," Computers & Fluids.167: 146-157.
[12] Dong H., Liu J., Chen Z., et al, 2018. "Numerical Investigation of Lateral Jet with Supersonic Reacting Flow. Journal of Spacecraft and Rockets," 55(4): 928-935.
[13] Guo Z. Liu J. and Chen G., 2008, "Numerical Analysis of Dynamic Properties of Safety Valve by Using Method of Fluid and Structure Interaction," CHINESE JOURNAL OF MECHANICAL ENGINEERING, 44(8), pp. 61-66.

Design of Power Supply System for High Power SAR Satellite

Yongwei Liu*, Yu Cheng, Kehai Bao, Houchun Li and Zhihao Wan

Shanghai Institute of Space Power-Sources, China
Email:lyw0209@hrbeu.edu.cn

Abstract. Synthetic Aperture Radar (SAR) satellite power system has the characteristics of high power, pulse power supply and fast response speed. Based on the combination of domestic and foreign SAR satellite power supply structures, a SAR satellite power supply system scheme with a composite bus architecture is proposed in this paper. By adding battery discharge regulators (BDR) between the platform bus and the SAR load bus as a grid-connected controller, the SAR load bus can supply power to the platform load, and the solar array and battery pack can be used efficiently while achieving dual bus output. Aiming at the composite bus, this paper also proposes an efficient high-power filter technology. Finally, the ground test results verify the correctness of the scheme, which can meet the different power supply requirements of the SAR satellite stable and pulse load equipment.

1. Introduction

Synthetic Aperture Radar (SAR) is an active microwave imaging sensor, which is not affected by weather and climate, and can observe all-day, all-weather, high-resolution, and large-area earth observations[1]. It has been widely used in military reconnaissance, activity monitoring, target recognition and other fields. As an important part of the satellite, the power supply system provides energy for the satellite, which is the basic guarantee for the normal operation of the satellite.

However, the SAR satellite power system usually has the characteristics of high power and pulsed power supply, and its demand can reach 5-30kW, which is much greater than the power demand of ordinary satellites[2]-[5]. The high-power pulsating load will have an adverse effect on the stability and reliability of the satellite, which in turn brings great challenges to the design of the entire satellite power system. Therefore, how to improve the reliability and electromagnetic compatibility of the SAR satellite power system to meet the stable power supply demand of the platform load while meeting the high-power pulse power supply demand of the SAR load has become the research focus of the SAR satellite power system design.

Aiming at the design requirements of SAR satellite power system, this paper firstly compares and analyzes the characteristics of domestic and foreign SAR satellite power system; then, a composite bus system SAR satellite power supply scheme is proposed, which can realize that the SAR load bus provides energy for the platform load through the grid-connected controller(BDR) when the platform bus energy is insufficient. Besides, by adding a high-efficiency and high-power filter to eliminate the impact of the SAR load bus pulsation caused by SAR work on the input of the grid-connected converter, and further improve the reliability and safety of the entire satellite power supply system. Finally, the ground test results prove the correctness of the scheme design.

2. Features of the SAR satellite power system

The typical SAR satellite mainly include India's Sky Eye 1 (RISAT-1), which adopts an unregulated single-bus power supply system based on MPPT, has a small number of system equipment, relatively low cost, and good dynamic characteristics[5]; limited to the platform load and the SAR pulse load sharing a bus, the serious bus pulse noise interference makes the platform load power scalability poor. The German Terrestrial Radar-X-band satellite (Terra SAR-X) provides stable power demand for the platform load through the DC/DC converter on the unregulated bus, the SAR load bus fluctuation caused by pulsed power supply also increases the DC/DC converter design difficulty[6]-[7]. China's Gaofen No.3 adopts a dual independent bus design, that is, the platform load uses a fully regulated bus, and the SAR load uses an unregulated bus; they are independently equipped with solar arrays and battery packs; although the two buses do not interfere with each other, this scheme requires a large number of equipment, relatively large volume and quality, and the technical scheme is not optimized enough. Through the comparative analysis of the topological structure and power adjustment control methods of the above mentioned domestic and foreign SAR satellite power supply systems, the characteristics are summarized as follows:

2.1. Single bus power supply system
The single bus power system is a primary power with single bus output, the power distribution and regulation of the satellite is realized through the power controller and battery, so this power supply system has a small number of equipment, small volume and mass, can make full use of energy, and has a relatively low cost[8]-[9]. Besides, the power supply bus is an unregulated bus and has extremely low output impedance due to its direct connection to the battery pack, which is beneficial to meet the short-term peak load and pulse load power requirements to the greatest extent. However, since the stable and the pulse load share the same bus and the stable power supply demand of the platform bus, the single-bus power supply system is often used for small and medium satellites with SAR load less than 3kW.

2.2. Dual bus power supply system
The dual bus power supply system is a primary power supply with dual independent bus output- the fully regulated bus and the unregulated bus[10]. Because the big power ratio between the stable and pulse load of the SAR satellite, the peak power of the pulse load can reach tens of thousands of watts, and the stable platform power is usually 1-3kW. Therefore, the satellite power system is equipped with stable bus and the unregulated bus for the stable platform load and the pulse load respectively, and each bus has its own solar array and battery pack and they are independent of each other, which can effectively avoid the interference of pulse load to stable load. However, considering the high stability characteristics of the platform equipment, the power system usually requires a DC/DC converter to realize the power distribution from the platform battery pack to the stable bus during the shadow period. Therefore, the advantages of this adjustment method are stable bus voltage accuracy, good EMC characteristics, and stable input voltage for various aerospace instruments, so the configuration is very flexible, and it is easier to achieve modularization and generalization; but its disadvantage is that the volume and heat consumption of the power supply controller are relatively large, and also has the problems of complex control and low energy utilization.

3. Design of SAR satellite power system based on composite bus
The SAR satellite power system provides energy for platform equipment and SAR loads. Most of the platform devices are stable and require high-quality bus to achieve precise satellite control; the SAR load is in a working state of high power, variable frequency, and repetitive pulses, which brings greater electromagnetic interference to the system, and increases the ripple of the power supply bus[11]-[12]. Therefore, the design of power system must take into account various power supply requirements.

3.1. Topological structure design
Combining the above analysis and comprehensively considering various technical characteristics, in order to avoid the interference of the SAR pulse load on the platform equipment and ensure the safety and reliability of the entire satellite power supply, this paper proposes a composite bus design scheme. Compared with the traditional dual independent bus configuration, this solution uses BDR as a grid-connected controller to realize that the SAR load bus provides energy for the platform equipment

when the fully regulated bus energy is insufficient. The main indexes and requirements of the power system are shown in table 1, and the topological structure is shown in figure 1.

Table 1. Main technical specification of the power supply system.

Category	Specific requirements
Bus configuration	A fully regulated bus for stable load, and an unregulated bus for the pulse load.
Bus voltage	The fully regulated bus voltage range is 30±1V, the unregulated bus voltage is 50~65V.
Solar array	The solar array adopts triple junction gallium arsenide cell, the average conversion efficiency is not less than 32%.
Battery Pack	Configure a battery pack with a rated capacity of 270AH, and the average depth of discharge does not exceed 30%.
Power controller	Platform power controller: shunt efficiency, discharge efficiency greater than 85%; SAR load power controller: shunt efficiency, charging efficiency is greater than 90%
Quality	Not more than 280kg

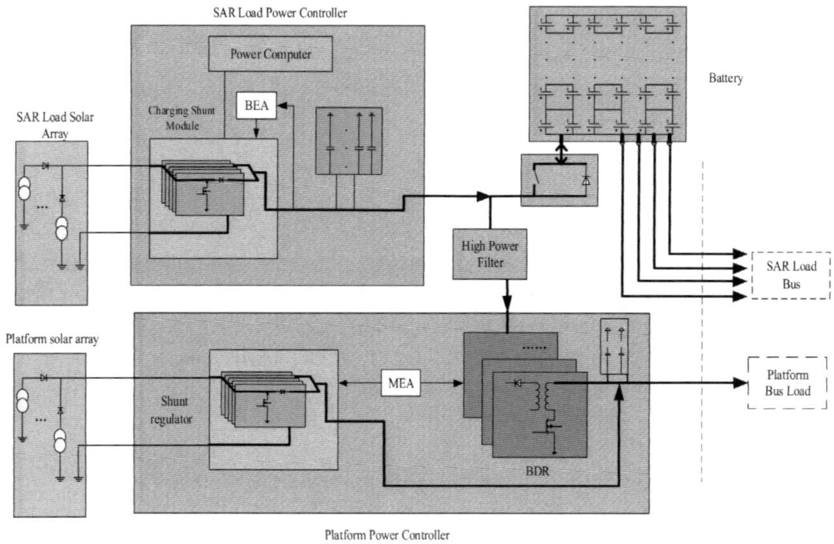

Figure 1. SAR satellite power system structure.

As can be seen from figure 1, the stable fully regulated bus is mainly composed of platform solar arrays, shut regulator and BDR, which adopts frequency-limiting sequential switch shunt adjustment technology to realize that the solar array provides 30V±1V and constant power consumption greater than 1050W for the platform load during the light period; during the shadow period, the grid-connected controller(BDR) provides a maximum peak consumption of 3000W for the platform load from the battery. The unstable bus for pulse load consists of the SAR load solar array, the battery pack and power controller. It is an unregulated bus with the voltage range of 50~65V and a maximum power of 25kW. Therefore, the SAR satellite power system topology in this paper has the following advantages[13]:

The platform based on fully regulated bus can provide a high-quality power supply bus for platform load. For equipment with higher input voltage accuracy, no secondary conversion is required, which reduces the complexity of power conversion; The SAR load adopts an unregulated bus, which is directly connected to the battery to meet the power supply requirements of the short-term peak load and pulse load to the greatest extent. They are still independent of each other under the composite bus power supply system, which can guarantee the reliability and electromagnetic compatibility of the entire satellite. Besides, the fully regulated bus is converted from the SAR load bus through the grid-connected controller BDR circuit when the platform bus energy is insufficient, so there is no need to configure two sets of battery packs, which can achieve efficient use of battery packs and substantially reduce the volume and quality of the power supply system. But at the same time, the effect of unregulated bus fluctuations on the input of the BDR circuits must be taken into account, and we will discuss this issue in Section 3.3.

3.2. The working principle of the power system

According to the requirements of power supply and distribution tasks, the entire satellite power supply system has three modes: external power mode, combined power supply mode and internal power mode, as follows:

External power mode: when the system load during the orbit sunlight period is less than the output power of the solar array of the platform, the solar array supplies power to the platform load and SAR load; combined power supply mode: when the effective load during the sunlight period is greater than the output power of the solar battery, the solar array and the battery pack are combined as the platform load power supply; internal power mode: during the orbit shadow period, the battery pack supplies power to the entire satellite load.

In order to ensure the stability of the satellite platform bus, the switching between various modes of the power system is executed according to the main error amplifier (MEA). The main error amplifier circuit mainly includes three parts: an error amplifier circuit, a two-out-of-three voting circuit, and an error voltage adjustment circuit. Its principle block diagram is shown in figure 2:

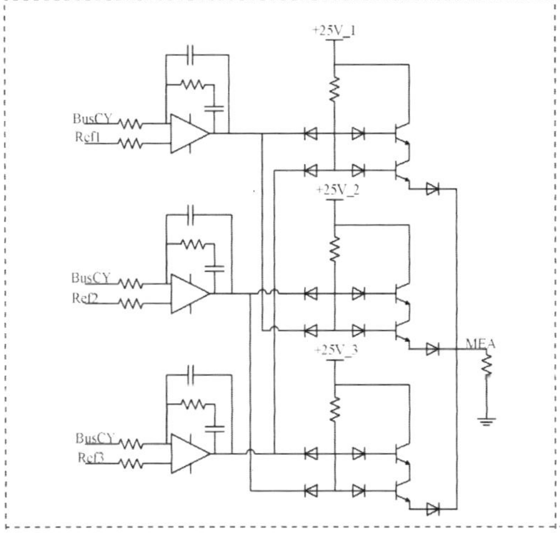

figure 2. Block diagram of the MEA.

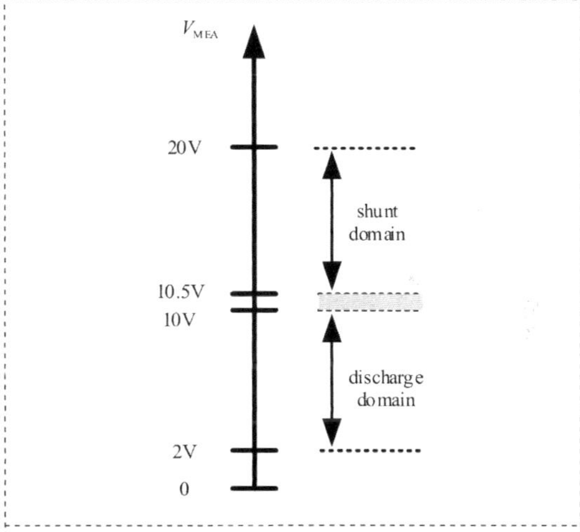

figure 3. MEA control domain division.

Two-domain control for MEA is adopted to schedule the BDR and the shunt regulator, and the purpose of stabilizing the bus voltage is achieved through the correct division of the MEA area. The MEA control domain voltage division is shown in Figure 3. The effective utilization range of the main error amplifier voltage is 2.0V-20V. When the output voltage of the main error amplifier is below 2V, the bus voltage may be less than 30V, and the BDR will output with the maximum capacity to meet the load requirements of the platform; When the output voltage of the main error amplifier is in the discharge domain of 2V-10V, the system works in the combined power supply mode, and as the output voltage of the MEA gradually decreases, the system changes from the combined power supply mode to the internal power mode; When the output voltage of the MEA is in the shunt domain of 10V-20V, the system works in the external power mode, and the platform bus mainly depends on the shunt regulator for shunt regulation and provide energy. In addition, the load charging shunt module realizes the charge and discharge control of the battery pack by sampling the battery terminal voltage, thereby enabling the battery pack to achieve energy balance.

3.3. High power filter design

According to the requirements of power supply tasks, in order to reduce the impact of bus pulsation caused by the SAR load on the input of the BDR, setting high power filter is necessary in the composite bus system, which can suppress the transient disturbance from the load bus and also avoid the noise generated by the BDR to the load bus. The traditional LC filter in figure 4 is a second-order underdamped system. Although it has good steady-state filtering effect, due to the low damping factor, transient resonance problems are prone to occur when the system is under light load. Compared with the LC filter, the LC damping filter adds a damping resistor R_d and a DC blocking capacitor C_d, as shown in figure 5. The damping resistor R_d can increase the system damping to suppress the resonance peak of the LC filter, and changing R_d can achieve controllable damping; The DC blocking capacitor C_d blocks the DC component of the power source to avoid power loss caused by the damping resistor.

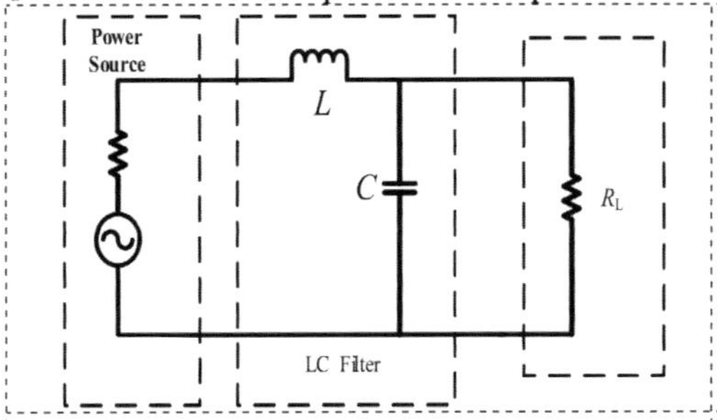

Figure 4. LC filter.

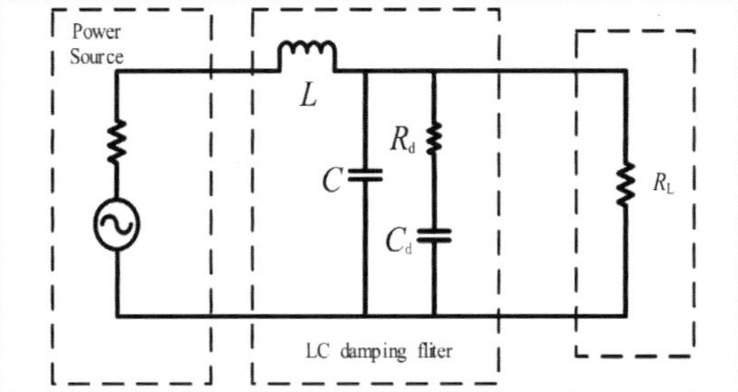

Figure 5. LC damping filter.

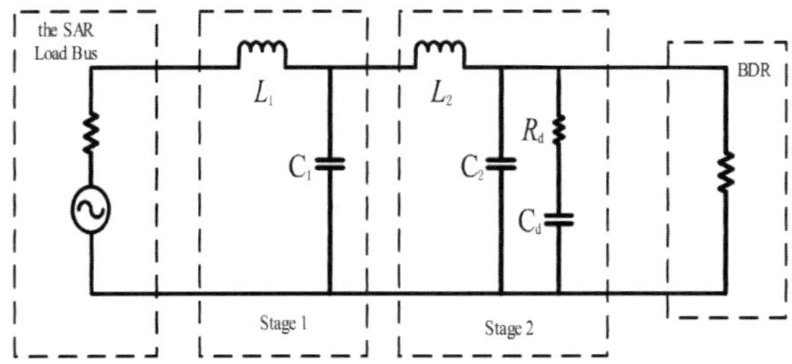

Figure 6. Two-stage LC damping filter.

Considering that BDR is composed of multiple DC/DC converters, and to limit transient pulses and ensure each converter input isolation, a low-loss controllable damping two-stage LC filter is proposed in this paper. As shown in figure 6, the stage1 (L_1, C_1) is used for high-power filter, which mainly eliminates the effect of power ripple caused by SAR load on each BDR input; the stage2 is used as the input filter of each DC/DC converter, which can effectively control the resonance peak of the filter and realize the isolation between the inputs of each DC/DC circuit. Therefore, the low-loss controllable damping two-stage LC filter used in this paper can not only limit the resonance peak of the transient process, but also ensure the steady-state filtering effect.

4. Experimental results

To verify the feasibility of the solution proposed in this paper, a product prototype was built as shown in figure 7, and related test results can be seen from figure 8. It can be seen from result that when the unregulated bus fluctuates due to the SAR load, the bus voltage ripple is significantly reduced after being filtered by the high-power filter, and the peak-to-peak value drops from 30.3V to about 630mV. At this time, the platform bus output remains stable, and the peak-to-peak value is less than 500mV, which meets the design requirements. It should be noted that due to the small capacity of the test battery during the ground test, the battery voltage fluctuation range when the SAR load is turned on is larger than the actual value, which further illustrates the feasibility of the SAR satellite power system based on the composite bus proposed in this paper.

Figure 7. Product prototype.

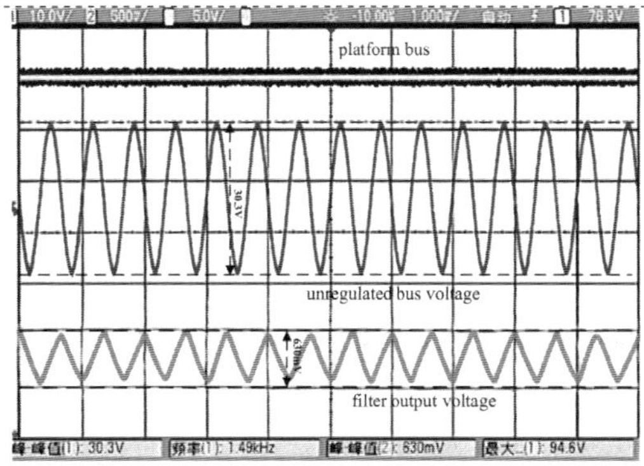

Figure 8. Text results.

5. Conclusion

On the basis of analyzing the mainstream SAR satellite power system, this paper proposes a SAR satellite power system based on the composite bus, and builds the principle prototype for experimental verification. The text results verify the feasibility and effectiveness of the proposed solution, and provide a practical idea for the development of SAR satellite power system.

6. References

[1] Chen Q. Spacecraft Power System Technologies[J], **15**(2020)

[2] Qiao M., Zhu L., Li X., et al. Design and Simulation on Power Supply System of SAR Satellite[J]. Spacecraft Engineering, **34**(2015)

[3] Yang S., Zhao C. Design of Power Supply System for High Power SAR Satellite[J], Spacecraft Engineering, (2017)

[4] Zhang X., Zhang W., Guo W., et al. Analysis and Enlightenment of Electrical Power System for SAR Satellite[J]. Spacecraft Engineering,(2015)

[5] Chen N., Li Q, et al. Modeling analysis and Design of Battery Chargers for Micro-nano SAR Satellites[C]. 2021 2nd China International SAR Symposium (CISS), (2021)

[6] Yan Z., Zhang Y., Zhang H. A Hybrid SAR/ISAR Approach for Refocusing Maritime Moving Targets with the GF-3 SAR Satellite[J], Sensors, **20**(2020)

[7] Sarafianos A, Steyaert M. A modelling and design approach for push/pull switched capacitor DC-DC converters[C]. 2016 IEEE 17th Workshop on Control and Modeling for Power Electronics (COMPEL). IEEE, (2016)

[8] Hendra A, Hamada F, Yusivar F. Voltage Control in Push-Pull DC-DC Converter using State Space Averaging PID with Soft-Start for Electric Vehicle Auxiliary System[C]. 2019 6th International Conference on Instrumentation Control and Automation (ICA), (2019)

[9] Li H., Zhang X., et al. A Fast Response Control Strategy for Power Converter of SAR Satellite[J]. Spacecraft Engineering, (2019)

[10] Qiao M., Li X., et al. Design and Simulation of SAR satellite power supply system[J] Spacecraft engineering, **24**(2015)

[11] Zhao C., Chen Q., et al. Design and verification of solar cell array for GF-3 satellite[J] Spacecraft engineering, **26**(2017)

[12] Zhang X., Li H., et al. A Fast-Dynamic Response Control Strategy for Power Supply Unit of SAR Satellites[J]. IOP Conference Series: Earth and Environmental Science, **603**(2020)

[13] Song J., Wang G., et al. Research on a Power Supply Scheme of Composite Bus for SAR Satellite[C]. 2021 2nd China International SAR Symposium (CISS), (2021)

Research on characteristic parameters for multi-factor aging of stator bar insulation of pumped storage generator

Xiangdong Liu[1], Jiaxuan Han[2], Liangliang Nie[1], Xingxiao Shao[2], Qing Li[1], Xiaofeng Huang[1], Penghu Li[2] and Haiyun Jin[2,*]

[1]Maintenance and Test Company, CSG Power Generation Co., Ltd, Guangzhou, 511400, Guangdong, China
[2]State Key Laboratory of Electrical Insulation and Power Equipment, Xi'an Jiaotong University, Xi'an, 710049, Shaanxi, China

Corresponding author: *hyjin@mail.xjtu.edu.cn

Abstract. The real stator bars of an 18kV pumped storage generator were taken as research objects. The multi-factor sequential aging tests of thermal cycling, electrothermal and vibration were carried out. The dielectric loss angle tangent (tanδ), capacitance (C), Δtanδ, ΔC and maximum partial discharge of the samples at different aging periods were measured. The variation of these parameters with aging time was summarized, and the aging mechanism was analysed. The results show that tanδ, Δtanδ, and ΔC increase overall with aging time, while the capacitance decreases slightly with aging time. The maximum partial discharge decreases gradually at beginning of aging, and then increases after aging of 4 periods. All of tanδ, Δtanδ, ΔC and maximum partial discharge increase obviously in the later stages of aging. They can be used as characteristic parameters for multi-factor aging of stator bars insulation of pumped storage generators, to evaluate the aging state of the insulation with epoxy mica.

1. Introduction

Pumped storage power station plays an important role on peak modulation and frequency modulation in power system. The machine units should change the operation conditions between generators and motors frequently [1,2]. The main insulation of stator bar of generators suffers a severer process of electrical, thermal and mechanical stress together with all kinds of environmental factors during long-term running, which accelerate the aging of main insulation. The electrical and mechanical properties of insulation degrade gradually and the main insulation fails finally, which seriously affects the security of the equipment and system [3,4]. Therefore, it is necessary to design and construct an experimental platform to test the insulation aging characteristics of stator bar of generators and do research on the aging laws. The work has great significance in improving the reliability of generators.

There are many researches about the aging of generator's bars insulation at home and abroad. Teresa H proposed that the aging state of the insulation could be characterized by the voltage of the first and second current surge points and the AC composite parameters [5]. Kimura K obtained the correlation between the residual breakdown voltage of the large motor stator insulation with the maximum partial discharge, the dielectric loss tangent and its increase, and the AC current surge rate. He believed that it was the most reliable to determine the residual breakdown voltage by the AC current surge rate [6]. Hao Yanpeng developed the equipment of ultrasonic and impact echo detection, and proposed relevant electrical parameters reflecting insulation aging and defects [7-9]. Yang Min analysed the influence of thermal aging time on dielectric loss and relative dielectric constant of stator bars, but this study only analysed the change of insulation characteristics under the effect of a single thermal field, which did not conform to the actual aging condition of stator bars [10]. Wang Xin studied the influence of thermal aging on partial discharge characteristics of insulation defects of stator bar, but this study focused on different insulation defects, and did not study the influence of different aging time on the results [11]. Wang Xiaojian tested the frequency domain dielectric spectroscopy of the insulation of stator bar with different electrothermal aging periods, and used the Havriliak-Negami model to extract the characteristic parameters that characterize the insulation aging degree. However, the characteristic parameters selected in this study were not intuitive enough, and only the electrothermal aging was carried out [12].

There are more and more characteristic parameters reflecting the aging state of the main insulation of stator bars with the development of modern testing technology. But the parameters that can truly reflect the aging state of the main insulation of stator bars are still under debate, which is a hot research topic in the field of insulation aging state evaluation. Therefore, it is significant to design a multi-factor aging platform for stator bar of pumped storage and select reasonable electrical parameters that characterize the aging state. In this paper, the new stator bars of a real pumped storage generator with rated voltage of 18kV were taken as research objects. The multi-factor aging tests were carried out according to the operation conditions. The electrical parameters of the stator bars were measured, and their variation during aging was analysed. This research has great significance in revealing the mechanism during the aging of stator bar insulation of pumped storage generators.

2. Experimental methods

2.1. Aging test

In order to simulate the real aging condition of the pumped storage generator stator bar, the aging test of stator bar must be carried out under the action of various factors. During the operation of pumped storage generator, the main stresses are thermal stress, thermal mechanical stress, electrical stress and mechanical stress. According to actual operating conditions and relevant standards [13-16], a multi-factor sequential aging test scheme was developed in this paper, as shown in Figure 1. Samples needed to be subjected to thermal cycling aging, electrothermal aging and vibration aging in turn. The above processes constituted an aging period.

The specific parameters of thermal cycling aging are as follow. Each aging period carried out 67 times of thermal cycling aging, with a temperature range of 40~155°C and a heating rate of 1.5°C/min. The cooling method was natural cooling. It took about 6h for a thermal cycling and about 400h for a period.

The specific parameters of electrothermal aging are as follow. Considered the fluctuation range of voltage and temperature in the actual operation of the real stator bars, the voltage of electrical aging was determined to be $1.2U_N$ (21.6kV), the temperature of thermal aging was 120°C, and the aging time of one period was 250h.

The specific parameters of vibration aging are as follow. Considered the accelerated aging test and the operating environment of the real stator, the vibration frequency f of vibration aging was 100Hz, the amplitude A was 0.5mm, and the aging time of one period was 144h.

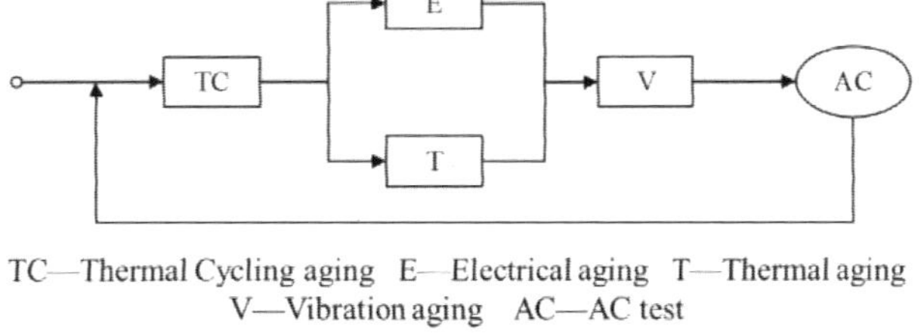

Figure 1. Multi-factor sequential aging test model.

Firstly, the characteristic electrical parameters of 21 original bars were measured, and 5 bars with similar insulation performance were selected as initial samples to carry out the above multi-factor sequential aging test. After each aging period, a series of non-destructive characteristic electrical parameters were tested before the next aging period. A total of 7 periods of aging were completed.

2.2. Characteristic parameters measurement

After each aging period, the non-destructive characteristic electrical parameters to be tested were tanδ, capacitance C, Δtanδ, ΔC and the maximum partial discharge.

In this paper, tanδ and capacitance C of the sample were measured by QS301 high precision high voltage Cilling bridge. According to relevant standards [16], tanδ and capacitance C at $0.2U_N$, $0.4U_N$, $0.6U_N$, $0.8U_N$ and U_N should be measured for each sample. Each data was averaged over the results of multiple samples.

According to relevant standards [16], Δtanδ and ΔC were the difference between the measured dielectric parameters at $0.6U_N$ and $0.2U_N$, which in this paper is the difference between the measured values at 10.8kV and 3.6kV.

Finally, two samples that have experienced 7 periods of aging were selected as sample A and sample B. Partial discharge was measured by the digital partial discharge test system of Doble Lemke. Since the rated voltage of the stator bar was 18kV and the corresponding rated phase voltage was 10.8kV, this paper studied the maximum partial discharge under these two voltages.

3. Results and discussion

3.1. Variation of tanδ and capacitance C with aging period

The variation of tanδ with aging period is shown in Figure 2. As can be seen from the figure, tanδ increases with aging. After the 5th period of aging, tanδ increases obviously. After 7 periods of aging, tanδ at 18kV increases by 185.7% (from 0.0126 to 0.0360) compared with that without aging. In addition, tanδ increases as the test voltage increases. In the early stage of aging, tanδ increases slowly with the increase of voltage. However, tanδ increases faster and faster with the increase of voltage in the middle and late stage of aging.

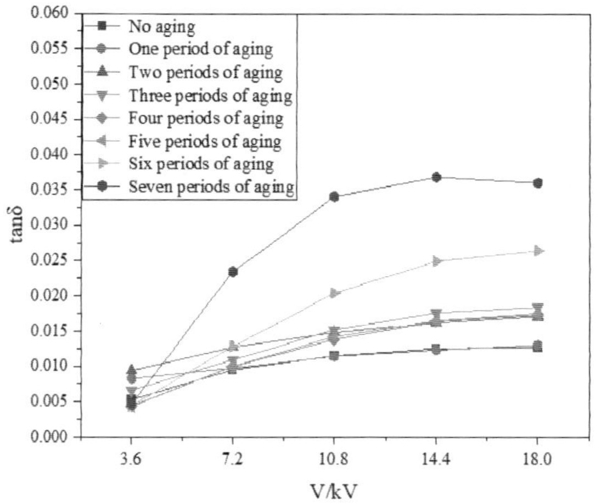

Figure 2. tanδ (mean value) of sample with different aging periods varies with the test voltage.

The reason why tanδ increases with aging is the generation of discharge by-products and the development of defects. And the reason why tanδ increases with the test voltage is partial discharge. At the early stage of aging, internal defects of insulation develop slowly, partial discharge is not obvious, and discharge by-products are less, so tanδ increases slowly with time. At this time, the partial discharge level is low because the air gap cracks in the insulation have not developed significantly. and partial discharge will not increase significantly even if the voltage increases. So tanδ

increases slowly with the test voltage. With the aging process, the air gap cracks in the insulation gradually develop, the defects increase, and the partial discharge level rises, the discharge by-products gradually accumulate, so tanδ is significantly increased by the aging time and test voltage.
The variation of capacitance C with aging period is shown in Figure 3. As can be seen from the figure, The capacitance of the sample decreases with aging. The effect of the third, fourth and fifth periods on capacitance change is not obvious. After 7 periods of aging, the capacitance value at 18kV voltage is reduced by 8.1% (from 1.23nF to 1.13nF) compared with that without aging. The slope of C-u curve increases slightly with aging time.

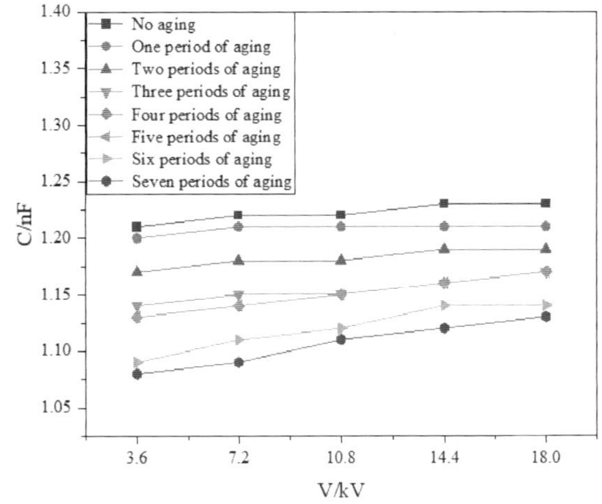

Figure 3. C(mean value) of sample with different aging periods varies with the test voltage.

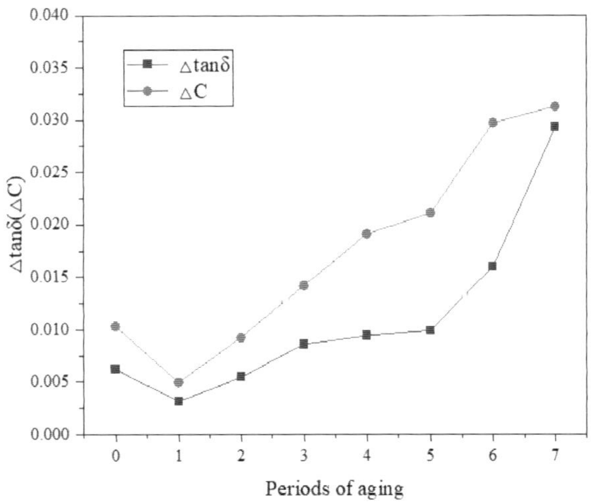

Figure 4. Variation of Δtanδ and ΔC(mean value) with aging period.

The capacitance of the sample can be regarded as distributed capacitance. When a new air gap is generated in the aging process, it is equivalent to connecting a new capacitor in series in the capacitor branch, and the total capacitance decreases. With the aging process, there are more and more air gap defects, and the capacitance of the sample decreases. When the sample is tested, if the test voltage is higher than the inception voltage of partial discharge, the air gap is short-circuited and the capacitance increases. As the test voltage continues to rise, the short-circuited air gap increases and the capacitance increases further. Thus, the capacitance of the sample increases with the test voltage. As the partial discharge level increases in the late stage of aging, the voltage has a greater influence, so the slope increases a little.

3.2. Variation of $\Delta \tan\delta$ and ΔC with aging period

The variation of $\Delta\tan\delta$ and ΔC with aging period are shown in Figure 4. Both $\Delta\tan\delta$ and ΔC show an obvious increasing trend. After 7 periods of aging, $\Delta\tan\delta$ increases by 372.6% (from 0.0062 to 0.0293) and ΔC increases by 203.9% (from 0.0103 to 0.0313) compared with that without aging.

Both $\Delta\tan\delta$ and ΔC reflect the difference of electrical parameters of samples at different voltages. The direct cause of this difference is partial discharge, and the factors affecting partial discharge are the air gap and cracks in the insulation. Compared with the samples in the early stage of aging, there are more defects in the insulation in the late stage of aging, and the partial discharge at high voltage is more intense than that at low voltage, so $\Delta\tan\delta$ and ΔC increase gradually with the aging time.

3.3. Variation of maximum partial discharge with aging period

Partial discharge test was conducted at 10.8kV and 18kV respectively for sample A and sample B that had experienced 7 periods of aging. The maximum partial discharge are shown in Figure 5 and Figure 6. The maximum partial discharge of samples at 10.8kV and 18kV both show a downward trend in the early stage of aging. After 4 periods of aging, the maximum partial discharge of the samples at different voltages begins to increase. The maximum partial discharge of sample A increases obviously after the sixth aging period. Taking the test results at 18kV as an example, the maximum partial discharge of sample A in the 7th period increased by 512% (from 925.3 to 5661.4) compared with that in the 4th period, and that of sample B increased by 105% (from 1083.9 to 2218.8).

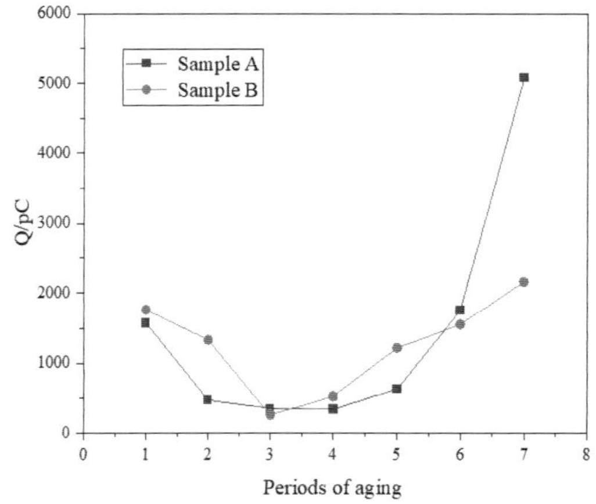

Figure 5. Variation of maximum partial discharge with aging period of samples at 10.8kV.

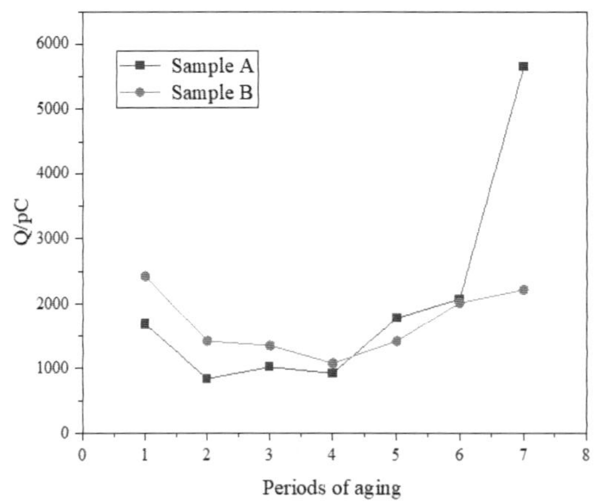

Figure 6. Variation of maximum partial discharge with aging period of samples at 18kV.

The reason for this phenomenon is that new bars were used in the aging test. In the early stage of aging, the electrothermal aging process further solidifies the epoxy of the main insulation, enhances the performance of the stator bar, and the maximum partial discharge of the sample gradually decreases. After three aging periods, new defects appear in the main insulation, leading to the deterioration of the performance of the stator bar and the gradual increase of maximum partial discharge.

4. Conclusions

1) With the aging process, tanδ, Δtanδ and ΔC increase significantly, while the capacitance value C decreases slightly. After 7 periods of aging, tanδ at 18kV voltage increases by 185.7%, and capacitance value at 18kV voltage decreases by 8.1%. Δtanδ increases by 372.6% and ΔC increases by 203.9% compared with that without aging.

2) The maximum partial discharge of the samples decrease in the early stage of aging. After 4 periods of aging, the maximum partial discharge of the samples begin to increase. Taking the test results at 18kV as an example, the maximum partial discharge of sample A in the 7th period increased by 512% compared with that in the 4th period, and that of sample B increased by 105%.

3) tanδ, Δtanδ, ΔC and the maximum partial discharge all increase significantly in the late stage of aging, which can be used as characteristic parameters for multi-factor aging of stator bar insulation of pumped storage generators, to evaluate the aging degree of insulation. The capacitance value C does not change obviously with aging time, so it is not suitable as the characteristic parameter to evaluate the aging state.

Acknowledgments

The work was financially supported by science and technology project of CSG Power Generation Co., Ltd (022200KK52190007).

References

[1] Zhang Shu 2008 *The research on strategy of pumped storage power station's development in China* (Harbin: Harbin Institute of Technology) p 1
[2] Pickard WF 2012 *Proc. IEEE* **100** 473-483
[3] He Yongxiu, Guan Lei, Cai Qi, Liu Xiaoli and Li Chengren 2004 *Power Syst. Techno.* **28** 54-57
[4] Wenqi F 2008 *Water Resour. Power* **26** **1**54
[5] Liu Yaonan 1994 *Electrical Insulation Testing Technology* (Beijing: China Machine Press) p 10
[6] Kimura K, Kaneda Y and Itoh K 1991 *Prop. Appl. Dielect. Mater.* 2 769-72
[7] Hao Yanpeng, Wang Guoli, Yan Bo, Jiang Xiongwei, Jia Zhidong and Xie Hengkun 2001 *High Volt. Eng.* **27** 15-36
[8] Wu Peng, Liu Libing, Chen Zhiyong and Gao Naikui 2004 *High Volt. Eng.* **30** 23-24
[9] Hao Yanpeng, Xie Hengkun, Wang Guoli and Gao Naikui 2004 *Proc. CSEE* **25** 76-82
[10] Yang Min, Wang Xin, Tong Xiaozhong, Jiang Tie, Xu Xing and Chen Xiangrong 2021 *High Volt. Appar.* **57** 94-99
[11] Wang Xin, Yang Min, Jin Yang, Tong Xiaozhong, Du Hao and Chen Xiangrong 2021 *Insul. Mater.***54** 58-63
[12] Wang Xiaojian, He Tianlei, Xu Junyuan, Chen Cong, Hu Lei and Shi Jie 2019 *Insul. Mater.* **52** 41-46
[13] IEEE STD 275 1992 Recommended practice for thermal evaluation of insulation systems for alternating - current electric machinery employing form - wound reinsulated stator coils for machines rated 6900 V and below Institute of Electrical and Electronics Engineers
[14] GB/T 17948.6-2007 2007 Rotating electrical machine - Functional evaluation of insulation systems - Test procedures for form-wound windings - Evaluation of thermomechanical endurance of insulation systems General Administration of Quality Supervision, Inspection and Quarantine of the People's Republic of China

[15] GB/T 29311-2012 2012 Electrical insulating materials - A.C. voltage endurance evaluation - Introduction General Administration of Quality Supervision, Inspection and Quarantine of the People's Republic of China
[16] DL/T 492-2009 2009 Guide for aging appraise of generator stator winding insulation with epoxy mica National Energy Administration of the People's Republic of China

Scheduling Scheme Optimization for Emergency Resources on a Deep-sea Considering Environmental Impacts

Ya-jie Wang, Jian-chun Fan*, Sheng-nan Wu
College of Safety and Ocean Engineering, China University of Petroleum-Beijing, Beijing, 102249, China

E-mail address:fjc6488@126.com (J.-c. Fan)

Abstract. An efficient and timely emergency response to a major accident is more challenging for the exploration and utilization of deep-sea oil and gas fields due to harsh external environment factors. An optimal solution is needed to tackle the external environment changes in the demand for emergency resources. To effectively balance emergency response time and the satisfaction of resource scheduling quantity, this paper proposes a multiple rescue points and multi-objective optimization model to address such challenges. Such model takes into account the environment factors involving wind speed and wave height caused by the variability of the deep-sea monsoon climate. Additionally, to handle the uncertainty of the resource demand at the accident sites, a linear programming and heuristic hybrid algorithm is proposed to help decision makers select the ideal routes and minimal scheduling time of deep-sea emergency resources, as well as the optimal operation season. A case study of a blowout accident is conducted to demonstrate the application of the proposed model and the real-world implications.

1. Introduction

As the demand for oil and gas grows continuously, the exploration and production of deep-sea oil and gas fields expand rapidly. Due to harsh external environments (including deep water, high waves, and strong winds) with complex processes and equipment, the development in deep-sea is far behind the inland counterparts. Meanwhile, sudden accidents on deep-sea oil and gas fields will bring geometrically larger damages than those on offshore fields. Once an accident occurs in the operation area, an efficient and timely supply of emergency resources is inevitably demanded. In this paper, resource storage terminal (RST) for deep-sea exploration are used to meet such demand. Specifically, the location of RSTs and the allocation of the required resources are two main challenges in this work.

Some complex mathematical models have been established to select the location to store resources. For example. [1] tried to maximize the efficiency of emergency resource distribution while minimizing the total construction cost of resource points and balancing the emergency demand of various demand points. With limited scheduling of static resources, they built an optimization model for resource locations under dynamic demands. [2] proposed a two-stage location-routing model with recourse for integrated preparedness and response planning under uncertainty. This model is used for risk management in disaster situations where there are uncertainties in demand and the state of the infrastructure. [3] proposed an index-based emergency response management system (IERMS) based on the Location Hazard Index (LHI) and the response time optimization model, which determines the locations of resource storage by solving the shortest response time and the optimal resource allocation in an emergency. [4] used a recent machine learning multiple layer perceptron (MLP) model and a Taguchi method to calibrate the MLP variables, combined an optomised pluvial flooding probability model with ideal location allocation methods on a geographic information system platform to construct the proposed

model for achieving accurate emergency response centres spatial planning. [5] took the maximum emergency efficiency as the objective function and constructed a location model of marine emergency material reserve based on timeliness, effectiveness, and differences in the demands for materials under accidents of various levels. [6] used a fully fuzzy-binary linear program to analyze the problem of maximum coverage locations. They proposed an augment-weighted Tchebycheff-based method to solve the problem. [7] proposed a Markov decision process (MDP) formulation for the stochastic road network recovery problem (SRNRP), and an approximate dynamic programming (ADP) approach to heuristically solve SRNRP. The authors considered a disaster scenario on a road infrastructure network that obstructs the flow of relief-aid commodities and search-and-rescue teams between critical service provided facilities and locations in need of these critical service.
Besides, demand-resource deployment has been comprehensively examined in the literature. For example. [8] discussed how the choice of different paths affects the efficiency and time uncertainty in scheduling. The problem of resource allocation and scheduling was solved using a decision support model and simulative analysis. [9] studied the Arctic region where there are few infrastructures with much difficulty in resource allocation. In their work, by allocating and scheduling public and private resources, a database and a dynamic network resource scheduling model were established to meet the needs for resources in the Arctic area. [10] used the spatial-temporal Poisson distribution to generate simulations of dynamic customers and analyzed their spatial and temporal characteristics in uncertain environments. In the end, the authors established a joint allocation model of multi-category resources that integrated the overall operation cost and vehicle fixed cost, which was proved to be better than the single category resource allocation model. [11] aimed at cutting down the response time toward events, as well as reduced the fleet operation costs and balanced the ship workload and emergency material dispatch. With the distribution plan of ships and materials under different target weights determined, an event type-based emergency model for material and ship distribution was established, and its effectiveness has been optimized. [12] judged the severity of oil spills by the established model of marine ships oil spill diffusion and created a resource scheduling model for different severity degrees of oil spills. The results revealed the earliest cleanse time for oil spill pollution and the optimal dispatch quantity of emergency materials. [13] reviewed the work done in D2D communication of resource allocation in the context of device to device communication, and the authors used the game theory to address the problem. Also provided the insight into the evolvement that has been made in the area of resource allocation and highlight various open issues that needs to be addressed. [14] proposed a resource scheduling method for cooperative spatial target monitoring when detected spatial situations. Based on the task modes, objective functions, and the model and solution algorithms of the problem, the scheduling and target capacities, as well as the response time of the resource scheduling system were calculated. [15] considered a problem of supporting resource allocation decisions affecting multiple beneficiaries, and incorporated fairness in the form of welfare dominance, ensured that the resultant distribution of benefits. The authors introduced a new approach based on the paradigm of maximizing efficiency subject to constraints to ensure that the decision was acceptably fair. [16] used the single-type resources were to be assigned in the tree nodes such that the total weight in the rooted path from each leaf to the root equals its demand, and the authors set the aim was to minimize total costs of the allocated resources. [17] proposed a new differential evolution (DE) algorithm, called DEMIDRA, which each individual represented the resource allocation of a participating miner and the resources allocations of all participating miners constitute the whole population. The DE was adopted to optimize the resource allocation as for the optimization of the mining decision.
These efforts provide very important foundations for the research about RST locations and demand resource scheduling. However, as the drilling operations move to the deep-sea area, the risk of accidents in operations rises dramatically. In summary, special attention should be paid to how the external environmental factors interfere with the location selection and resource scheduling in emergency decision-making, and they are crucial to optimize decision-making and reduce risks. Therefore, in this paper, the main contributions are as follows:

1. A multi-objective optimization model for the deep-sea emergency resource scheduling is proposed to effectively balance emergency response time and the satisfaction of resource scheduling quantity.
2. The environment factors involving wind speed and wave height caused by the variability of the

deep-sea monsoon climate are taken into account.

3. A linear programming and heuristic hybrid algorithm is proposed to handle the uncertainty of the resource demand at the accident sites.

The remainder of this article is structured as follows. Section 2 a model to calculate the ship speed is introduced, and the multi-objective optimization model for RSTs location and scheduling in the deep-sea are established. This model can achieve the minimum emergency response time and satisfy the demand for emergency dispatching resources to the utmost extent. Then using a solution to the proposed model is given by LPHAL. Section 3 we are analyze the information about the wind field and wave heights in the deep-sea and using a case study of a blowout accident to demonstrate the application of the proposed model and the real-world implications. Section 4 the case study is discussed and some valuable features are summarized. Section 5 concludes the paper.

2. Methodology

2.1 Velocity formula

Deep-sea oil fields are especially susceptible to expansion and escalation. Besides, when an incident occurs, the meteorological, hydrographic, and sea conditions will interfere with the timeliness and reliability of emergency resource delivery by affecting the speed of rescue vehicles. When experiencing bad weather, the ship speed becomes subject to factors such as wind, windward angle, wave height, wavelength, wave-front angle, and the flow speed of wind-generated currents. Until now, the scientific field has offered a lot of findings of the speed of a ship sailing in wind and waves, such as the empirical ship stall equations proposed by James and the Central Institute of Shipping in the former Soviet Union. [18]

Specifically, the equation given by the Qingdao Meteorological Navigation Consortium was established based on that proposed by the former Soviet Maritime Research Institute in 1967, which includes the wave height quadratic coefficient k_2 and the wave direction coefficient k_3. Besides, the Qingdao equation also refers to the B·C Clashock stall nomogram and an empirical coefficient G based on the actual ship measurements to calculate the stall amount [19]. Therefore, this paper applies the Qingdao equation to solve the actual speed of a ships in deep-sea, and Table 1 shows the description of each variable in the Eq. (1).

$$V_s = V - (k_1 h + k_2 h^2 - k_3 \alpha_1 h)(G - k_4 D_t V) \tag{1}$$

Table. 1 Scalar descriptions in the ship stall equation

Variable	Description	Variable	Description
V	speed of ships in still water	V_s	ship actual speed
h	wave height	G	empirical coefficient
D_t	actual displacement of the ship	k	each performance factor of the ship
α_1	the angle between the bow of the ship and the direction of the incoming waves		

2.2 Optimization model

When scheduling resources, a balanced resource distribution can prevent unnecessary losses to the emergency response. Therefore, the function of fuzzy affiliation in the triangular fuzzy $\tilde{a}_{jk}$ is invoked in this paper, and the importance of different resources categories to the emergency point E_j, namely g_{jk}, was implemented to satisfy the demand for each type of resources to the utmost extent.

2.2.1 Model assumption. The contingency dispatch model for multiple resources and demand points in the deep-sea are set up with the following assumptions.

1) To find out the maximum emergency response time, each vessel will only distribute resources for one demand point in one voyage.

2) The carriers are all capable of meeting the voyage demand, and there are no halfway failures throughout any voyage.

3) RSTs have sufficient reserves of all types of resources to meet the demands at multiple

emergency points.

4) The ships travel safely during their voyages without ship accidents or piracy terrorism incidents.

5) All the ships share identical parameters.

2.2.2 Model. In this paper, the multistage decision-making model for the multi-objective optimization problem in the deep-sea is the primary concern. This model would seek solutions over a set of possible choices to optimize certain criteria. The minimum emergency response time $minT$ and the resource demand scheduling satisfaction $maxQ$ are expressed as follows.

Sets/indices	
R	set of RSTs, $i \in R$
E	set of emergency points, $j \in E$
K	set of resource types, $k \in K$
R_{ik}	storage capacity of *k* resources in R_i
g_{jk}	importance of *k* resources to emergency point E_j
q_{ijk}	quantity of *k* resources distributed by R_i to emergency point E_j
c_{jk}	number of *k* resources received by emergency point E_j
$\tilde{a}_{jk}$	emergency point E_j triangular fuzzy number of resource requirement for category *k*
a^l_{jk}	emergency point E_j minimum value of demand for category *k*
a^m_{jk}	maximum demand of emergency point E_j for *k* resources
a^r_{jk}	emergency point E_j most likely value of demand for category *k* resources
$\mu(c_{jk})$	the demand satisfaction corresponding to c_{jk} is expressed by the fuzzy membership function of $\tilde{a}_{jk}$
S_{ij}	traveling distance between resource storage terminals R_i and emergency point E_j

Mathematical formulations Eq. (2) and (3) are the objective functions of *T* and *Q*, respectively. Objective function (2): Minimize the value emergency response time from R_i to E_j in case of an accident. Objective function (3): Maximize the demand satisfaction of the dispatched resource *k* when resource distribution is carried out.

$$minT = \sum_{j=1}^{E}(max \sum_{i=1}^{R} \frac{S_{ij}}{V_s} X_{ij}) \tag{2}$$

$$maxQ = \sum_{j=1}^{E} \sum_{k=1}^{K}(\mu(c_{jk}) g_{jk}) \tag{3}$$

$$\text{s.t. } X_{ij} = \begin{cases} 1, & \sum_{k=1}^{K} q_{ijk} > 0, \ R_i \text{ distribute } k \text{ resources to } E_j \\ 0, & \sum_{k=1}^{K} q_{ijk} = 0, \ R_i \text{ doesn't distribute } k \text{ resources to } E_j \end{cases} \tag{4}$$

$$\sum_{j=1}^{E} q_{ijk} \le f_{ik} \tag{5}$$

$$c_{jk} = \sum_{i=1}^{R} q_{ijk} \ge a^l_{jk} \tag{6}$$

$$\tilde{a}_{jk} = (a^l_{jk}, a^r_{jk}, a^m_{jk}) \tag{7}$$

$$\mu(c_{jk}) = \begin{cases} \frac{c_{jk}-a^l_{jk}}{a^r_{jk}-a^l_{jk}}, & c_{jk} \in [a^l_{jk}, a^r_{jk}] \\ \frac{c_{jk}-a^m_{jk}}{a^r_{jk}-a^m_{jk}}, & c_{jk} \in [a^r_{jk}, a^m_{jk}] \\ 0, & else \end{cases} \tag{8}$$

$$g_{jk} \epsilon (0,1) \tag{9}$$

Constraint (4) is used to control the resource transportation from R_i to E_j. When E_j is asking for resources from R_i, $X_{ij} = 1$; Otherwise, $X_{ij} = 0$.

Constraint (5) is used to constrain the number of resources dispatched by each RST. In other words, the actual supply of each type of material from any R_i cannot exceed the storage capacity of that material.

Constraints (6) to (8) are the triangular fuzzy numbers of resources required by emergency point E_j. When $c_{jk} < a^l_{jk}$, the number of class *k* resources from R_i to E_j cannot meet the emergency demands, making the whole schemes not feasible. However, if $c_{jk} > a^m_{jk}$, the amount of class *k* resources supplied by R_i to E_j will exceed the demand at E_j, resulting in a waste of resources and failure to obtain the

decision maker's utmost satisfaction. Therefore, constraint (8) is used to calculate the demand satisfaction of c_{jk}, and it is a constraint for the scheme feasibility.
Constraint (9) is a set of values to define the importance of category k resources to E_j.
This multi-objective optimization problem is an NP-Hard (Nondeterministic polynomial-Hard), namely, the solution will not be optimal for all objective functions at the same time. Therefore, the total objective function of the model is obtained by weighting each sub-objective linearly in this paper. However, since each sub-objective has a different magnitude, all of them must be dimensionless before subsequent operations. $T_{max}$and$T_{min}$are assumed the maximum and minimum values of objective function (2), while Q_{max} and Q_{min} are assumed the maximum and minimum values of objective function (3). After normalization, the two objective functions became:

$$f_T = (T_{max-}T)/(T_{max} - T_{min}) \quad (10)$$

$$f_Q = (Q_{max} - Q)/(Q_{max} - Q_{min}) \quad (11)$$

Where f_T and f_Q are the objective functions after the dimensionless processing of sub-objectives (2) and (3), respectively. Subsequently, assume the weight coefficients of emergency response time and resource dispatch satisfaction during deep-sea emergency rescue operations to be λ_T and λ_Q, respectively. These two coefficients must reside within (0,1), and $\lambda_T + \lambda_Q = 1$. With the two coefficients, the total objective function can be obtained.

$$max\, F = \lambda_T f_T - \lambda_Q f_Q \quad (12)$$

2.3 Computational Procedure
Based on the features of deep-sea emergencies and the situation that the number of RST and emergency points remains low, the model is solved using a linear programming and heuristic algorithm language (LPHAL), and the specific steps are described below.
The shortest navigation range and route from R_i to E_j are found with the Dijkstra algorithm. Under constraints (4) to (9), a linear programming algorithm is used to find the extremums of emergency response time and resource dispatch satisfaction of Eq. (2) and Eq. (3), namely,T_{max}, T_{min}, Q_{max}, Q_{min}. These extremums are taken into Eq. (10) and (11) to obtain the values of f_T and f_Q. Subsequently, the expert evaluation method is used to find λ_T and λ_Q in the total objective function, which are then taken into Eq. (12).

3. Case study
The response time and the effectiveness of resource dispatch are two key indicators of emergency response at deep-sea. By analyzing the changes of wind speed and wave height caused by monsoon changed in the past 30 years as the interference factors of deep-sea emergency response, we can analyze the impact of different seasons on emergency response time and resource scheduling.

3.1 Hydrological information
Based on the WRF (weather research and forecasting) mesoscale climate model, Mike 21 software is used to simulate and analyze the wind field data of the deep-sea exploration area in the past 30 years. The simulation and analysis results are taken to calculate the wave field distribution in the area by the wave models of WAVEWARCH III and Mike21 SW, and the workflow is shown in Fig.1.

3.1.1 Monsoon. The monsoon is a frequent condition in any deep-sea operation area, and the changes in wind speeds and directions affect the selection of RST locations. The wind field model is calculated within the region of 0.6°N, 103.2°E~26.1°N, 126.2°E. Fig.2 shows the specific roadmap of RST candidate points (A ~ G) and emergency points (Accident Site1-3) during deep-sea oil and gas exploration. WRF simulation is used to get the changes of wind speed and direction at these points in different seasons, and the results are shown in Fig. 3.

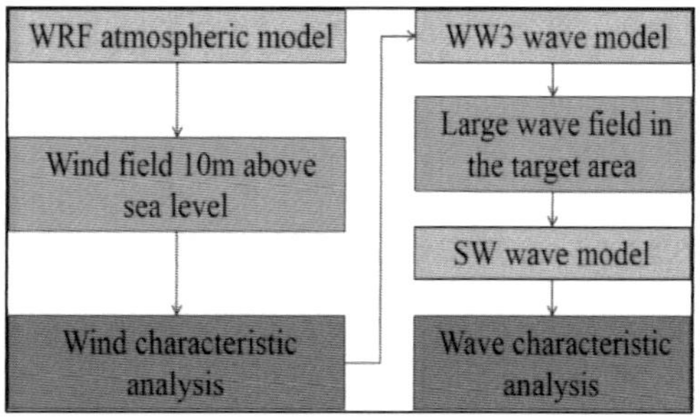

Fig.1 Flow chart of the calculation of wind and wave fields with the mathematical model

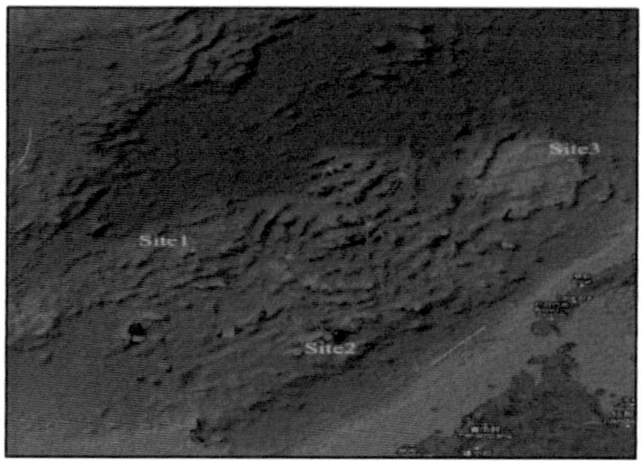

Fig.2 Regional road network of RST and emergency points

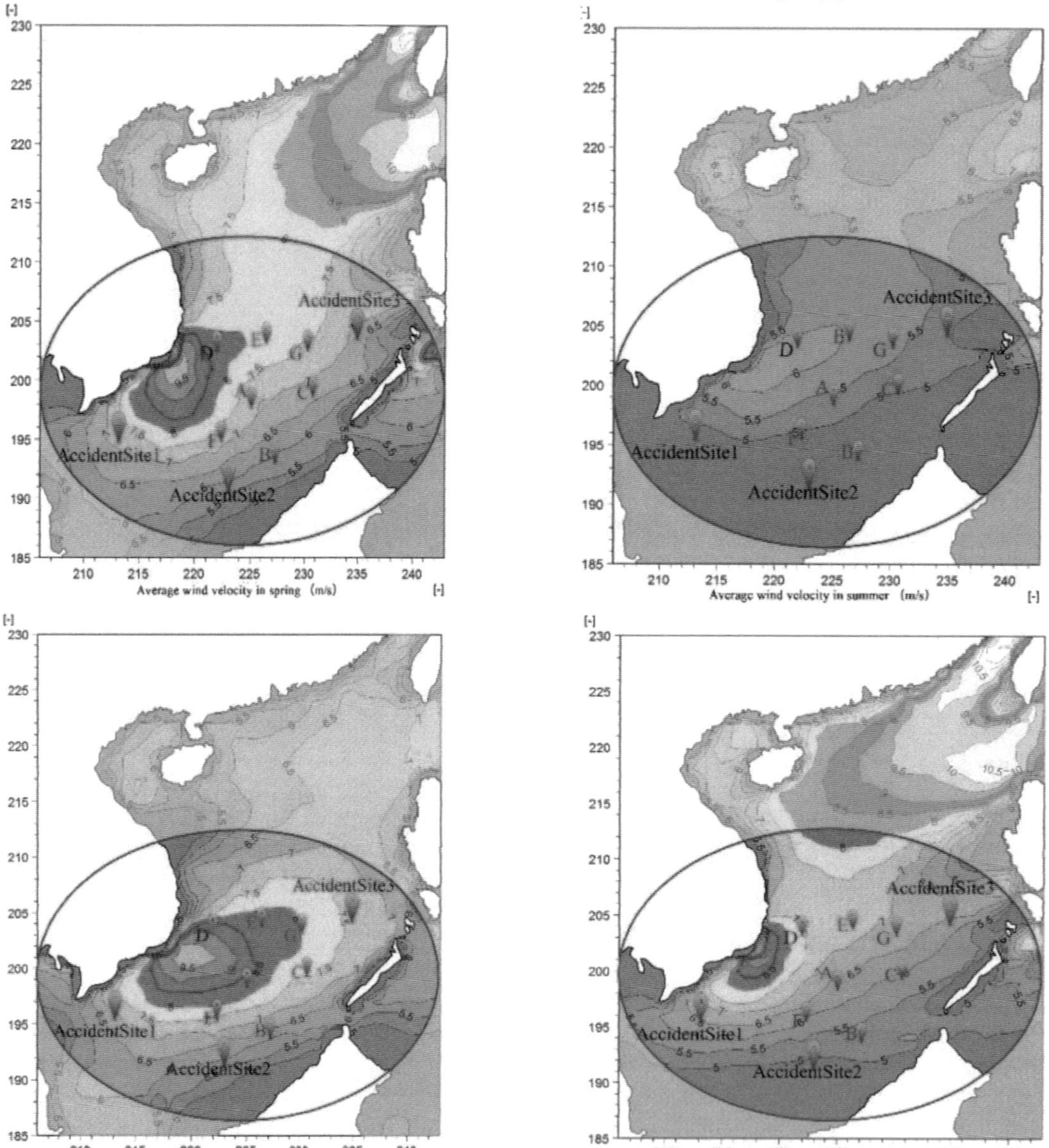

Fig.3 Wind speed and direction at the RST and emergency points in different seasons

From Fig. 3, it can be seen that the target exploration area is the least affected by the monsoon in summer, while in the other three seasons, the monsoon can bring significant impacts on the target area, especially on the RST candidate points in the northwest of the region of interest. Moreover, in spring, autumn, and winter, the monsoon usually comes from the northeast, while the direction of monsoon origin shifts to the southwest in summer. Tables 2-5 show the information about the angles between the forward waves and bows at each node in different seasons.

Table.2 Angle between forward wave and bow in spring (α_1)

	A	B	C	D	E	F	G	Site1	Site2	Site3
A		165°	13°	63°	7°	170°	25°			
B	52°		35°			114°			150°	
C	135°	175°			65°		13°			
D	120°				79°	155°				
E	165°		115°	105°			45°			
F	23°	35°		5°				95°	145°	
G	150°		145°		125°					25°

Table.3 Angle between forward wave and bow in summer (α_1)

	A	B	C	D	E	F	G	Site1	Site2	Site3
A		25°	105°	105°	175°	15°	170°			
B	135°		170°			35°			5°	
C	55°	5°			95°		150°			
D	65°				105°	12°				
E	7°		45°	50°			120°			
F	170°	120°		175°				35°	30°	
G	3°		10°		40°					155°

Table.4 Angle between forward wave and bow in autumn (α_1)

	A	B	C	D	E	F	G	Site1	Site2	Site3
A		160°	12°	60°	10°	160°	20°			
B	47°		33°			105°			143°	
C	130°	170°			60°		10°			
D	123°				83°	160°				
E	150°		105°	100°			40°			
F	25°	35°		5°				90°	150°	
G	145°		145°		120°					30°

Table.5 Angle between forward wave and bow in winter (α_1)

	A	B	C	D	E	F	G	Site1	Site2	Site3
A		155°	10°	55°	5°	150°	20°			
B	45°		30°			100°			140°	
C	140°	170°			65°		15°			
D	125°				85°	140°				
E	160°		110°	95°			50°			
F	30°	40°		15°				90°	150°	
G	130°		125°		135°					20°

3.1.2 Wave. Monsoon is the main effective factor for RST and waves around the emergency points. When the northeast monsoon is prevailing, the NE (northeast) wave is dominant in October. From November to the next April, NE waves appeared at a percentage of more than 40%. Also, the wave heights showed a significant variation across months. Specifically, the maximum wave height was 1.8m in December, which dropped to 0.7m in April. When the southwest monsoon is on the way, S (south) and SW (southwest) waves prevail. In the data, S and SW waves appeared with a frequency of 40% - 50% in summer, and the monthly average wave height was 1.0-1.2m. During monsoon alternation, wave directions fluctuate, and no obvious prevailing waves are observed. The minimum wave heights are found from April to May, and the wave direction varied.

The WAVEWATCH III model is applied to the region of 98.0～135.0°E, 8°S~27.0°N. Fig.4 shows the effective distribution of seasonal mean wave heights in that region after simulation, and Table 6 shows the wave height distribution of RST and emergency points in each season. The results suggest that wave height is obvious in spring, autumn, and winter (when NE monsoon is dominant) while being low in summer (when SW monsoon is dominant).

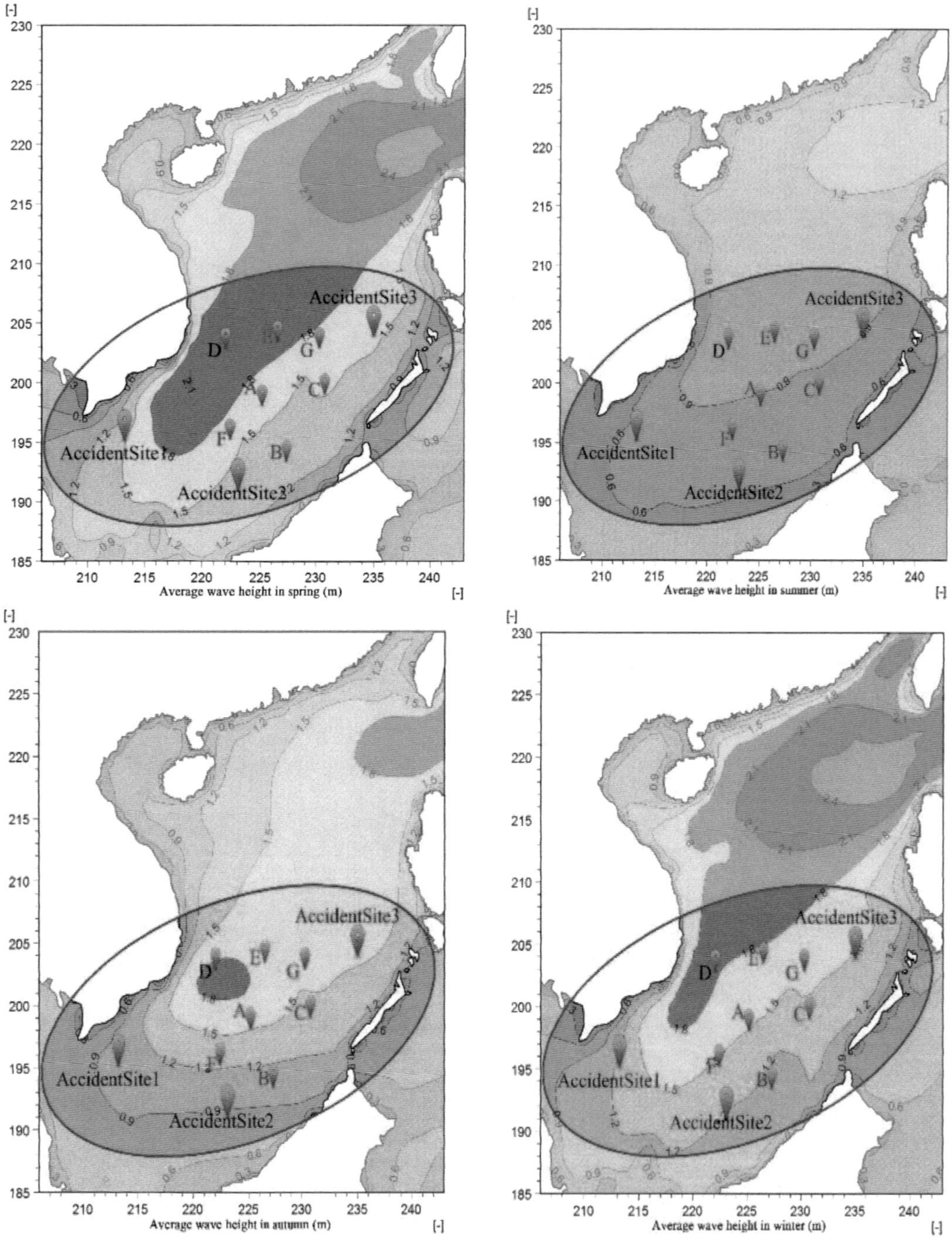

Fig.4 Effective distribution of mean wave heights by season

Table.6 Wave height distribution of RST and emergency points by season

h/m	A	B	C	D	E	F	G	Site1	Site2	Site3

Spring	1.6	1.3	1.5	2.4	2.2	1.6	1.7	1.5	1.4	1.6
Summer	0.85	0.65	0.8	1.1	1.2	0.7	1.0	0.65	0.6	0.9
Autumn	1.5	1.1	1.4	1.9	1.7	1.25	1.65	1.1	0.8	1.6
Winter	1.5	1.1	1.4	1.85	1.7	1.45	1.6	1.3	1.2	1.5

3.2 Example verification

This part aims to compare the changes in the sailing times and routes under different monsoons and wave conditions and analyze how these changes affect the distribution of RSTs for different materials. In this paper, a blowout accident is taken as an example, and the emergency relief materials belong to eight categories, namely, bop, barite, mud materials, cementing materials, crane, cutting machine, oil containment booms, and oil spill dispersants. Table 7 shows the importance of each emergency resource demonstrated by their expert scoring (g_{jk}). Table 8 shows the fuzzy number of resource demands at each emergency point based on the empirical knowledge ($\tilde{a}_{jk}$).

Table. 7 Importance of various types of emergency resources to the emergency points

Number	Name	Importance/g_{jk}	Number	Name	Importance/g_{jk}
1	bop	0.149	5	crane	0.114
2	barite	0.142	6	cutting machine	0.108
3	mud materials	0.129	7	oil containment boom	0.114
4	cementing material	0.136	8	oil spill dispersant	0.108

Table. 8 Fuzzy number of resource demands at the incident site ($\tilde{a}_{jk}$)

	AccidentSite1	AccidentSite2	AccidentSite3
bop	(0,1,2)	(0,1,2)	(0,1,2)
barite	(70,100,120)	(60,85,100)	(50,95,120)
mud materials	(80,110,150)	(75,100,140)	(100,120,150)
cementing material	(95,120,145)	(80,100,130)	(90,120,150)
crane	(1,2,3)	(1,2,3)	(1,2,3)
cutting machine	(1,2,3)	(1,2,3)	(1,2,3)
oil containment boom	(800,1000,1500)	(700,900,1500)	(900,1200,1700)
oil spill dispersant	(5,10,15)	(6,10,13)	(7,12,16)

The ship speed is set to 14 knots and other parameters are Dt =15197t,$k_1 = 0.745, k_2 = 0.05015, k_3 = 4.5 \times 10^{-3}, k_4 = 1.35 \times 10^{-6}$, G = 1.09. $\lambda_T = 0.6, \lambda_Q = 0.4$. Both numbers are derived from expert evaluations. The verification is performed at Visual Studio 2012, Lingo 18.0, and MatlabR2016b under a Windows 7 computer with the host configuration of Intel corei5, CPU 2.5GHz, and a RAM of 12G.

The actual ship speed in different seasons are solved based on Table 2 - 6 and Eq. (1). Fig.5 shows the sailing speed from RSTs to each node in different seasons. Among all four seasons, the ship speed in summer is the closest to that in still water, which is more conducive to emergency rescue.

With the LPHAL method described in Section 4.1, the minimum navigation time from RSTs to the emergency points, as well as the route variations are calculated for each season (as shown in Table 9). Besides, the extremums of T and Q are also calculated for each season under the constraints mentioned above being met (as shown in Table 10).

Based on these results, the number of resources of category k, namely,q_{ijk}, dispatched from R_i to E_j can be obtained, which is shown in Fig.6. The dispatch amounts of eight emergency resources in different seasons are reflected on the links between any two nodes, e.g. the amount of *barite* from point E to *Site3* in *spring* is 40t. A wireless segment link between two points means no resource transportation from R_i to E_j. Then the c_{jk} and X_{ij} can be found by q_{ijk} following a similar procedure, and the calculations are not repeated here.

Table. 9 Minimum sailing times and routes from $R_i$to E_j in different seasons

		Spring		Summer		Autumn		Winter	
		T_{min}/h	Route	T_{min}/h	Route	T_{min}/h	Route	T_{min}/h	Route
Site1	A	4.53	A-F-Site1	3.54	A-F-Site1	3.24	A-F-Site1	4.63	A-F-Site1
	B	4.29	B-F-Site1	3.88	B-F-Site1	3.88	B-F-Site1	4.49	B-F-Site1
	C	6.52	C-B-F-Site1	5.83	C-B-F-Site1	5.80	C-B-F-Site1	6.61	C-B-F-Site1
	D	5.31	D-F-Site1	4.92	D-F-Site1	4.69	D-F-Site1	5.41	D-F-Site1
	E	7.40	E-D-F-Site1	6.10	E-D-F-Site1	6.38	E-D-F-Site1	7.29	E-D-F-Site1
	F	2.54	F-Site1	2.03	F-Site1	2.03	F-Site1	2.54	F-Site1
	G	8.12	G-A-F-Site1	7.76	G-A-F-Site1	7.51	G-A-F-Site1	8.13	G-A-F-Site1
Site2	A	4.26	A-B-Site2	3.28	A-B-Site2	3.25	A-B-Site2	4.46	A-B-Site2
	B	2.01	B-Site2	1.62	B-Site2	1.51	B-Site2	2.31	B-Site2
	C	4.14	C-B-Site2	3.27	C-B-Site2	3.53	C-B-Site2	4.14	C-B-Site2
	D	6.65	D-F-B-Site2	5.65	D-F-B-Site2	5.63	D-F-B-Site2	6.65	D-F-B-Site2
	E	6.79	E-C-B-Site2	5.81	E-C-B-Site2	5.78	E-C-B-Site2	6.59	E-C-B-Site2
	F	3.78	F-B-Site2	2.77	F-B-Site2	2.57	F-B-Site2	3.88	F-B-Site2
	G	6.35	G-C-B-Site2	5.40	G-C-B-Site2	5.34	G-C-B-Site2	6.65	G-C-B-Site2
Site3	A	5.38	A-G-Site3	4.25	A-G-Site3	4.38	A-G-Site3	5.63	A-G-Site3
	B	6.05	B-C-G-Site3	5.61	B-C-G-Site3	5.85	B-C-G-Site3	6.47	B-C-G-Site3
	C	3.88	C-G-Site3	2.79	C-G-Site3	2.88	C-G-Site3	4.08	C-G-Site3
	D	4.90	D-E-G-Site3	3.81	D-E-G-Site3	3.89	D-E-G-Site3	4.88	D-E-G-Site3
	E	3.20	E-G-Site3	2.74	E-G-Site3	2.90	E-G-Site3	3.19	E-G-Site3
	F	7.32	F-A-G-Site3	6.84	F-A-G-Site3	6.62	F-A-G-Site3	7.61	F-A-G-Site3
	G	1.62	G-Site3	1.09	G-Site3	1.32	G-Site3	1.72	G-Site3

Table. 10 Maximum and minimum values of each sub-objective function

	Target value	Response time /h	Demand satisfaction
Spring	Max.	38.31	1.848
	Min.	5.5	0
Summer	Max.	38.36	1.848
	Min.	5.38	0
Autumn	Max.	38.03	1.848
	Min.	5.5	0
Winter	Max.	38.3	1.848
	Min.	5.5	0

The deep-sea scheduling scheme of emergency resources obtained by this model can achieve the shortest emergency response time and the highest demand satisfaction. The target values of each seasonal sub-target and program running times are shown in Fig7, which could well meet the requirements of the emergency environment for rapid decision-making.
It should be noted that T_{obj} derived from this model is the sum of the time required to collect all supplies when incidents occur at all three incident points simultaneously. Table 11 gives the resource dispatch results when an incident happened at a single incident point. The dispatch of *cementing* materials for a sudden blowout at Site3 in autumn operations was used as an example to illustrate this model.

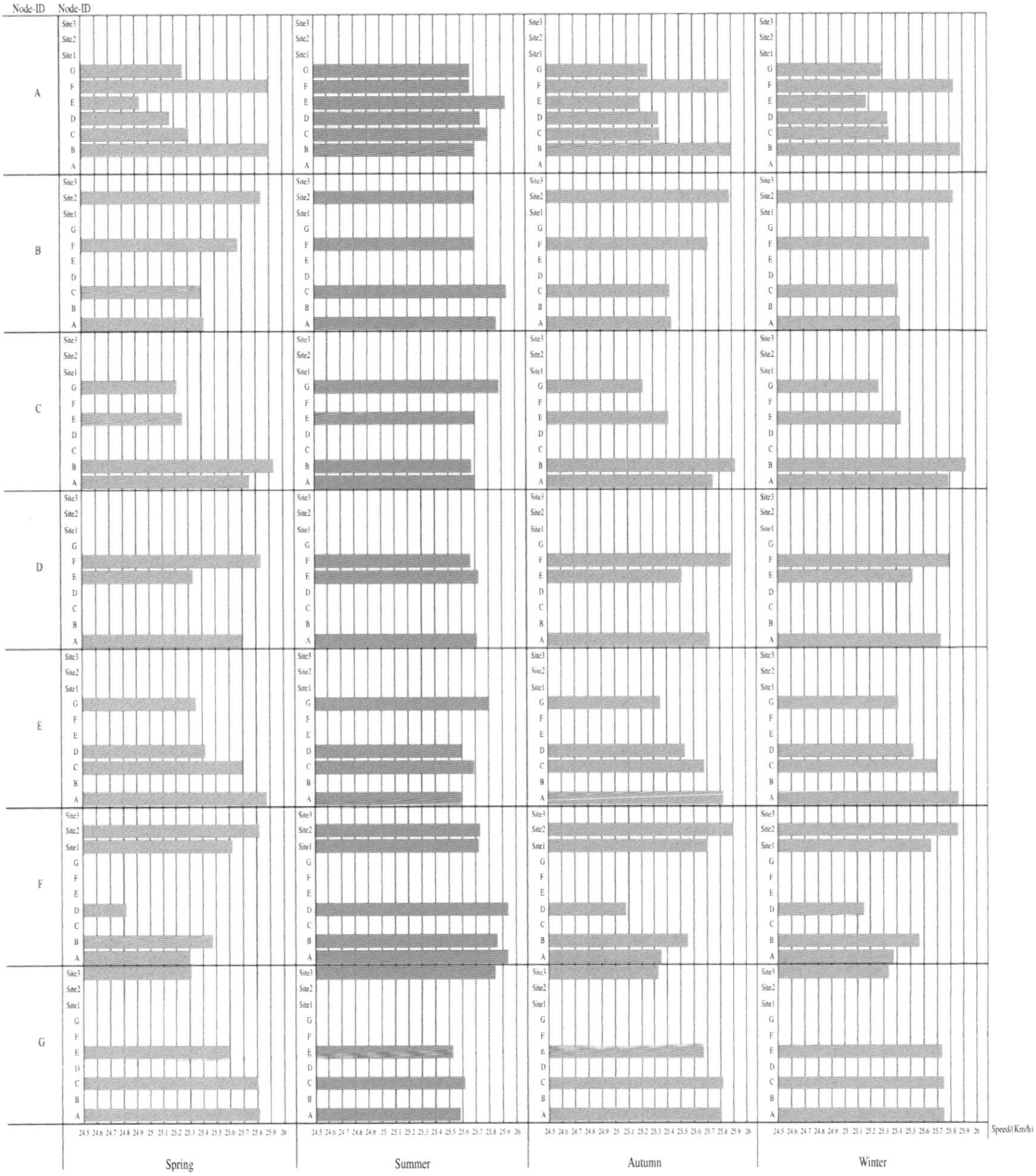

Fig.5 Variations in sailing speeds from RST to each node in different seasons

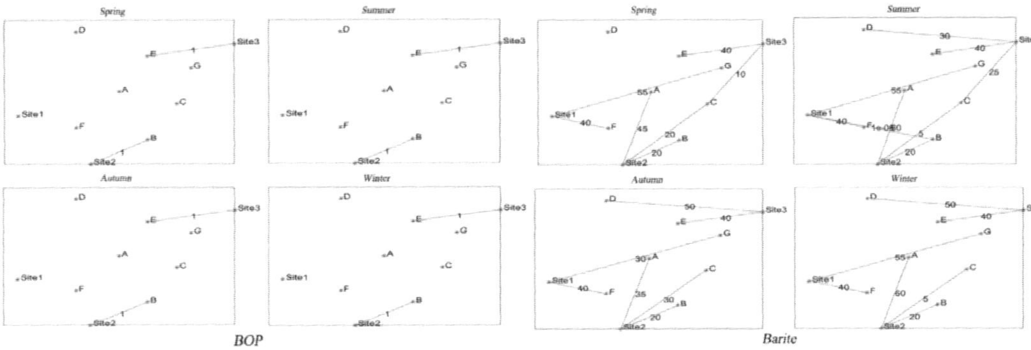

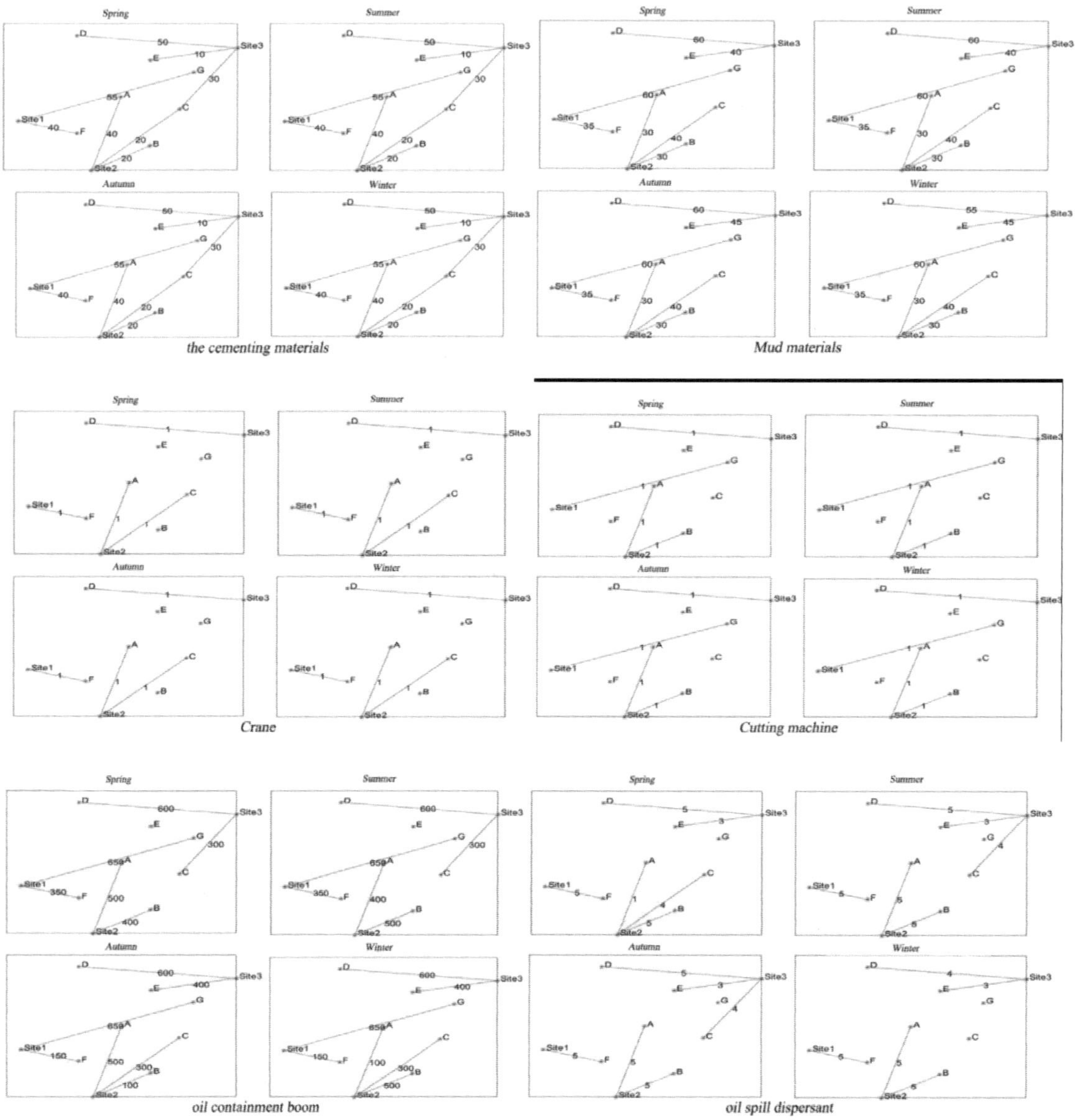

Fig.6 Comparison of resource scheduling options for blowout incidents by season in deep-sea

4. Discussion

According to the analysis in Section 4, the speed change from RSTs to each node in different seasons is shown in Fig. 4, which is determined by variables α_1 and h. The change of V_s can be used to determine the resource scheduling and the selection of RST in different seasons, as shown in Fig.5 and Table 10. In Table 10, the routes selection from RSTs to emergency points are not affected by seasonal factors, and the sailing time is shorter in summer and autumn. The resource allocation in Fig.5 shows that for the resource with small demands, the selection and scheduling quantity of RST are almost not affected by seasonal changes. For resources with large demands, the selection and scheduling quantity of RSTs are greatly affected by seasonal changes, and the scheduling results in autumn and winter are almost the same. To sum up, the operation time window from April to September every year are more suitable for deep-sea oil and gas field exploration and development.

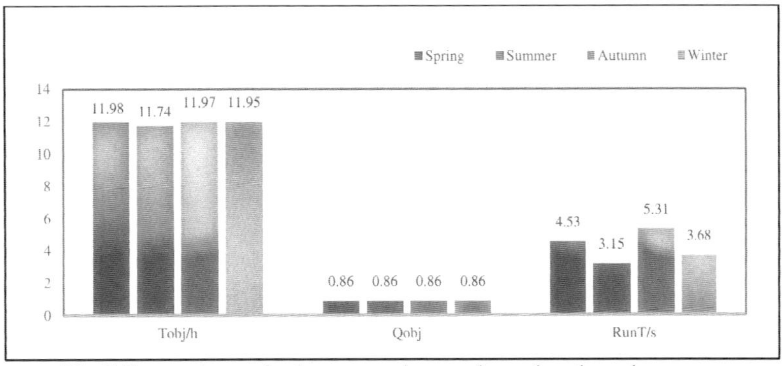

Fig.7 Comparison of sub-target values and running times by season

Table.11 an example of the resource dispatch

RST	Route	Time/h	Dispatch quantity/t
R_C	C-G-Site3	2.88	30
R_D	D-E-G-Site3	3.89	50
R_E	E-G-Site3	2.90	10

5. Conclusions

In this research, the selection of emergency routes and the dispatch of emergency resources in the deep-sea are studied based on the variations of wind speed and wave height in different seasons. Considering the changes of monsoon, the ship stall equation proposed by the Qingdao Meteorological Navigation Consortium of China is cited here. Based on the triangular fuzzy function, a deep-sea emergency resource dispatch model is proposed for a small number of RSTs to achieve minimum emergency response time and maximum resource dispatch satisfaction. The proposed model is validated with the LPHAL algorithm in an assumed blowout incident. Our model accomplished three objective: (1) Addressing the fuzzy uncertainty of the resource demands at the accident points, and the supply of each point are different due to the change of season. (2) The time and routes to dispatch resources in deep-sea emergencies at multiple rescue and accident points are determined. (3) According to the case study, the response time of summer is the shortest among the four seasons, and the resource dispatch satisfaction is the highest. Therefore, summer is the most preferred season for drilling operations.

The study provides a tool for analyzing and refining the siting and configuration of deep-sea emergency RSTs, making it important for marine systems in exploration areas that are vulnerable to natural factors, such as the South China Sea and North Atlantic Ocean.

Three directions are recommended for future work: (1) Based on the response time models, the effects of monsoons will be replaced by those of typhoons. Typhoons of different levels should be analyzed to better describe the impact of weather factors on the dispatch of deep-sea emergency resources. (2) Change the transport means, such as helicopters. We will collected more field data to diversify the choice of transportation means and make the model produce more practical results. (3) Improvements to the model are still to be made. In the future, data about the world's oceans should be included, and the user only needs to ensure the accuracy and reliability of the input parameters to start intelligent analysis and calculation, which is more conducive to the application of the model in the field.

Acknowledgements

Funding: The authors are grateful to China National Deep-sea Oil Corporation (CNOOC) for financial support to this work.

Reference

[1] Gao, X, 2012. Research on the optimization methods of large-scale emergency rescue resources layout and scheduling. Jilin Univ. (in Chinese).
[2] Akil M. Caunhye, Zhang, Y., Li, M., Nie, X.,2016. A location-routing model for prepositioning and distributing emergency supplies. Transp. Res. Part E. 90,161-176.

[3] Yousuf Rebeeh, Shaligram Pokharel, Galal M. Abdella, Abdelmagid Hammuda, 2019. A framework based on location hazard index for optimizing operational performance of emergency response strategies: The case of petrochemical industrial cities. Saf. Sci. 117, 33-42.
[4] Hossein Mojaddadi Rizeei, Biswajeet Pradhan, Maryam Adel Saharkhiz,2019. Allocation of emergency response centres in response to pluvial flooding-prone demand points using integrated multiple layer perceptron and maximum coverage location problem models. Int. J. Disaster Risk Reduct. 38,101205.
[5] Zhang, L., Lu, J., 2019. Location of maritime emergency supplies repertories considering risk uncertainty. Chin. Saf. Sci. J. 9, 173-180.
[6] Manuel Arana-Jiménez, Víctor Blanco, Elena Fernández, 2020. On the fuzzy maximal covering location problem. Eur. J. Oper. Res. 283,692-705.
[7] Aybike Ulusan, Özlem Ergun, 2021. Approximate dynamic programming for network recovery problems with stochastic demand. Transp. Res. Part E. 151,102358.
[8] Duncan T. Wilson, Glenn I. Hawe, Graham Coates, Roger S. Crouch, 2014. Evaluation of centralized and autonomous routing strategies in major incident response. Saf. Sci. 79, 80-88.
[9] Martha Grabowski, Christopher Rizzo, Travis Graig, 2016. Data challenges in dynamic, large-scale resource allocation in remote regions. Saf. Sci. 87, 76-86.
[10] Ge, X., Xue, G., 2019. Multi-products joint distribution vehicle routing optimization with uncertain circumstance. Comput. Eng. Appl.55, 264-270.
[11] Zhu, X., Chen, C., 2019. Multi-objective optimization of marine emergency resource dispatching. Navig. Chin. 8,46-51.
[12] Chen, Z., 2020. Research on the optimization of emergency materials scheduling in maritime oil spill accident. Dalian Marit. Univ. (in Chinese).
[13] Roopsi Rathi, Neeraj Gupta, 2020. Game theoretic and non-game theoretic resource allocation approaches for D2D communication. Ain Shams Eng. J.
[14] Zhuang, H., Pan, T., He, Z., 2020. Method of resource scheduling for space object cooperative surveillance using Tabu Genetic Algorithm. Spacecr. Eng. 8, 46-51
[15] Nikos Argyris, Özlem Karsu, Mirel Yavuz, 2021. Fair Resource Allocation:using Welfare-based dominance constraints. Eur. J. Oper. Res. EOR,17212.
[16] Nir Halman, Shmuel Wimer, 2021. Resource allocation in rooted trees subject to sum constraints and nonlinear cost functions. Inf. Process. Lett. 170,106114.
[17] Yong Wang, Chun-Rong Chen, Pei-Qiu Huang, Kezhi Wang, 2021. A new differential evolution algorithm for joint mining decision and resource allocation in a MEC-enabled wireless blockchain network. Comput. Ind. Eng. 155,107186.
[18] Zhang, Y.S., Lu, Y.Q., Cai, F., Shi, A.G., 2005. The comparison of several computing methods of ship's loss. Mar. Technol. 1, 7-9.
[19] Wang, M.R., 2010. The study on the scheduling of rescuing resources for marine perils. Dalian Marit. Univ. (in Chinese).

Journal of Physics: Conference Series **2361** (2022) 012019 doi:10.1088/1742-6596/2361/1/012019

MSCNN-BLSTM based Prediction of the Remaining Useful Life of Aeroengine

Lei Wang[1], Dongrun Chang[2], Zongshuai Li[2]

[1] Engineering Technical Training Center, Civil Aviation University of China, Tianjin
[2] College of Electronic Information and Automation, Civil Aviation University of China, Tianjin

695220661@qq.com; 18234777561@163.com; zsli@cauc.edu.cn

Abstract: Accurate prediction of the aeroengine remaining useful life (RUL) is essential to improve engine availability and reliability. Aiming at the reliable prediction of residual life of aeroengine system, an engine residual life prediction model based on the fusion of multiscale fusion two-dimensional convolutional neural network and bidirectional long and short term memory (MSCNN-BLSTM) is proposed. Based on the fusion of two-dimensional convolutional neural network and bidirectional long and short time memory (BLSTM) network, the engine medium and advanced features extracted by the convolutional neural network are integrated to make residual life prediction. Finally, C-MAPSS dataset provided by NASA was used for validation. It is shown that the proposed multiscale hybrid model, compared with other model predictions, reduces the performance index score and root mean square error by 32.2% and 14.7% respectively. It can be seen that the data-driven model can effectively extract the information from the degradation data, which improves the prediction performance of aeroengine remaining life.

1. Introduction

The residual life prediction techniques can be grouped into three varieties[1]: physical model-based methods, reliability-based methods, and data-driven methods. These three mentioned residual life prediction methods all have their own disadvantages and advantages[2]. To predict the remaining life based on physical models[3], the development of the physical models is a complicated process. In addition, in complex systems such as nuclear power plants[4], it is difficult to use physical models to understand the degradation mechanism. Besides this, many parameters need to be ascertained by concrete experiments, thereby restricting the use of this method. The reliability-based residual[5] life prediction needs to fit the observed data using probability theory and mathematical statistics, with the most extensive applicability. However, these methods need to assume the prior probability and periodic distributions.

No limitations for two other methods have been found in data-driven methods[6]. Deep learning, as a new subfield of machine learning, develops deep complex nonlinear relationships in feature and state monitoring (Condition Monitoring, CM) datasets[7]. In technologies based on deep learning, researchers concentrate on Recurrent Neural Networks (RNN), using it to learn features deeply and capture the temporal information of device state monitoring data for residual lifetime prediction. J. Chen et al.[8] Proposed a RNN-based health indicator of bearing residual lifetime. LeCun Y et al.[9] came up with a RNN-based Auto-encoder (AE). However, there is always a flaw: during the backpropagation of network training, RNN are prone to gradient extinction or gradient explosion, which limits their ability to capture the long-term dependence on time[10].

To solve the shortcomings that RNN model cannot solve the long-term data dependence due to the disappearance of gradient, Long Short Term Memory Memory (LSTM) networks are derived on the change of RNN structure. However, T. Xia et al.[11] Proposed a time-window-based Bi-directional LSTM (BLSTM) network framework. The BLSTM framework can handle sequential information in two forward and reverse directions. C. -G. Huang et al.[12] put forward estimation of the remaining lifetime with the BLSTM-based prediction framework. In addition to the RNN and its variant networks, Convolutional Neural Networks (CNN) have been lead into the field of engine residual life prediction in recent years, and have achieved considerable results in processing sequential data. H.Liu et al.[13] achieved the prediction of residual engine life using the BLSTM and CNN combination.

The CNN and BLSTM neural fusion network can both replace the process of manually extracting the data features and solve the long-term dependencies and time-series prediction problems between the data. However, as the neural network feature extraction proceeds, the above method generally only uses the extracted features (advanced global features) at the last level to solve the problems such as classification. In this way, some of the low-level local features generated by the middle layer will be ignored, losing the key degraded information in the data, and seriously affecting the accuracy of the remaining life prediction results. To address the shortcomings of the above methods, the paper cited a idea of residual network solving the network degradation of deep neural networks caused by excessive network layers proposed by He K et al.[14] A new reconciled neural network prediction model, Multi-scale fusion CNN and Bidirectional LSTM (MSCNN-BLSTM), is proposed using a constant mapping cascade. This method first processes the raw sensor monitoring data through min-max standardization and sliding time window; then the MSCNN model learns the processed data characteristics on multiple scales, and BLSTM enhances the dependence on time series during equipment degradation and implements RUL prediction. Finally, the availability of this proposed method is demonstrated by employing the Commercial Modular Aero-propulsion System Simulation (C-MAPSS) dataset.

2. Basic knowledge of neural networks

2.1 Two-dimensional convolutional neural networks(2DCNN)

In each layer of 2DCNN, a two-dimensional input array and a two-dimensional convolutional kernel output a two-dimensional array containing a large number of local features through convolutional operations. Convolution operation are detailed in (1).

$$p_i = relu(H * k_i + b_i) \tag{1}$$

The p_i in the formula represents the resulting i-th feature map; the relu is the activation function applied to the convolutional layer; H is the input matrix; $*$represents the convolutional operation; k_i delegates the i-th kernel of the current convolutional layer; and b_i represents the bias value. The m feature maps obtained by the filter are indicated as follows:

$$P = [p_1 \cdots p_i \cdots p_m] \tag{2}$$

2.2 BLSTM

Each LSTM unit receives the state vectors h_{t-1} and c_{t-1} of the previous time node while receiving x_t through internal operations to obtain two new state vectors h_t and c_t. Three gates (input, output, forgetting gates) are used within the LSTM to control the inflow and outflow of state vectors. Thus keeping long-and short-term memory in the cell state vector to prevent the gradient from vanishing or exploding. The specific LSTM unit structure is shown in Figure 1.

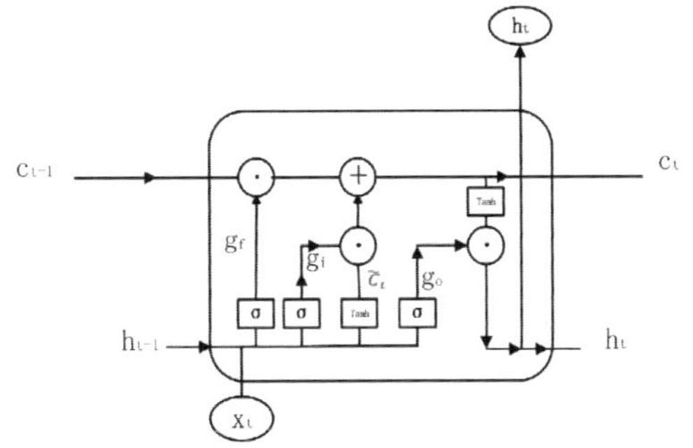

Figure 1. LSTM cell structure

G_f, g_i and g_o are control variables for forgetting gate, input gate and output gate respectively; (W_f, b_f), (W_i, b_i), (W_o, b_o) are parameter variables corresponding to three gates; σand Tanh is activation function. The interval of g is [0,1], and when the value is 1, the gating unit is fully opened for the accepted data connected in the graph. ⊙represents the multiplication between each element of a matrix.

To wipe out the problem that it only considers information about the past sequences of the input data and ignores information in future sequences, the BLSTM network is used in this method. The basic formation of BLSTM layer is shown in Figure 2:

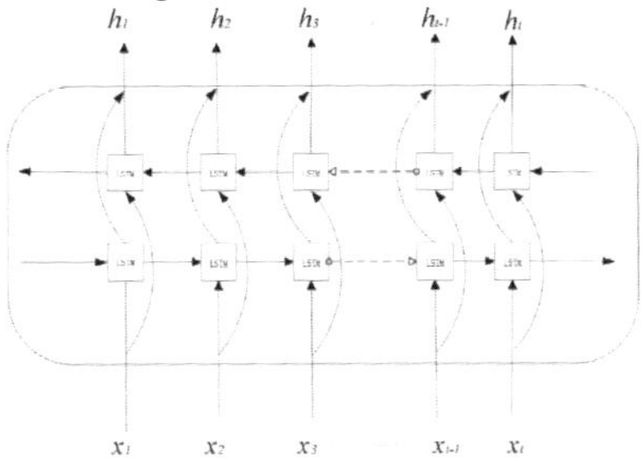

Figure 2. Basic structure of BLSTM layer

One layer flows in the positive direction, calculates h_1^+, h_2^+, ..., h_t^+; the second layer network calculates h_1^-, h_2^-, ..., h_3^-, in the opposite direction; and finally combines the saved forward and backward output, namely $ht{=}h_t^+ \oplus h_t^-$.

3. Method

3.1 MSCNN

Assuming the sample is X0 to be a two-dimensional matrix with T time nodes and F features.

$$X_0 = [x_1, x_2 \cdots x_t \cdots x_T]^T \in R^{T \times F} \tag{3}$$

$$x_t = \left[x_t^1, x_t^2 \cdots x_t^f \cdots x_t^F\right] \in R^{1 \times F},\ t \in [1,\ T] \tag{4}$$

In (3), x_t represents a time series. x_t^f means the data characteristics of the f-th sensor at the t-th time point in (4). Convolutional neural networks can handle problems in a complex environment and context, using the characteristics of weight sharing to integrate local features of data into a multilayer perceptron, with good performance for fault identification and diagnosis performance. After features are extracted by a multilayer convolutional neural network, advanced degradation features are cascades with the original input features, so that advanced degradation features can be used as input to the next stage of the model without losing key degradation information. This detail is shown in Figure 3.

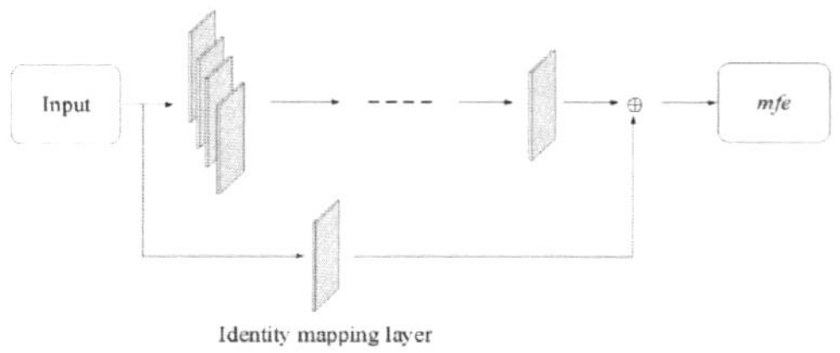

Figure 3. MSCNN model

The input data in the figure are cascade after the multilayer 2DCNN and the input through the constant mapping layer to generate multiscale feature extraction(*mfe*)with the formula as follows,

$$mfe = X_t \oplus X_0 \tag{5}$$

In (5), X_t represents the eigenvector after t convolution and X_0 represents the input matrix vector.

3.2 MSCNN-BLSTM

The model is mainly divided into two stages: encoding and decoding. As shown in Figure 4.

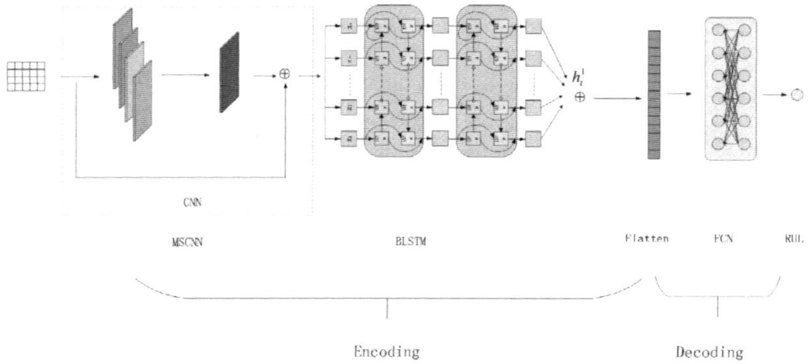

Figure 4. MSCNN-BLSTM Model

The encoding phase consists of the MSCNN and the BLSTM networks. After the original data of the engine was preprocessed, the deep features were extracted as the input to the 2DCNN. After the CNN output of each layer was batch normalized by the BN layer, the nonlinear feature mapping was performed using the ReLU activation function; subsequently, the multiscale feature mfe extracted by MSCNN serves as input to the BLSTM network, and each layer of network output passes through the Dropout layer to prevent data overfitting, thus resolving long time dependencies between degraded data. Finally, multiple temporal output features of the second layer BLSTM network are cascades as the output of the encoding phase.

Fully-connected Network(FCN) was selected in decoding stage was used to analyze the encoding. The FCN is often used in combination with other neural networks in the field of deep learning lifetime prediction. First, to convert the encoder output to one-dimensional form using a flatten layer; then, with a full connection layer of three layers.

During the training procedure, the raw training set was randomly classified as two groups: training set and verification set. Random 1/3 of the raw training data was token as the verification set, and the parameters were regulated according to the performance of the verification set. In addition, the Adaptive Moment Estimation(Adam) algorithm is selected to optimize the model parameters. The Adam algorithm obtains the advantages of the adaptive gradient algorithm(Adagrad) and the root mean square propagation(RMSProp) algorithm. The proposed method calculates the adaptive learning rate for different parameters through the values of the first and second moments of the gradient.

3.3 Data Preprocessing

Units and values generally vary in the range of state information data measured by different aviation engine sensors. If the raw data is straightforwardly used for data analysis, some important degradation features with small values are easily ignored, making the neural network difficult to converge, thus affecting the accuracy of the final prediction. Data normalization can scale the state data to a very small interval, and it can be converted into dimensionless data, helping to improve the neural network convergence rate and prediction accuracy. The input data can be scaled to the interval [0,1].

A sliding window can be used to capture the long-term dependencies between time series data, and it can be expanded by the data. The samples in each time window belong to one and the same engine unit, and the remaining life of the last one data point in each window is taken as the target tag. The sizes and number of engine time windows for different subsets in the C-MAPSS dataset are shown in Table 1.

Table 1. The information about time windows

Dataset	FD001	FD002	FD003	FD004
Time window size	30	20	30	18
Training set samples	17731	48819	21820	57016
Test set samples	10196	29070	13696	36998

In the data driven prediction model, the goal is to get the expected remaining lifetime value and output from each input sample. However, its degradation was almost negligible in the early stages of engine operation. Instead, more engine degradation information exists in the later stages. Therefore, we use a piecewise linear descent function to imitate the degradation process of the actual RUL of an aeroengine. The maximum value of the remaining lifetime tag is 125.

3.4 Prevent overfitting

The BN and Dropout layers were used to avoid model overfitting.The BN layer uses the average value and standard deviation of batch samples to make the export of each layer of neural network more stable, thus accelerating the convergence rate of the model. As mentioned above, each BLSTM layer and the last FCN layer output passes through the Droupout layer and then is input to the next layer to achieve the purpose of enhancing the model generalization ability.

4. Results and validation

4.1 Proof technique

To test and verify the significance of the proposed model, this paper uses a practically degraded dataset (C-MAPSS dataset) of a simulated turbofan engine created by NASA. The C-MAPSS dataset can be classified as four different consonant sets based on the failure mode and running conditions; Each subset consists of a training set and a test set. The engine in the training set runs from one point to failure and provides the relevant remaining life value after each engine cycle.

There are two indicators used to evaluate the performance of the prediction method, namely, root mean square error(RMSE) and Score. RMSE, as shown in Formulas (6) and (7), is widely used in regression tasks.

$$d_i = RUL_i^{pre} - RUL_i^{act} \tag{6}$$

$$RMSE = \sqrt{\frac{1}{n}\sum_{i=1}^{n} {d_i}^2} \tag{7}$$

$$\text{Score}=\begin{cases}\sum_{i=1}^{n} e^{-\frac{d_i}{13}}-1, & d_i \leq 0\\ \sum_{i=1}^{n} e^{\frac{d_i}{10}}-1, & d_i \geq 0\end{cases} \quad (8)$$

In the expression (7), n expresses the sum number of engines in the used data set; d_i refers to the difference between the forecast RUL value and the actual RUL value of the i-th engine. As shown in (8), Score is a performance scoring function proposed by the supplier of the C-MPASS dataset. It is more stringent for predicting higher than below the actual remaining life value.

4.2 Experimental result

The computer equipment performing the experiments was configured with NVIDIA GeForce GTX 2080Ti, Intel Core i9-9900k, and 32GRAM.The programming tool used in the experiment was Python3.7, the deep learning library Keras2.2.4 and Tensorflow2.1.0.

As the experiment progressed, four independent engine units in the dataset were randomly selected to compare their predicted remaining life with the true values over the full period of engine life, as shown in Figure 5.

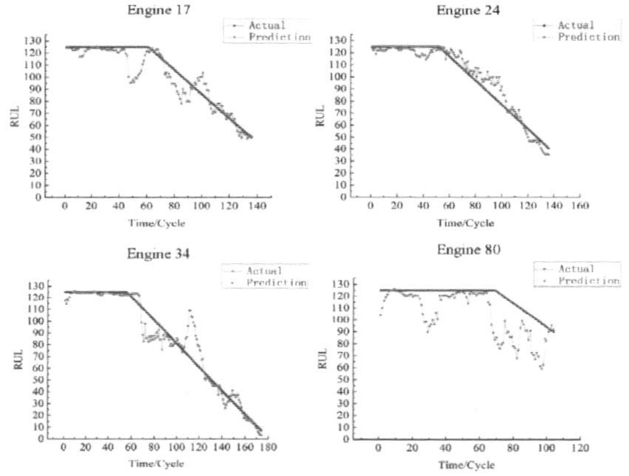

Figure 5. RUL prediction of random four engine units

Figure 5 shows the detailed comparison between the neural network predicted residual life values over the full run-to-failure period of the four random engine units in the FD001 dataset test set. The results show that because there is almost no effective degradation information in the early stage of engine operation, the low model accuracy leads to a relatively large error in the early prediction. But considering that there is basically no failure in the early stage of engine operation, the initial error can be negligible. With the increase of engine operation cycle, its later failure stage contains more abundant degradation information. The prediction model can precisely calculate the remaining life of engine unit by learning the engine failure characteristics.

4.3 Performance comparison with else advanced methods

To validate the predictive superiority of this model, it was compared with some other and more excellent methods. Figure 6 shown the life prediction results of the corresponding engine after performing the actual engine ranking according to the remaining life size of the engine unit.

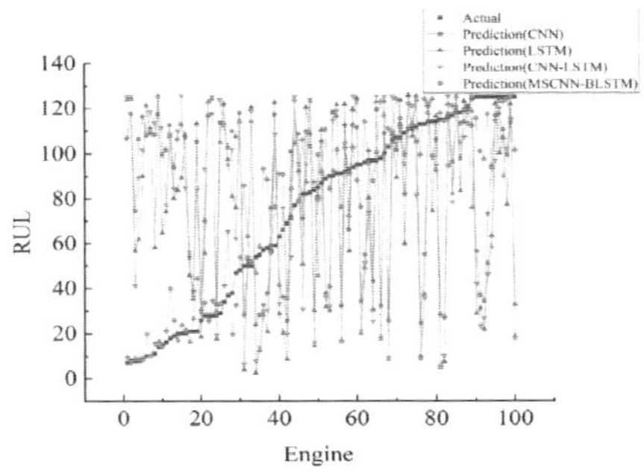

Figure 6. RUL prediction of 100 independent aeroengine units

The horizontal axis is the sorted 100 independent engine units, and the vertical axis represents the remaining life value corresponding to each engine unit. 'Actual' represents the actual surplus life value of all engine units, and the other discounts indicate the remaining life value of other advanced algorithms. The closer to the 'actual' curve, the better the algorithm predicts; at the same distance, below the blue curve indicates better performance. It is evident from the figure that the MSCNN-BLSTM model put forward by this paper significantly outperformed other methods.

Table 2. Contrast of experimental results of different advanced methods[15]

Method	RMSE	Score	R^2
SVR	2096	1381	N/A
DLSTM	18.33	655	N/A
CNN	18.42	906	0.80
LSTM	17.85	686	0.83
MDBGRU	17.08	N/A	N/A
CNN-LSTM	16.36	443	0.86
MSCNN-BLSTM	13.95	300	0.89

The results obtained from the experiments were compared with those obtained by the currently publicly available state-of-the-art methods, see Tables 2 and Figure 6, and N /A represents the missing data. The advanced methods for this comparison mainly include multi-layer CNN, LSTM, multi-scale deep bidirectional GRU Network (MDBGRU), one-dimensional Convolution and Long and Short Time Memory Fusion Network (CNN-LSTM[15]). According to the comparison results, the method scoring Score and RMSE are the lowest among the listed algorithms, and the Score and RMSE decrease by 32.2% and 14.7% compared with the previous optimal results, respectively.

5. Conclusions

This paper mainly proposes and uses a new neural network model based on multi-scale 2DCNN and BLSTM network to solve the issue of aeroengine residual life prediction and validated using the C-MPASS dataset, yielding the following conclusions.

Data-driven equipment residual life prediction efforts overcome the defects of conventional machine learning methods in building degradation models. And it can be applied to aeroengines to solve prediction problems in large-scale industrial fields outside of engines. In future work, it may be necessary to consider the limited applicability of prediction data and health control data, and attempt to introduce specialized knowledge into prognosis models to further enhance its predictive precision.

6. References

[1] Gouriveau R, Medjaher K, Zerhouni N. From prognostics and health systems management to predictive maintenance 1: Monitoring and prognostics[M]. *John Wiley & Sons*, 2016.

[2] Kundu P, Darpe A K, Kulkarni M S. Weibull accelerated failure time regression model for remaining useful life prediction of bearing working under multiple operating conditions[J]. *Mechanical Systems and Signal Processing*, 2019, 134: 106302.
[3] Downey A, Lui Y H, Hu C, et al. Physics-based prognostics of lithium-ion battery using non-linear least squares with dynamic bounds[J]. *Reliability Engineering & System Safety*, 2019, 182: 1-12.
[4] Sato M, Moura L S, Galvis A F, et al. Analysis of two-dimensional fatigue crack propagation in thin aluminum plates using the Paris law modified by a closure concept[J]. *Engineering Analysis with Boundary Elements*, 2019, 106: 513-527.
[5] Lee C, Kwon D. A similarity based prognostics approach for real time health management of electronics using impedance analysis and SVM regression[J]. *Microelectronics Reliability*, 2018, 83: 77-83.
[6] Li X, Ding Q, Sun J Q. Remaining useful life estimation in prognostics using deep convolution neural networks[J]. *Reliability Engineering & System Safety*, 2018, 172: 1-11.
[7] Liu H, Zhou J, Zheng Y, et al. Fault diagnosis of rolling bearings with recurrent neural network-based autoencoders[J]. *ISA transactions*, 2018, 77: 167-178.
[8] Chen J, Jing H, Chang Y, et al. Gated recurrent unit based recurrent neural network for remaining useful life prediction of nonlinear deterioration process[J]. *Reliability Engineering & System Safety*, 2019, 185: 372-382.
[9] LeCun Y, Bengio Y, Hinton G. Deep learning[J]. *nature*, 2015, 521(7553): 436-444.
[10] Yuan M, Wu Y, Lin L. Fault diagnosis and remaining useful life estimation of aero engine using LSTM neural network[C]. *2016 IEEE international conference on aircraft utility systems (AUS). IEEE*, 2016: 135-140.
[11] Xia T, Song Y, Zheng Y, et al. An ensemble framework based on convolutional bi-directional LSTM with multiple time windows for remaining useful life estimation[J]. *Computers in Industry*, 2020, 115: 103182.
[12] Huang C G, Huang H Z, Li Y F. A bidirectional LSTM prognostics method under multiple operational conditions[J]. *IEEE Transactions on Industrial Electronics*, 2019, 66(11): 8792-8802.
[13] Liu H, Liu Z, Jia W, et al. A novel deep learning-based encoder-decoder model for remaining useful life prediction[C]. *2019 International Joint Conference on Neural Networks (IJCNN). IEEE*, 2019: 1-8.
[14] He K, Zhang X, Ren S, et al. Deep residual learning for image recognition[C]//Proceedings of the IEEE conference on computer vision and pattern recognition. 2016: 770-778.
[15] LI Jie, JIA Yuan-jie, ZHANG Zhi-xin, LI Run-ran. Remaining Useful Life Prediction of Aeroengine Based on Fusion Neural Network[J]. *Journal of Propulsion Technology*, 2021, 1-10

Acknowledgments
Supported by the Basic Scientific Research Business of Central University Special Fund Project of Civil Aviation University of China (3122020025).

A Smooth Nadir-Pointing Attitude Guidance law for Satellites Constrained by the Deviation of Sun-Pointing

Zhibo Yang[1], Yang Qin[2], Jia Liu[1], Ningbo Sun[1], Da Huo[1], Haoran Ji[3], Qiang Cong[1], Yanzhang Li[1]

1. DFH Satellite Co., Ltd, Beijing, China;
2. China Aerospace Science and Technology Corporation, Beijing, China;
3. College of Aerospace Science and Engineering, National University of Defense Technology, Hunan Province, China

yangzhibo212@163.com

Abstract. This paper proposes a novel nadir-pointing guidance method for satellites constrained by the deviation of sun-pointing, to ensure the attitude of the satellite steering smoothly. The main idea of establishing the expected attitude contains two steps. Firstly, establishing a temporary attitude according to the nadir direction and the normal direction of the ecliptic. Secondly, in order to reduce the deviation between the expected Sun-pointing axis and the Sun-pointing direction, the temporary attitude will be rotated along one Euler axis, and ensure that the expected nadir-pointing axis fixed in the nadir direction at the same time. The nadir-pointing method provided in this paper overcomes the singular phenomenon that the expected attitude of the satellite flips, without serious loss of Sun-pointing. In this way, the peak expected angular velocity and the peak power consumption will be greatly reduced.

1. Introduction

Nadir orientation is one of the commonly used attitude guidance modes of satellites[1]. By setting the expected attitude of the satellite, it can effectively keep the payloads, like data transmission antenna and optical camera, aligned in the nadir direction to ensure their normal operation[2,3]. Meanwhile, appropriate constraints can be set during nadir-pointing, to meet the needs of satellite solar cell array for daily charging[4]. The nadir-pointing-aligned sun-pointing-constrained mode is a typical nadir-pointing choice for satellites. In this mode, the one axis of the satellite will be strictly pointed to the direction from the satellite to the center of the Earth, and makes the angle between another axis, which is expected to point to the Sun, and the direction from the satellite to the Sun relatively small[5-7].

The traditional nadir-pointing-aligned Sun-pointing-constrained guidance method requires that the expected nadir-pointing axis points to the Earth center, while the expected sun-pointing axis be arranged in the plane determined by the Sun, the Earth and the satellite itself, to minimize the deviation angle of sun-pointing[8,9]. However, when utilizing this kind of traditional method, the expected attitude solve by the guidance law cannot change smoothly near the short time period that the satellite-Sun connection and the satellite-Earth connection become parallel, which means the attitude would have to flip greatly. This singular phenomenon leads to the lack of safety and stability to the attitude control system, and also makes the attitude control system work frequently and with high power consumption, which will damage the service life of the satellite.

To overcome the disadvantage of traditional methods, this paper proposes a novel smooth method for nadir pointing by keep the satellite rotating smoothly along a preferred temporary attitude. In this way, the peak expected angular velocity and the peak power consumption will be greatly reduced, without any serious sacrifice of Sun-pointing.

2. Traditional Nadir-Pointing-Aligned Sun-Pointing-Constrained Guidance Law

Without loss of generality, for a satellite, it is supposed that one camera is installed in the z_b axis of the satellite body, while the normal of the solar array is the the $-y_b$ direction of the body. Thus, when utilizing the traditional nadir-pointing-aligned Sun-pointing-constrained method, three axes should be calculated under the inertial frame as,

$$\begin{aligned} \boldsymbol{x}_{bi} &= \frac{\boldsymbol{r}_e \times \boldsymbol{r}_s}{\|\boldsymbol{r}_e \times \boldsymbol{r}_s\|} \\ \boldsymbol{y}_{bi} &= \boldsymbol{r}_e \times \frac{\boldsymbol{r}_e \times \boldsymbol{r}_s}{\|\boldsymbol{r}_e \times \boldsymbol{r}_s\|} \\ \boldsymbol{z}_{bi} &= \boldsymbol{r}_e \end{aligned} \tag{1}$$

Wherein, $\boldsymbol{r}_s$ is the position of the satellited measured in the inertial frame, $\boldsymbol{r}_e$ is the unit vector from the satellite to the center of the Earth.

As can be seen from Eq. (1), the singularity might happen because of the calculation of $\boldsymbol{r}_e \times \boldsymbol{r}_s$. Also, the change of order between $\boldsymbol{r}_e$ and $\boldsymbol{r}_s$ will make the attitude flips greatly in a short time.

Supposing that the satellite operates in a 480km 35°-inclination orbit, and taking J2 perturbation into consideration, one can find the deviation between $-y_b$ axis and $\boldsymbol{r}_s$ varies in 1.5 years as Fig. 1.

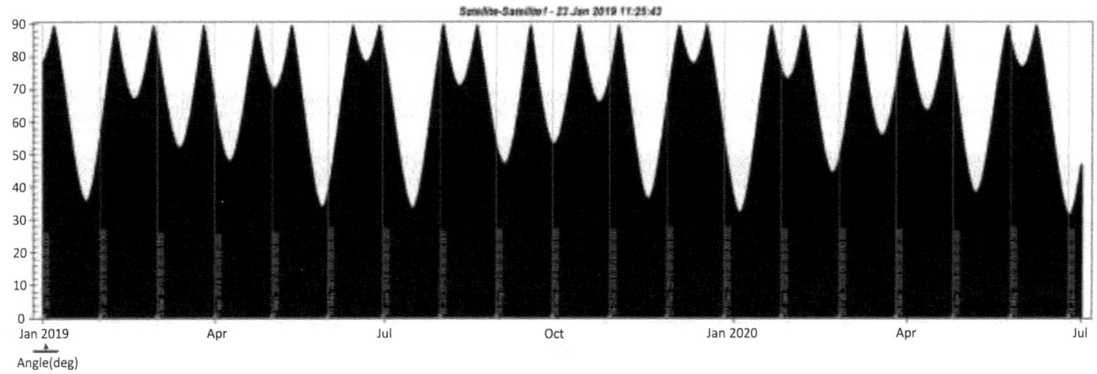

Figure 1. The deviation of the Sun-pointing under traditional method

Also, under the traditional method, the angular velocity of the satellite relative to the inertial frame in all three axes are shown in Fig. 2 to Fig. 4.

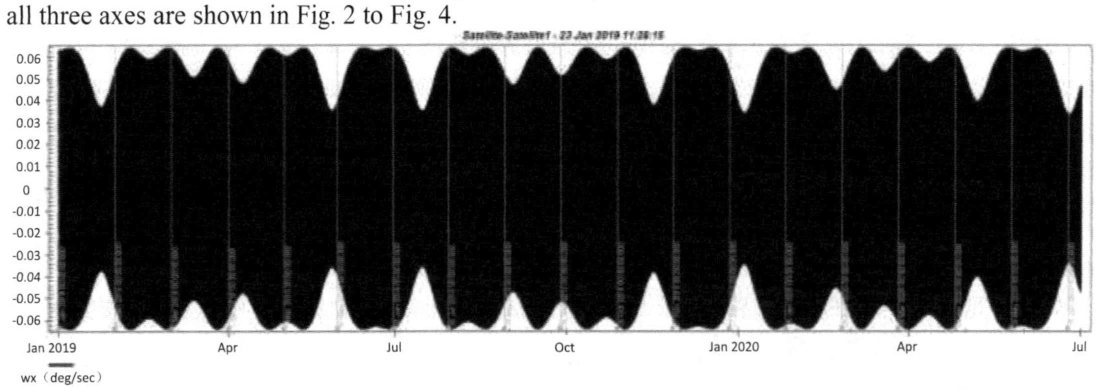

Figure 2. The angular velocity in the x direction under traditional method

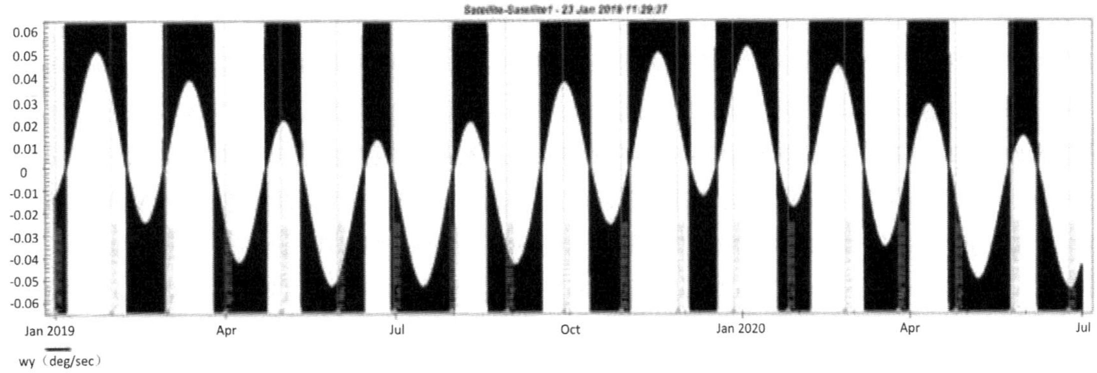

Figure 3. The angular velocity in the y direction under traditional method

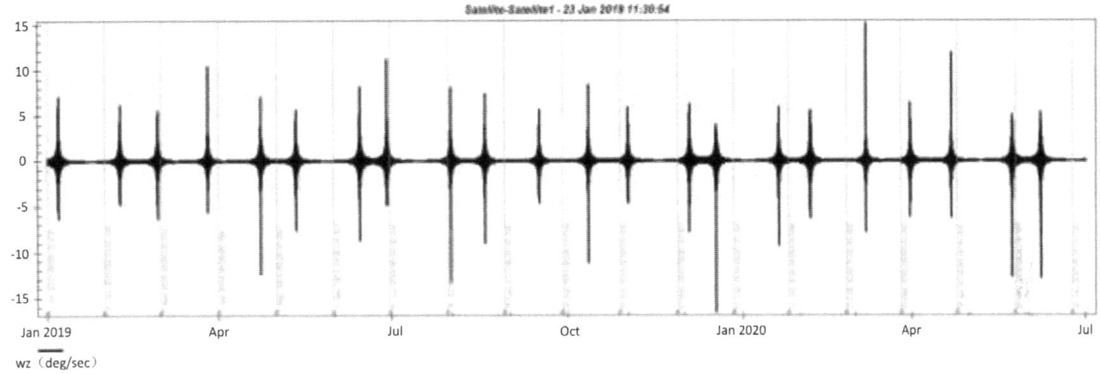

Figure 4. The angular velocity in the z direction under traditional method

As can be seen from Fig. to Fig. , under the traditional guidance method, the deviation between $-y_b$ axis and $\boldsymbol{r}_s$ will always smaller than 90°. Also, the peak angular velocity in both the x direction and the y direction will be up to $0.06\ ^\circ/s$. However, because the attitude flipped sometimes, the angular velocity in the z direction might have sudden changes, with a peak near $5\ ^\circ/s$.
To explain the sudden change of the attitude, the deviation between $\boldsymbol{r}_e$ and $\boldsymbol{r}_s$ is calculated as Fig. 5.

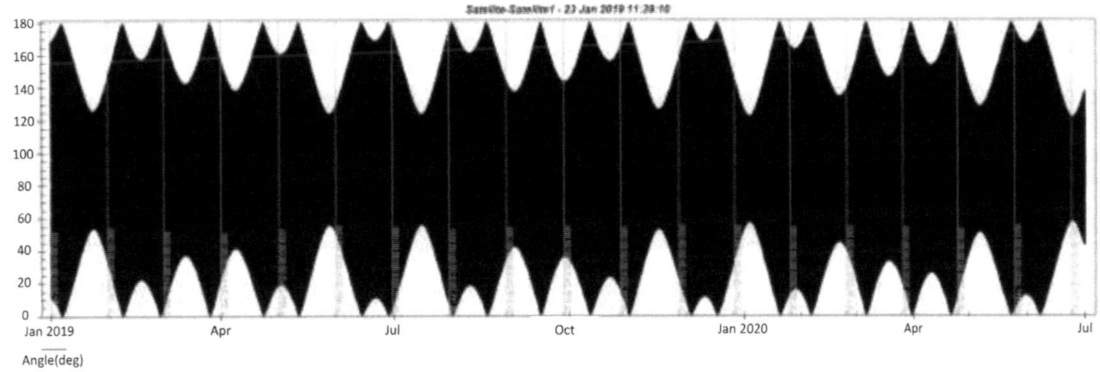

Figure 5. The deviation angler between nadir direction and Sun-pointing direction

By comparing Fig. 4 and Fig. 5, we can find that the attitude meets a sudden change as long as $\boldsymbol{r}_e$ and $\boldsymbol{r}_s$ get parallel to each other, as predicted.

3. Smooth Nadir-Pointing-Aligned Sun-Pointing-Constrained Guidance Law

In this section, the satellite is assumed with the same body axes as in section 2. The z_b axis is asked to be aligned in nadir direction, while the $-y_b$ axis should as close to $\boldsymbol{r}_s$ as possible. In the novel guidance method, the satellite is set to rotate along the z_b axis, and its rotation period is same with the changing period between $\boldsymbol{r}_e$ and $\boldsymbol{r}_s$.

The temporary attitude is set as B_0, under B_0, z_{b0} is coincide with $\boldsymbol{r}_e$, while the $-y_b$ axis is arranged in the plane determined by the normal of ecliptic $\boldsymbol{n}_{sun}$ and $\boldsymbol{r}_e$, which means,

$$\begin{aligned} \boldsymbol{x}_{b_0 i} &= \frac{\boldsymbol{n}_{sun} \times \boldsymbol{r}_s}{\left\| \boldsymbol{n}_{sun} \times \boldsymbol{r}_s \right\|} \\ \boldsymbol{y}_{b_0 i} &= \boldsymbol{z}_{b_0 i} \times \boldsymbol{x}_{b_0 i} \\ \boldsymbol{z}_{b_0 i} &= \boldsymbol{r}_e \end{aligned} \tag{2}$$

as shown in Fig. 6.

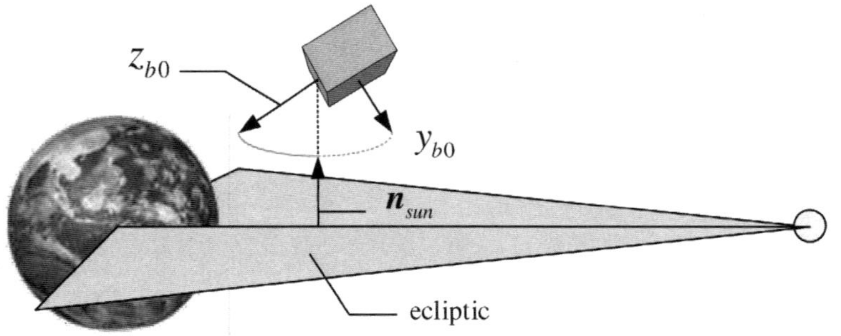

Figure 6. The temporary attitude

The transform matrix from the inertial frame to the temporary attitude B_0 can be calculated as,

$$\mathbf{M}_1 = \left[\boldsymbol{x}_{b_0 i}^T, \boldsymbol{y}_{b_0 i}^T, \boldsymbol{z}_{b_0 i}^T \right]^T \tag{3}$$

Rotate attitude B_0 along the direction z_{b_0} with angle θ to reach the final expected attitude $B_{\exp}$. To ensure the attitude changing smoothly, the value of θ should be consecutive in the range $[0, 2\pi)$. Meanwhile, to make the deviation between $-y_b$ and $\boldsymbol{r}_s$ meets 0 twice in one period of θ, θ is set to be equal to an angle that if one rotate the projection of $\boldsymbol{r}_e$ in the ecliptic, marked as $\boldsymbol{r}_{en}$, along $\boldsymbol{n}_{sun}$ direction with this angle, the vector will be parallel to $\boldsymbol{r}_s$, as shown in Fig. 7.

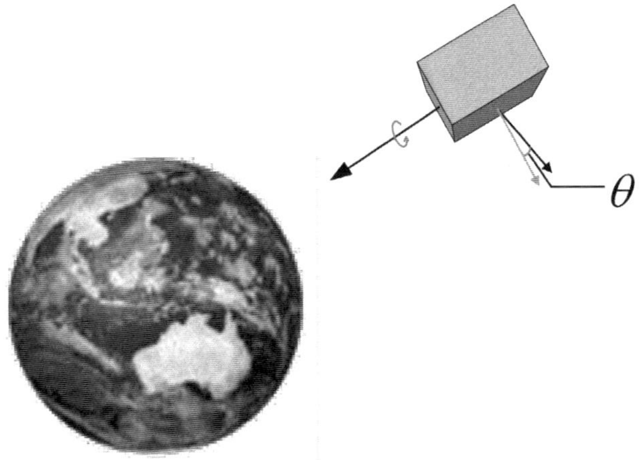

Figure 7. The rotation method of the temporary attitude

Thus,

$$\sin\theta = \operatorname{sgn}\left(\left(\boldsymbol{r}_{en} \times \boldsymbol{r}_s\right)\cdot \boldsymbol{n}_{sun}\right)\frac{\left\|\boldsymbol{r}_{en} \times \boldsymbol{r}_s\right\|}{\left\|\boldsymbol{r}_{en}\right\|\cdot\left\|\boldsymbol{r}_s\right\|} = \operatorname{sgn}\left(\left(\boldsymbol{r}_{en} \times \boldsymbol{r}_s\right)\cdot \boldsymbol{n}_{sun}\right)\left\|\boldsymbol{r}_{en} \times \boldsymbol{r}_s\right\| \tag{4}$$

$$\cos\theta = \frac{\boldsymbol{r}_{en}\cdot\boldsymbol{r}_s}{\left\|\boldsymbol{r}_{en}\right\|\cdot\left\|\boldsymbol{r}_s\right\|} = \boldsymbol{r}_{en}\cdot\boldsymbol{r}_s \tag{5}$$

where,

$$\boldsymbol{r}_{en} = \frac{\boldsymbol{r}_e - \left(\boldsymbol{r}_e\cdot\boldsymbol{n}_{sun}\right)\boldsymbol{n}_{sun}}{\left\|\boldsymbol{r}_e - \left(\boldsymbol{r}_e\cdot\boldsymbol{n}_{sun}\right)\boldsymbol{n}_{sun}\right\|} \tag{6}$$

Therefore, the transformation matrix from attitude B_0 to attitude $B_{\exp}$ can be calculated as,

$$\mathbf{M}_2 = \mathbf{R}_z(\theta) \tag{7}$$

Finally, the transform matrix from the inertial frame to the expected body fixed frame can be solved as,

$$\mathbf{M} = \mathbf{M}_2\mathbf{M}_1 \tag{8}$$

The transformation matrix from inertial system to the body fixed frame uniquely determines the desired attitude of the satellite. Through the transfer matrix $\mathbf{M}$ the quaternion $\bar{q}^*$ of the desired attitude of the satellite can be obtained. $\dot{\bar{q}}^*$ can be obtained through the difference of $\bar{q}^*$, and then the desired angular velocity ω^* can be obtained through the kinematic equation.
The smooth attitude guidance law is tested in the same condition as in section 2. Under the novel smooth law, the deviation between $-y_b$ axis and $\boldsymbol{r}_s$ varies in 1.5 years as Fig. 8.

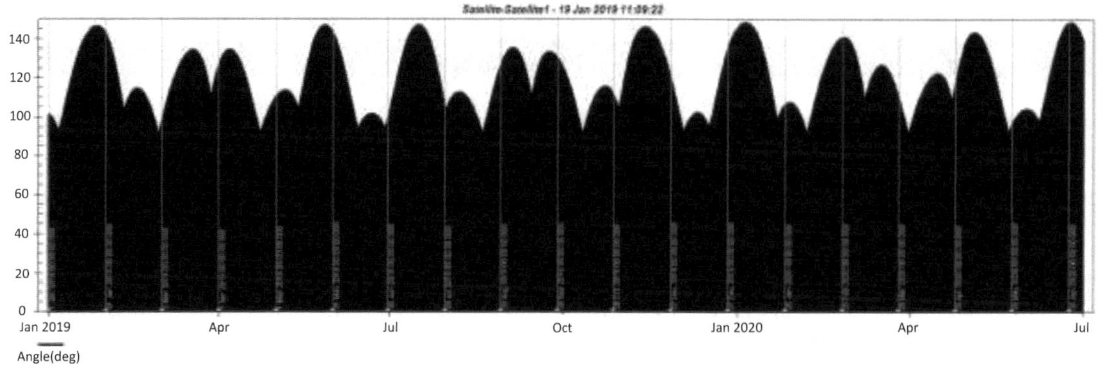

Figure 8. The deviation of the Sun-pointing under smooth method

Also, under the novel method, the angular velocity of the satellite relative to the inertial frame in all three axes are shown in Fig. 9 to Fig. 11.

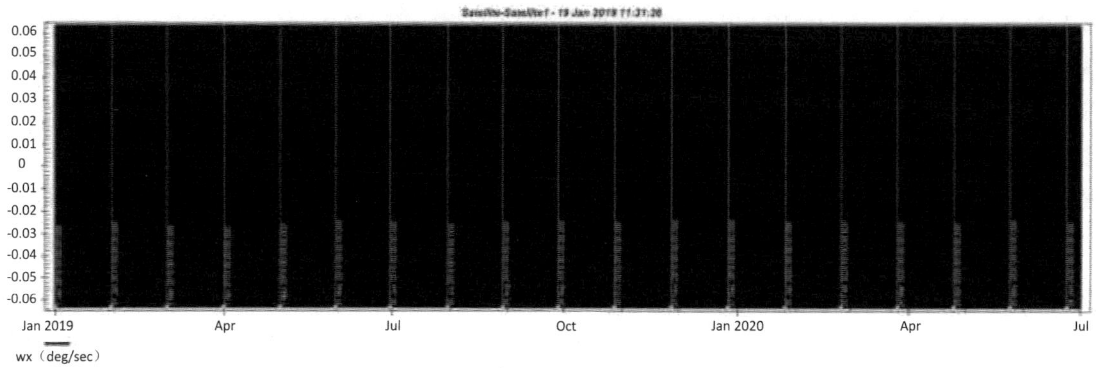

Figure 9. The angular velocity in the x direction under smooth method

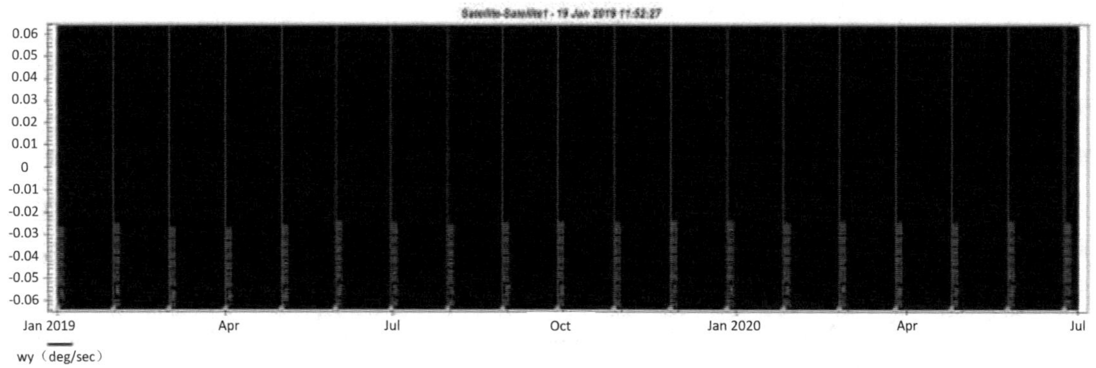

Figure 10. The angular velocity in the x direction under smooth method

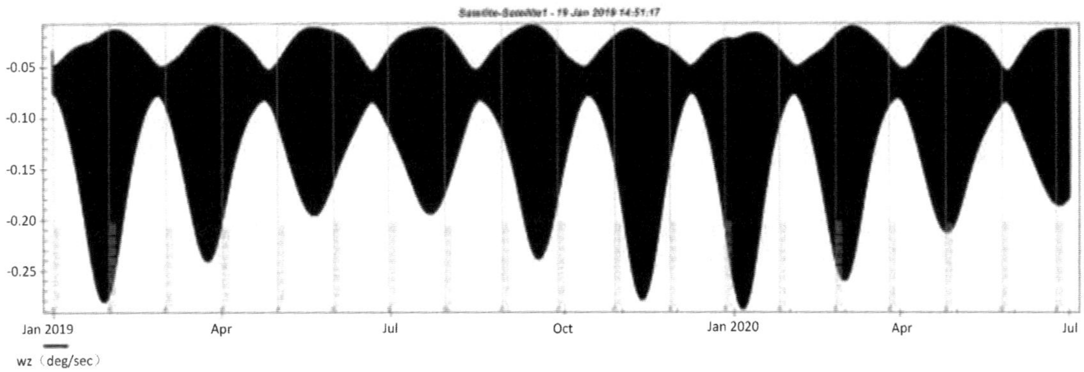

Figure 11. The angular velocity in the x direction under smooth method

As can be seen from Fig. 9 to Fig. 11, under the novel smooth guidance method, the deviation between $-y_b$ axis and $\boldsymbol{r}_s$ will be greater than the traditional way, with an average peak about 100°. During the 1.5 year, the maximum Sun-pointing deviation angle gets 150°. Also, the peak angular velocity in both the x direction and the y direction change periodically and be up to $0.06\ ^\circ/s$. The angular velocity in the z direction is greatly reduced compared with the traditional method, with a peak near $0.3\ ^\circ/s$.

4. Conclusion

The smooth nadir-pointing-aligned Sun-pointing-constrained guidance law base on the periodical rotation of the temporary attitude proposed in this paper is able to reduce the peak angular velocity of the satellite, and overcome the singular phenomenon of flipping. It is efficient for satellites to operating and charging at the same time.

5. Reference

[1] Teofilatto P, Testani P, Celani F, et al. A nadir-pointing magnetic attitude control system for TigriSat nanosatellite[C]. Proceedings of the International Astronautical Congress, IAC. 2013, 6: 4818-4829.
[2] JI H R, ZENG G Q. A method for spatial effective coverage analysis in space-based optical observation[J]. Optik, 2018, 166:116-126.
[3] Kalweit C C. Optimum yaw motion for satellites with a nadir-pointing payload[J]. Journal of Guidance, Control, and Dynamics, 1983, 6(1): 47-52.
[4] STARIN S, BOURKLAND K. Persistent attitude error in a sun-pointing controller due to nonlinear dynamics[C]. AIAA Guidance, Navigation & Control Conference & Exhibit. 2013.
[5] Holben B N, Justice C O. The topographic effect on spectral response from nadir-pointing sensors[J]. Photogrammetric Engineering and Remote Sensing, 1980, 46(9): 1191-1200.
[6] Yang Y. Quaternion based model for momentum biased nadir pointing spacecraft[J]. Aerospace Science and Technology, 2010, 14(3): 199-202.
[7] Reyhanoglu M, Ton C, Drakunov S. Attitude stabilization of a nadir-pointing small satellite using only magnetic actuators[J]. IFAC Proceedings Volumes, 2009, 42(19): 292-297.
[8] Ribah A Z, Ramayanti S. Power produced analysis of solar arrays in nadir pointing mode for low-earth equatorial micro-satellite conceptual design[C]. IOP Conference Series: Earth and Environmental Science. IOP Publishing, 2019, 284(1): 012048.
[9] Meneghini R, Atlas D. Simultaneous ocean cross-section and rainfall measurements from space with a nadir-pointing radar[M]. National Aeronautics and Space Administration, Goddard Space Flight Center, 1984.

Control Method for the Deorbit of Geostationary Satellite in the State of Gas-Liquid Mixing Variable Thrust at the End of its Service Life

ShuHua Zhang[1]*, JianPing Li[1], Heng Yao[1], ZhongMou Lei[1]

[1] Beijing Institute of Control Engineering, Beijing,100094, P.R. China

Email address: shzbice99@126.com

Abstract. By analyzing the gas-liquid mixing phenomenon in the thruster pipeline at the end of the service life of geostationary satellite, it is found that the gas in the pipeline leads to insufficient propellant combustion and unstable thrust output, which brings great difficulties to the orbit control of satellite. Aiming at the problem of thrust instability caused by gas-liquid mixing in the process of geostationary satellite orbit control, a joint orbit control method of normal mode + station keeping mode is proposed, which effectively solves the problem of deorbit at the end of the life of geostationary satellite in actual on orbit operation.

1. Introduction

Geostationary orbit resources are very valuable. When geostationary orbit satellites are unable to continue their business at the end of their service life or in case of serious failure, they need to carry out deorbit operation to raise the satellite orbit by hundreds of kilometers. The satellite deorbit operation can leave the geostationary orbit position for the use of subsequent satellites, and can also avoid the threat to other satellites in geostationary orbit after satellite orbit drift [1-5]. Usually, the satellite's orbit departure will be carried out in station keeping mode, using the same direction double pulse orbit maneuver with an interval of half an orbit period [6]. However, when the propellant is exhausted, the instability of thrust makes it very difficult to control the attitude of the satellite in the process of orbit departure, so this method is no longer applicable.

At the end of the service life of geostationary satellite, when the propellant (including oxidant and incendiary agent) is nearly exhausted, the working condition of the thruster is very complex, and will experience a variety of states such as pure liquid, gas-liquid mixing and pure gas. When liquid propellant is ejected from the channel under the condition of pure liquid, the two propellants can have chemical reaction and large thrust. In the case of gas-liquid mixing, when the gas working medium is ejected from the channel, only one liquid propellant can not cause chemical reaction, and can only work in the mode of single component cold air, the thrust will drop sharply and fluctuate violently, and the thruster temperature will drop due to the evaporation and heat absorption of liquid propellant. When all propellant channels are discharged by gas, the thruster will work in single component cold air mode, and the output thrust will be greatly reduced to less than 1N. The above three states will occur repeatedly during the deorbit until the pure gas state is maintained.

* Corresponding author: ShuHua Zhang

In this paper, a new method of combined normal mode and station keeping mode (need to modify the controller parameters) is proposed for the mission of deorbit operation when the propellant is exhausted. Aiming at the situation that the propellant of an on orbit satellite is exhausted, this method is used to successfully complete the deorbit mission of the satellite.

Closely combined with the working characteristics of the thruster when the propellant is exhausted, the satellite's deorbit control is carried out in stages: (1) in the normal mode, give full play to the control ability of the momentum wheel to carry out eastward orbit control and consume the propellant in the propulsion pipeline. (2) In the station keeping mode, modify the controller parameters, widen the attitude control threshold, and carry out the eastward station keeping.

2. Attitude control analysis with propellant exhausted

2.1. Satellite profile

Taking a satellite as an example, the basic situation of the control object is introduced. The control system adopts the full orbit three-axis stability mode, the two-component unified propulsion system and the V + L momentum wheel as the actuator. Use the control computer to control the satellite attitude and orbit [7-10].

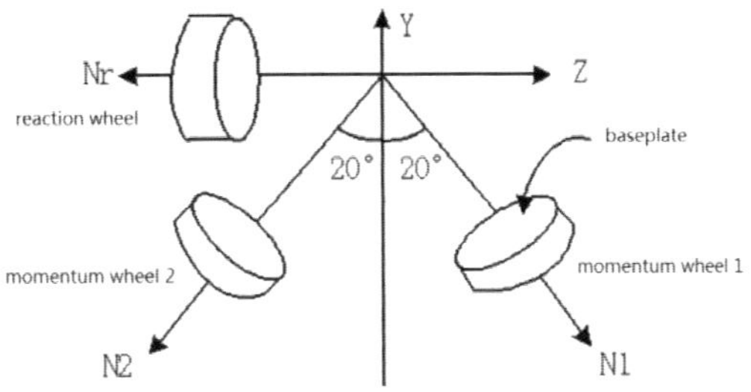

Figure 1. Installation diagram of momentum wheel

The normal mode is in the synchronous orbit as the daily working mode to ensure the satellite attitude required by the payload. The pitch and roll axes in the normal mode are controlled by the V-mounted momentum wheel or the offset momentum wheel and reaction wheel with L-shaped structure, in which the reaction wheel is the backup wheel. At the same time, accelerate and decelerate the momentum wheel to realize the attitude control of y-axis. The roll error is continuously controlled by differentially changing the angular momentum of the momentum wheel on the z-axis to meet the accuracy requirements. The yaw observer is used to estimate the angular momentum of the satellite, and the 10N thruster completes the management of the angular momentum. The installation diagram of v-wheel is shown in Figure 1.

The station keeping mode is used for the fixed-point acquisition and inclination correction of the quasi synchronous orbit, as well as the east-west and north-south position maintenance orbit control after reaching the fixed-point, and makes the satellite maintain the three-axis attitude of earth orientation.

The propulsion system consists of 14 10N thrusters, which are divided into two branches A and B, 7 each, to provide control torque and thrust for attitude and orbit control. Among them, 2A, 3A, 2B and 3B are combined with 4A, 5A, 4B and 5B respectively to control pitch or yaw, and 6A, 7A, 6B and 7B work separately to control roll. 2B + 3B or 4A + 5A can complete the westward station keeping, 2A + 3A or 4B + 5B can complete the eastward station keeping, and 6A + 7A and 6B + 7B can complete the northward station keeping and southward station keeping respectively. In the process of station keeping, the orbit control thruster works in the off modulation mode [11].

2.2. Analysis of propellant depletion in station keeping mode

Under normal conditions, the attitude of the satellite remains stable and the rolling attitude is near 0 ° during the position maintenance in the station keeping mode. As shown in Figure 2, the ignition flag in the figure is high, indicating that the station keeping ignition is in progress.

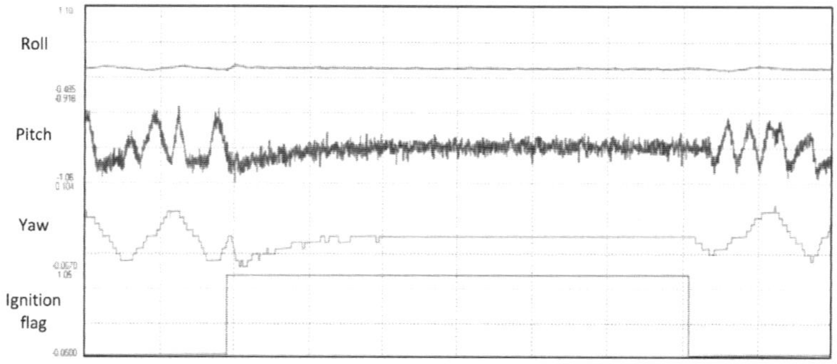

Figure 2. Station keeping attitude curve under normal conditions

Figure 3 shows the attitude curve of 7A thruster in the paired two thrusters (6a + 7a) when the propellant is suddenly exhausted in the process of station keeping ignition in the North position. The rolling attitude has a maximum attitude overshoot of about 1.025°, and then gradually returns to near 0°. In this process, the pitch attitude and yaw attitude remain stable. According to the data analysis, a few seconds before the attitude overshoot process, the pressure of propellant tank suddenly decreases, and the temperature of 7a thruster decreases at the same time of attitude overshoot. Therefore, it can be inferred that the propellant is exhausted, resulting in air mixing in the propulsion pipeline, resulting in unstable thrust during the operation of the two thrusters in the later stage of station keeping and attitude angle overshoot.

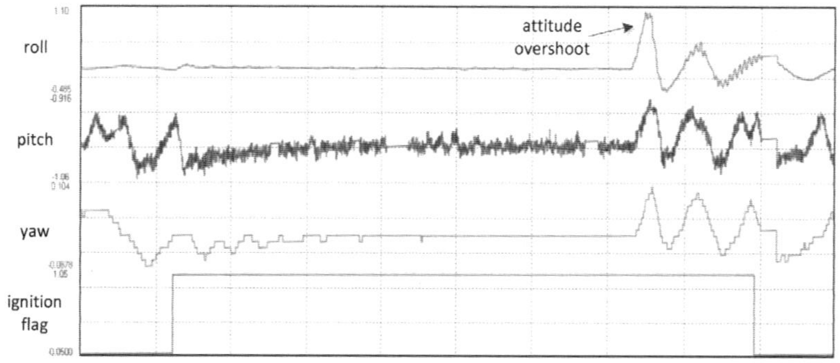

Figure 3. Station keeping attitude curve under propellant depletion

3. Joint orbit control method of normal mode + station keeping mode

This method is divided into two stages. In the first stage, the eastward station keeping is carried out in the normal mode, and the momentum wheel is used to overcome the interference torque generated by the mismatched orbit control thruster of the satellite. At the same time, the asymmetry of the orbit control thruster is evaluated. With the consumption of the remaining propellant in the pipeline, the second stage emergency orbit control is carried out when the symmetry of the orbit control thruster is significantly improved. That is, in the station keeping mode, by gradually increasing the jet pulse width and shortening the jet interval (the jet interval can be appropriately shortened according to the attitude control ability), the emergency orbit control task can be completed when the satellite attitude is basically stable.

The specific implementation steps are as follows (described in Figure 4):

(1) Inject the ignition pulse sequence in the normal mode, and select the pulse sequence with narrow pulse width and long period at the beginning.

(2) Start the station keeping ignition and monitor the range of relevant telemetry parameters (Including the attitude, the thruster temperature, the pressure of storage tank, and so on). If the telemetry parameters exceed the range, send a command to stop the ignition, wait for the satellite attitude to

stabilize and the relevant telemetry parameters to return to normal. After all telemetry parameters are normal, continue to inject the ignition pulse sequence and restart the ignition.
(3) After the station keeping ignition, adjust the pulse width and cycle of the ignition pulse sequence according to the ignition situation. If the relevant telemetry parameters are stable, gradually increase the pulse width or reduce the cycle. Otherwise, gradually reduce the pulse width or increase the cycle. Then continue the station keeping ignition according to step (2).
(4) Judge whether the fuel in the propellant pipeline is exhausted according to the ignition condition of the station keeping. The judgment condition of depletion is: ignition for 100 pulses continuously, and the telemetry parameters in step (2) can be maintained within the normal range. If not exhausted, continue with steps (2) and (3). If it has been exhausted, switch to the station keeping mode for the second stage of work.
(5) In the station keeping mode, modify the PID parameters of the controller and the parameters of the pseudorate modulator, widen the attitude control threshold to $\pm 0.5^{\circ}$, turn off the momentum wheel and reduce its speed to 0.
(6) Inject ignition pulse sequence in station keeping mode, start ignition and monitor relevant telemetry parameters. If telemetry parameters are abnormal, stop ignition, wait for attitude and relevant parameters to be normal, and then inject ignition parameters again to continue station keeping ignition.
(7) After the ignition, adjust the pulse width or cycle according to the telemetry parameters in the station keeping process and continue the station keeping ignition according to step (6).
(8) Judge whether the orbit control effect meets the expectation according to the orbit measurement results. If it does not meet the expectation, continue the station keeping ignition according to steps (6) and (7). If the expectation has been reached, the orbit control ends.

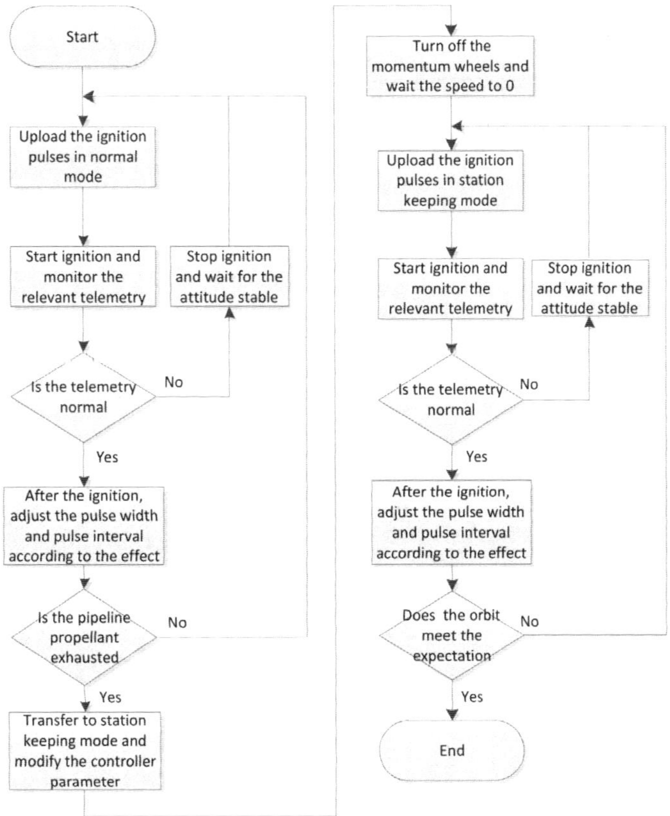

Figure 4. Implementation flow chart

4. Orbit control effect

Using the joint orbit control method of normal mode and station keeping mode, the orbit control of a geostationary satellite at the end of its service life is carried out.

In normal mode, the pulse sequence with pulse width of 64ms and interval of 100s is used for orbit control. The orbit control thrusters are 4B and 5B to generate eastward thrust. Normally, when 4B and 5B thrusters work at the same time, interference torque will be generated in the positive direction of the pitch axis, resulting in positive error in the attitude of the pitch axis. It can be seen from the pitch angle in Figure 5 during the orbit control process, the pitch angle is sometimes large and sometimes small, and the initial positive deviation becomes negative deviation, which reflects that the working condition of the thruster is very unstable under the condition of gas-liquid mixing.

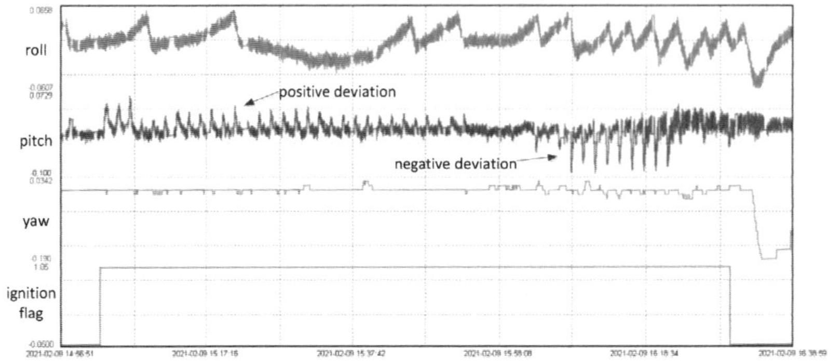

Figure 5. Attitude curve during normal mode ignition

The initial ignition pulse width is 64ms with an interval of 10s under the station keeping mode. With the gradual reduction of interference, continuously adjust the pulse width and interval. As shown in Figure 6, the curve is a three-axis attitude curve with a pulse width of 1.28s and an interval of 10s for orbit control. It can be seen from the curve that after early consumption, the satellite basically works in the pure gas state, the interference torque gradually decreases, and the control system can work under the condition of long pulse width.

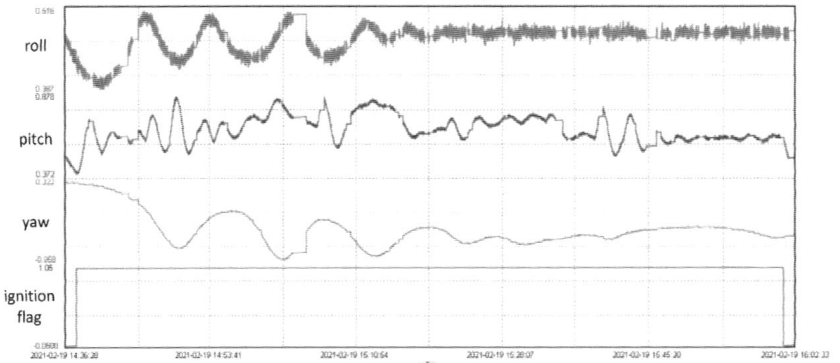

Figure 6. Attitude curve during ignition in station keeping mode

The working condition statistics in the whole deorbit process are shown in Table 1. It can be seen from the table that by making full use of the remaining propellant in the pipeline and the high-pressure gas in the tank at the end of the service life of the geostationary satellite to implement the orbit lifting control, the semi major axis of the orbit can be lifted by about 130km（the sum of the last column in table 1）, which basically meets the requirements of the orbit control at the end of the service life of the satellite.

Table 1. Statistics of a satellite deorbit process

time	Number of pulses	control model	Ignition duration (s)	Semi major axis lifting (km)
Day 1	15	normal mode	0.96	0.24
Day 2	101	normal mode	6.46	1.56
Day 3	151	normal mode	9.66	2.61
Day 4	164	normal mode	10.50	2.33
Day 5	410	normal mode station keeping mode	94.14	6.256
Day 6	1605	station keeping mode	514.56	19.273
Day 7	1618	station keeping mode	414.208	15.771
Day 8	1424	station keeping mode	473.088	40.544
Day 9	2267	station keeping mode	5038.4	39.377

5. Conclusion

Aiming at the problem of large thrust change caused by gas-liquid mixing in the propulsion pipeline at the end of the service life of geostationary satellite, a joint orbit control method of normal mode + station keeping mode is proposed to carry out orbit lifting. Through the practical verification and application of on orbit satellites, this method can effectively improve the satellite orbit height and meet the orbit control requirements at the end of satellite life.

6. References

[1] Geng Yuling, Zhao Jinbo, Li Dong, Liu Jingyong. Research on the control flow of geostationary satellite deorbit. Proceedings of the 13th Annual Conference on satellite communications, 2017, v47 v55:130-133(Chinese).
[2] Li Xin, Zhao Yue, Li Juan. Research on the method of geostationary communication satellite deorbit. Proceedings of the 12th Annual Conference on satellite communications, 2016, v47, v44:427-436(Chinese).
[3] Luo Gangqiao. Study on the end of life deorbiting scheme of geostationary satellites. Spacecraft environmental engineering, 2005, 22 (2): 73-76 (Chinese).
[4] Jing Yaoxiang, Yang Hui. The stability analysis of discarding orbit and deorbit strategy optimization for GEO satellite. Spacecraft engineering, 2013, 01:21-25 (Chinese).
[5] Zheng Jun, Chen Hong. Deorbit control strategy of GEO satellite in the end of life. Shanghai aerospace, 2007, 03:38-41(Chinese).
[6] Eric Mattias Soop. Handbook of geostationary orbits. Wang Zhengcai, Translated. Beijing: National Defense Industry Press,1999 (Chinese).
[7] Zhang Renwei. Satellite orbit and attitude dynamics and control. Beijing: Beihang Universiity Press,2006 (Chinese).
[8] Tu Shancheng. Satellite attitude dynamics and control. Beijing: China Astronautics Press,2001(Chinese).
[9] Liu Yuxin, Shang Haibin, Wang Shuai. Research on GEO satellite station keeping method using electric propulsion. Journal of Deep Space Exploration, 2015, 1(4):35-42 (Chinese).
[10] Yin Quan, Gao Yijun. Optimization method of east-west station keeping control parameters for geostationary satellites. Aerospace control and application, 2014, 40(5):48-51 (Chinese).
[11] Qiu Rui, Zuo Zijin, Dai xueyang. Efficiency compensation method and application of geostationary satellite north south position keeping thruster. Spacecraft engineering, 2018, 27 (3): 17-21 (Chinese).

Numerical Analysis of the Effect of Ducted Propeller Configuration Parameters on Aerodynamic Performance

Zeyu Li, Jianghao Wu, Haiying Lin

Beihang University, Beijing 100191, P.R. China

haiying@buaa.edu.cn

Abstract. In this paper, a numerical calculation method is applied to investigate the influences of ducted propeller aerodynamic configuration parameters. The thrust, torque and power of the propeller in the duct and the decoupled propeller are compared under identical conditions. The performance with equal input power is analysed considering the additional thrust of the duct. The relationships of system thrust, power and efficiency with the axial position of the propeller disk in the duct, the propeller disk diameter and the propeller-duct clearance are investigated.

1. Introduction

A ducted propeller consists of an annular duct surrounding one or multiple propellers, which provides vertical take-off and landing capability and excellent hovering performance for the aircraft [1]. The airflow generates a low-pressure area at the duct lip, providing additional thrust to the system. The enclosure of the duct effectively suppresses the propeller tip vortex [2], reduces losses [3], and provides higher aerodynamic efficiency as a power system [4].

The principle of the additional thrust of the duct was studied [5]. With experimental methods, the variations of the thrust characteristics were summarized with the radius of the duct lip, the duct height, and the duct expansion angle [6]. The Blade-element theory [7] and the lamellar strip theory [8] have been applied to numerical calculation methods [9]. Nested mesh techniques [10] and slip mesh models [11] have been validated for ducted propeller calculations.

The existing research on the ducted propeller system [12] primarily concentrates on the study of the flow field characteristics of the duct in the axial flow and forward flight states and the study of the influence of the duct geometric shape parameters on the overall performance of the ducted propeller [13].

A numerical solution method for the ducted propeller system is developed in this paper to solve the flow field for this configuration and analyse the effect of aerodynamic configuration parameters on the aerodynamic performance of the ducted propeller.

2. Numerical Calculation and Validation Methods

The aerodynamic modelling dealing with the relative motion between the duct and the propeller is performed using a dynamic region computational model. The multiple reference frame (MRF) model uses a steady-state solution method with an interface at the regional intersection for data transfer and has a computational accuracy sufficient to meet the demand for calculating the maximum load capacity and a computational volume smaller than that of transient calculations. The numerical calculation parameters are demonstrated in Table 1.

Table 1. Numerical calculation parameters.

Atmospheric temperature	Atmospheric pressure	Atmospheric density	Viscosity coefficient
288.15 K	1.01325×10^5 Pa	1.225 kg/m^3	1.7894×10^{-5}Ns/m^2

2.1. Numerical validation

This paper implements computational fluid dynamics (CFD) using the commercial solver ANSYS Fluent 19R1. For the aerodynamic calculation of the relative motion between the duct and the propeller, an MRF model is applied for the steady-state solution, and the k-ω SST turbulence model is selected to solve the turbulence calculation of the rotating flow field. Test experiments are carried out based on a standalone propeller to verify the computational model. The aerodynamic characteristics tests of this system were conducted in the Department of Airworthiness Engineering laboratory, Beihang University, with an open test area. Figure 1 shows the installation of the test model. A six-component box-type strain gauge balance was used to measure the thrust and torque, and the motor controller captured the Hall coil signal built into the motor to obtain the rotational speed. A PID speed control algorithm running in the Arduino development board was utilized to maintain the test rotation speed stability with a 1% speed error. The Amass DC power supplier was used to control and measure the input voltage and current to calculate the power. The motor system efficiency is 95%. Numerical results of power and thrust are displayed in Figure 1, compared with the obtained experimental data. The calculated thrust is underestimated by about 5% and the power by 10%, as shown in Figure 2, but within the measurement uncertainty for all operating conditions.

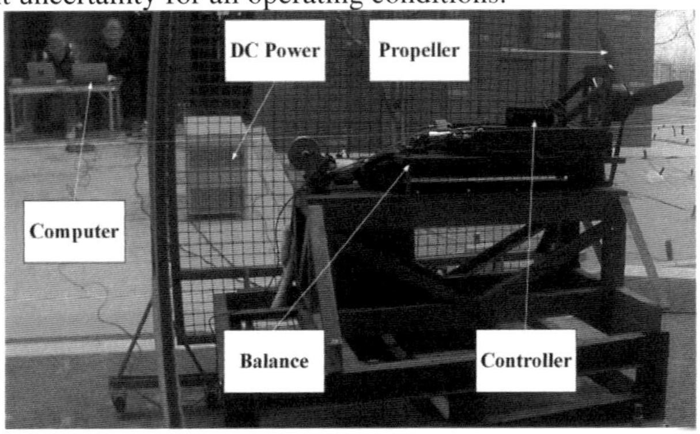

Figure 1. Aerodynamic test experiment of a propeller.

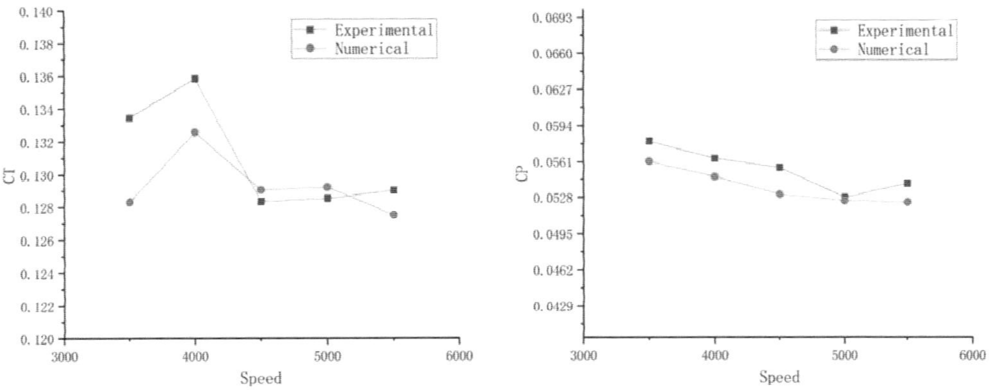

Figure 2. Comparisons of aerodynamic performances between numerical and experimental results.

2.2. Computational Models

A ducted propeller consisting of a duct and a propeller is utilized as the power system of the UAV, as shown in Figure 3. The propeller diameter is 546mm, the inner-diameter of the duct is 550mm, the outer-diameter is 671mm, the height is 334mm, the propeller blade is three blades, and the clearance between the blade tip and the inner duct wall is about 2mm. The propeller centre is 98mm from the entrance plane and 200mm from the exit plane of the duct. The propeller adopts a radially occupied variable airfoil design with ARA-D6%, ARA-D10% and ARA-D13% airfoils, with the pitch angle rangeing from 52° to 25° from 25% radial position to the tip.

The duct is a geometry formed by an airfoil rotated around an axis. A non-uniform rational B-sample curve combined with the characteristic parameter description method is considered to realize the geometric parameterization of the duct section airfoil.

Figure 4 shows that the duct section airfoil is characterized by straight-line segments and non-uniform rational B spline curves. C, G, and K are fixed lip boundary points. The horizontal coordinates of the beginning b point of the expansion section, the vertical coordinates of the trailing edge boundary A and the lip spline control points D, E, F, H, I, and J are used as design variables to control the duct exit radius, the length of the exit expansion section, and the shape and radius of the leading edge.

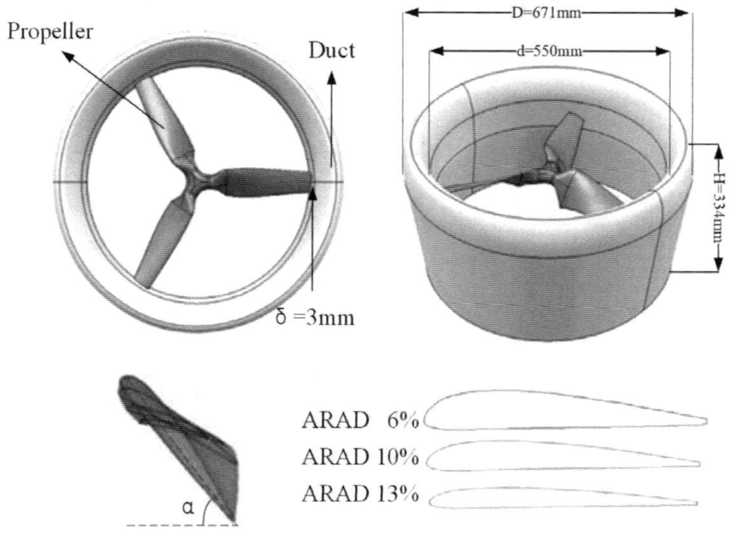

Figure 3. Ducted propeller model.

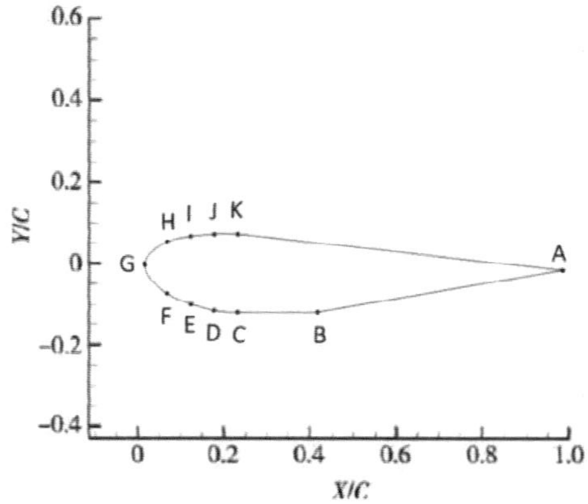

Figure 4. Duct section airfoil.

2.3. Boundary condition and mesh

The entire calculation basin is shown in Figure 5. The propeller diameter is denoted as D. The basin consists of two cylinders. The outer flow field extends radially to 4D. The rotating region is axially 6D away from the inlet and 10D away from the outlet.

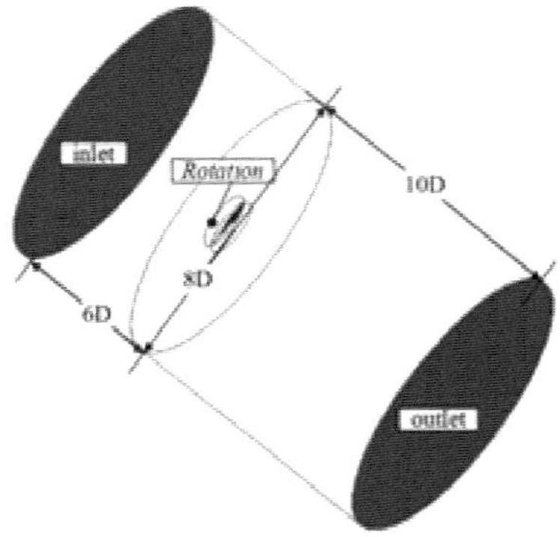

Figure 5. Calculation basin of ducted propeller.

The meshes of stationary and rotating regions are generated separately by ANSYS ICEM CFD 19R1. They are densified at the leading edge and trailing edge of the duct, propeller blade tip and propeller blade, and inner duct wall to improve the calculation accuracy. The meshing of the propeller and the outflow field is shown in Figure 6.

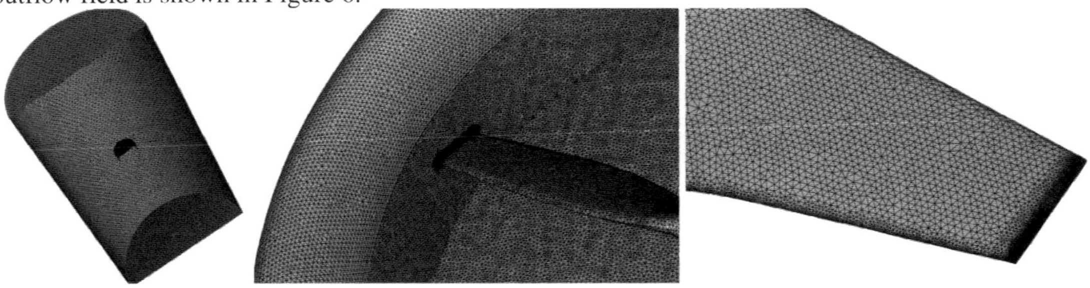

Figure 6. Unstructured mesh for stationery and rotation regions of the ducted propeller.

2.4. Aerodynamic performance parameters

The ducted propeller thrust direction is aligned with the incoming flow direction. Concerning the propeller aerodynamic evaluation, its aerodynamic performance is judged by considering its dimensionless thrust, torque, power, and efficiency parameters.

Thrust coefficient:

$$C_T = \frac{T}{\rho n_s^2 D^4} \tag{1}$$

Torque coefficient:

$$C_M = \frac{M}{\rho n_s^2 D^5} \tag{2}$$

Propeller power:

$$P = 2\pi M \times n_s \tag{3}$$

Power coefficient:

$$C_P = \frac{P}{\rho n_s^3 D^5} \tag{4}$$

The figure of merit (FM):

$$FM = \frac{P_i}{P} = \frac{P_i}{\kappa P_i + P_0} = \frac{C_T^{\frac{3}{2}}}{\sqrt{2}C_P} \tag{5}$$

Disk load (DL):

$$DL = \frac{T}{A} \tag{6}$$

Power load:

$$PL = \frac{T}{P} = \sqrt{\frac{2\rho}{DL}} \times FM \tag{7}$$

FM was generally considered to reflect the propeller efficiency but only applicable to different propellers with the duplicate DL. In the case of κ and P0 being the same, the greater the propeller with the more significant disc load, the greater the *FM*, at which point the two propellers produce the same pull, while the more efficient propeller consumes more of the total power. For a propeller with the same thrust coefficient and power coefficient, the greater the speed, the higher the thrust (proportional to the speed squared), and the higher the propeller disc load, but its power load is lower because the power is proportional to the speed cube.

Therefore, the efficiency can be expressed by the thrust ratio, i.e., the thrust, to the power that a ducted propeller can generate at a given power. The balance of the thrust coefficient to the power coefficient, *CT*/*CP*, a dimensionless quantity, is considered for judging the propeller efficiency.

2.5. Mesh independence study

The mesh volumes of 4.7 million, 6.2 million, 9.5 million and 11.7 million meshes were tested for the calculation, and the results of the aerodynamic characteristics of the four scale meshes were calculated as shown in Table 2. As the mesh density increases, the thrust coefficient of the ducted propeller increases, the torque coefficient decreases, and the efficiency increases. When the number of mesh nodes is more significant than 9.5 million, the ducted propeller system's aerodynamic coefficient and efficiency tend to be stable. Therefore, the mesh density chosen in this paper is 9.5 million.

Table 2. Comparison of aerodynamic results with different mesh sizes.

Mesh number	CT	CM	CP	Efficiency
4.7 million	0.45352	0.03876	0.24298	1.86648
6.2 million	0.45684	0.03832	0.24079	1.89726
9.5 million	0.46218	0.03785	0.23784	1.94326
11.7 million	0.46271	0.03773	0.23708	1.95165

3. Numerical analysis of the ducted propeller shape parameters

3.1. Influence of the axial position of the propeller

When considering the effect of the propeller disc axial position on the aerodynamic characteristics, the impact of the propeller-duct clearance is removed using the flat section of the inner duct wall, and the diameter of the propeller disc is used to measure the length of the propeller disc from the lip. H/d=0.232, 0.277, 0.322, 0.367 and 0.412. The corresponding propeller disc-lip distances are 77.5 mm, 92.5 mm, 107.5 mm, 122.5 mm and 137.5 mm, respectively. Propeller speed is set to 6000 rpm.

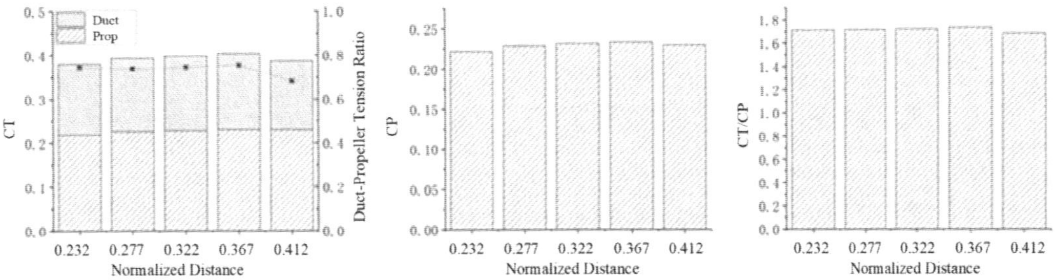

Figure 7. Aerodynamic characteristics of ducted propellers with different axial positions.

Figure 7 demonstrates the changes in the ducted propeller system and each component's thrust coefficient, power coefficient, and efficiency at 6000 rpm for five different axial positions of the propeller disc. As the propeller disc position moves from the duct inlet to the duct outlet, the propeller thrust in the duct first increases and then stays unchanged. The trend of changes in the additional duct thrust, total duct propeller system thrust, power coefficient, and efficiency are similar. The duct thrust coefficient, power coefficient, and efficiency decrease due to the propeller disc surface's overly backward position making the duct's inner wall as the rectification section excessively long. The flow separation occurs before the airflow enters the propeller, thus causing a significant decrease in the additional thrust and propeller power of the duct, and the reduction of the thrust coefficient is higher than that of the power coefficient.

3.2. Effect of propeller diameter

When considering the effect of different propeller diameters on the aerodynamic characteristics of the ducted propeller, the duct height is taken to characterize the different propeller diameters, d/H=1.635, 1.798, 1.962, and 2.452, corresponding to 546mm, 600.6mm, 655.2mm and 819mm. Numerical simulations are carried out for different propeller diameters of the ducted propeller at 6000rpm with the propeller-duct clearance as 2mm.

We compared the thrust coefficient, power coefficient and efficiency distribution, as shown in Figure 8. The total thrust coefficient of the ducted propeller improves with the increase of the propeller disc diameter, and the propeller thrust coefficient and the power coefficient of the ducted propeller are almost unchanged. The expansion of the total thrust coefficient of the ducted propeller mainly comes from the increase of the additional thrust of the duct, and the diameter of the propeller disc increases from the initial value to 1.5 times, and the duct-propeller thrust ratio also increases from 0.941 to 1.010. The variation trend of efficiency is similar to the variation trend of duct propeller thrust ratio, so we can assume that the change in total thrust due to the shift in propeller thrust leads to the exact trend change of power. In contrast, the evolution of total thrust due to the change of additional duct thrust does not result in the shift of power. Therefore, the key to the efficiency of a high ducted propeller system is to improve the extra duct thrust.

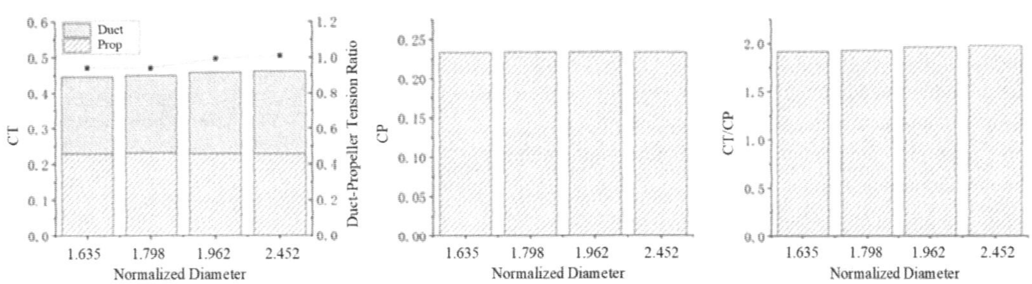

Figure 8. Aerodynamic characteristics of ducted propellers with different propeller diameters.

3.3. Influence of propeller-duct clearance

When considering the influences of different propeller clearances, the propeller diameter is maintained at 546, and numerical simulations are carried out for four extra ducted propellers with 2mm, 3mm, 4mm and 5mm propeller-duct clearances at 6000rpm.

The ducted propeller system's total thrust coefficient, power coefficient, and efficiency decreased significantly as the propeller-duct clearance increased. The maximum efficiency of 1.943 was achieved at 2 mm for the propeller duct clearance. At the same time, the total thrust coefficient decreased by about 10% for a propeller-duct clearance of 5 mm, and the efficiency declined by 4%. The trend of the duct-propeller thrust ratio indicates that the decrease in the total thrust coefficient is primarily caused by the reduction of the additional thrust in the duct. Figure 9 reflects the various values of each coefficient.

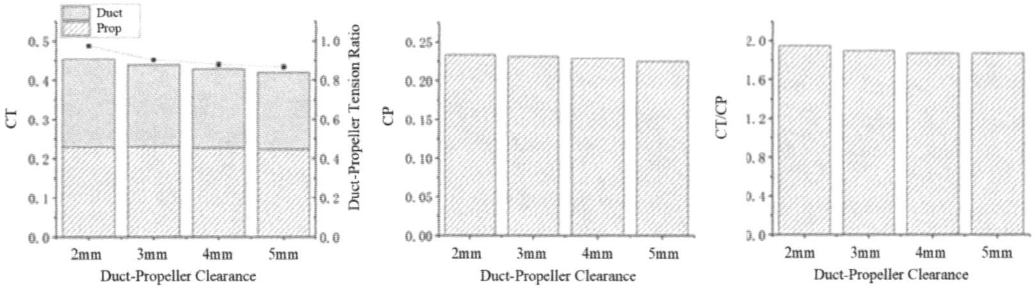

Figure 9. Aerodynamic characteristics of ducted propellers with different propeller-duct clearances.

Figure 10 (a)-(d) presents the local diagrams of the flow field at the propeller tip position with the clearance of 2 mm, 3 mm, 4 mm and 5 mm, respectively. As the clearance increases, the vortex at the propeller tip increases, the airflow separation along the inner wall of the duct intensifies, and the negative pressure in the low-pressure area at the lip of the duct is weakened, a decrease in the additional thrust generated by the duct. At the same time, the low-pressure area above the propeller disc and the high-pressure area below the propeller disc are both reduced, the pressure difference is decreased, and the thrust generated by the propeller is reduced, which leads to a decrease in the power coefficient. Since the propeller tip vortex dissipates the system energy, the larger the clearance leads to increased propeller tip vortex, which increases the energy loss, and the ducted propeller's efficiency decreases.

4. Conclusions

In this paper, the numerical simulation method is applied to investigate the principle of thrust increase of the ducted propeller by evaluating the flow field distribution of the decoupled propeller and the ducted propeller. The effects of the aerodynamic position parameters such as the axial position of the propeller disk, the diameter of the propeller disk and the propeller-duct clearance on the aerodynamic performance of the ducted propeller are investigated. The conclusions of the study are as follows.

The system total thrust coefficient, power coefficient and efficiency of the duct propeller increase and decrease as the propeller disc moves toward the duct outlet in the axial interval with constant propeller-duct clearance.

The total system thrust coefficient of the ducted propeller increases, the power coefficient remains constant, and the aerodynamic efficiency increases with a continual propeller-duct clearance and an increased disc diameter.

The total system thrust coefficient, power coefficient and efficiency of the ducted propeller decrease with the enhancement of the propeller-duct clearance.

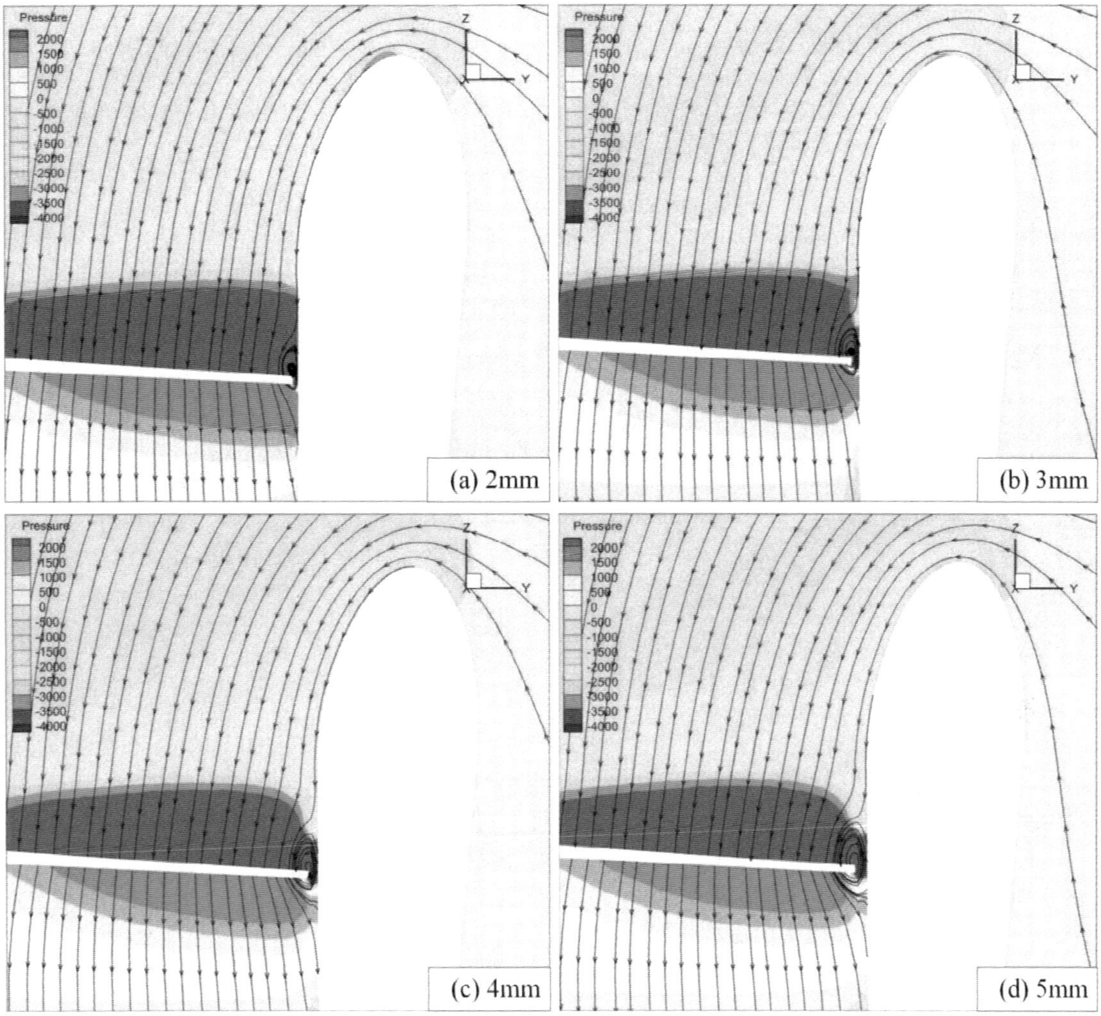

Figure 10. Pressure contours and streamlines of ducted propellers with different propeller-duct clearances.

5. References

[1] Fleming J, Langford M, Tweedie S, Ng W and Goossen E 2012 Development of an air data system for ducted fan unmanned aircraft. In: *50th AIAA Aerospace Sciences Meeting including the New Horizons Forum and Aerospace Exposition,* p 668

[2] GRAY R and Wright T 1969 Determination of the design parameters for optimum heavily loaded ducted fans. In: *VTOL Research, Design, and Operations Meeting,* p 222

[3] Deng S, Wang S and Zhang Z 2020 Aerodynamic performance assessment of a ducted fan UAV for VTOL applications *AEROSP SCI TECHNOL* **103** 105895

[4] Zhang T, Qiao G, Smith D A, Barakos G N and Kusyumov A 2021 Parametric study of aerodynamic performance of equivalent ducted/un-ducted rotors *AEROSP SCI TECHNOL* **117** 106984

[5] Yilmaz S, Erdem D and Kavsaoglu M 2013 Effects of duct shape on a ducted propeller performance. In: *51st AIAA aerospace sciences meeting including the new horizons forum and aerospace exposition,* p 803

[6] Li L, Huang G and Chen J 2019 Aerodynamic characteristics of a tip-jet fan with a large blade pitch angle *AEROSP SCI TECHNOL* **91** 49-58

[7] Misiorowski M P, Gandhi F S and Oberai A A 2019 Computational Study of Diffuser Length on Ducted Rotor Performance in Edgewise Flight *AIAA J* **57** 796-808
[8] Fu J and Zhou Z 2012 Research on the Characteristics of Computational Methods of Ducted Fan System of VTOL UAV *Science Technology and Engineering* **20** 1294-300
[9] Xu H and Ye Z 2011 Numerical simulation and comparison of aerodynamic characteristics between ducted and isolated propellers *Journal of Aerospace Power* **26** 2820-5
[10] Toleos Jr L R, Luna N, Manuel M C E, Chua J M R, Sangalang E M A and So P C 2020 Feasibility study for Fused Deposition Modeling (FDM) 3D-printed propellers for unmanned aerial vehicles *International Journal of Mechanical Engineering and Robotics Research* **9** 548-58
[11] Li X, Guo Z, Liu Z and Chen Q 2016 Influence of ducted fan profile parameters on aerodynamic characteristics *Journal of National University of Defense Technology* **38** 28-33
[12] Bich V N and Hoa N T N 2022 Numerical Evaluation of Roughness Influences on Open Water Propeller Characteristics Using RANSE Method *International Journal of Mechanical Engineering and Robotics Research* **11** 15-21
[13] Ottens B, Dimitrakakis C and Faltings B 2017 DUCT: An upper confidence bound approach to distributed constraint optimization problems *ACM Transactions on Intelligent Systems and Technology (TIST)* **8** 1-27

Optimization of Cooling Channel for Spark Plug Based on Conjugate Heat Transfer Simulation

Yongchao Liu[1], Yadong Li[1], Tingting Wang[2], Xiaoming Zhou[1,3]

[1]School of Mechanical and Electrical Engineering, Northwestern Polytechnical University, Chengdu 611731, China
[2]AVIC Sichuan Fanhua Aviation Instrument Electric Appliance Co., Ltd, Chengdu 610050, China

corresponding author's e-mail address :zhouxm@uestc.edu.cn

Abstract. Air cooling channel is very important to avoid the ablation of ignition spark plug by reducing the temperature on the end face, and thus increase the service life of the spark plug. For a typical aero-engine ignition spark plug, coupled simulation model describing the flow in the air cooling channel as well as the heat transfer overall the spark plug has been established with a CFD software. Orthogonal experiments have been performed to analyze the influence of the design parameters on the air flowrate and end surface temperature On the basis of orthogonal experiment, the cooling channel is optimized by multi-factor analysis of variance The results demonstrate that: (1) The most important issue that influence the overall flowrate in cooling channel and the maximum temperature on the ignition end face is the number of holes on end face, and second one is the outlet diameter, while the influence of inlet diameter is not significant. (2) There is a better structural design for the cooling channel of the spark plug. The optimized air cooling channel increases the volume flow rate by 52.21% compared to the original scheme, which helps maintain the maximum temperature of the ignition end face of the spark plug below 368.2°C.

1. Introduction

In the aero-engine ignition system, the spark plug is an accessory used to generate electric sparks, which is a special form of discharge device [1]. Its working principle is to conduct short-pulse high-voltage current to the side wall of the combustion chamber to ignite the mixed gas in the combustion chamber. The spark plugs used in aero-engines mainly include ordinary spark plugs, electro-erosion spark plugs and semiconductor spark plugs [2].

At present, the research on the engine spark plug mainly focuses on the following three aspects: (1) The research on the fault of the spark plug and ignition system. Yan Jun[3], Yang Yufeng[4], and Li Luoyang[5] summed up the failures of the spark plugs as excessive temperature of the spark plug, excessive clearance, damage to the insulating sleeve, lead accumulation, carbon accumulation, moisture and other aspects, and give corresponding maintenance and selection suggestions. Dong Ying [6] introduced the application of laser welding technology in the repair of spark plug wires. Ma Hongwei [7] developed an ignition fault detection system by analyzing the failure mechanism of the aero-engine ignition system. (2) Research on spark plug materials. Xiao Zunwen [8-12] did extensive research on semiconductor spark plug materials for high-energy ignition systems, and explored the possibility of using semiconductor ceramics such as TiO_2, SnO_2, Cu_2O, and SiC in spark plugs. Hironori Osamura et al [13] developed a new type of iridium alloy with both anti-oxidation and high melting point, and used it in the spark plug electrode. Zhao Jun [14] studied the formula and sintering process of semiconductor ceramics, and the experiment showed the influence of SiC and Al_2O_3 content on product performance. (3) Research on the structure of the spark plug. Over the past few decades, numerous studies have described the effect of spark plug structure on engine performance. Design features at the tip of the spark plug, such as gap bulge, gap size, electrode size, and tip configuration, all have an impact on engine performance. Based on actual test data, Shaoyue et al [15] analyzed the relationship between electrode gap, energy storage capacity and minimum firing voltage, which provided a basis for the rational selection of electrode gap for semiconductor spark plugs. Abdel-Rehim [16] selected a set of spark plugs with different numbers of ground electrodes and explored the effect of the number of ground electrodes on engine performance. Tawfik Badawy[17] conducted optical and thermal experiments on a constant volume combustion chamber and a single-cylinder gasoline direct injection engine to explore the effect of spark plug gap on flame nucleus propagation and engine performance.

In the development of aero-engines, in order to improve fuel efficiency, the inlet temperature of the compressor and the combustion temperature of the combustion chamber will be increased. Correspondingly, the working environment temperature of the ignition tip must also be improved, and the maximum working temperature of the ignition tip will even exceed 1300 °C in the future [18]. The spark plug works under high temperature conditions for a long time, and the corrosive gas formed by combustion and high-energy flowers will erode the electrode and cause the spark plug to ablate. When the spark plug is in the ablated state, faults such as increased breakdown voltage of the spark plug and advanced ignition will occur, and the spark plug with severe ablation must be replaced, thus reducing the service life of the spark plug. Designing the cooling channel at the firing end of the spark plug to reduce the temperature of the firing end is an important method to avoid ablation of the spark plug and improve the service life of the spark plug.

In this paper, for a specific semiconductor spark plug model, under the condition of high temperature and high pressure, the coupled simulation model of flow and heat transfer in the cooling channel of the ignition spark plug is established by Fluent, and orthogonal experiments and variance analysis are used to optimize the design, and the effects of side cooling aperture, end cooling aperture, and the number of end cooling holes on the gas flow rate and the temperature of the ignition end face are studied. The purpose is to reduce the temperature of the ignition end of the spark plug at high temperature, so as to improve the high temperature environmental performance of the spark plug, reduce the failure rate and prolong the life of the ignition spark plug.

2. Materials and Methods

2.1. Numerical modeling of spark plug

The complete geometric model of the spark plug is shown in figure 1. On the right side of plane2, the spark plug is located outside the engine, and the outer wall of the spark plug is exposed to the natural environment outside the engine room. Between plane1 and plane2, the spark plug is in a low-temperature and high-pressure gas environment (temperature about 300°C, pressure about 0.5 MPa), on the left side of plane 1, the outer wall of the spark plug is exposed to the combustion chamber of the engine, which is a high temperature and low pressure gas environment (temperature about 1100℃, pressure about 0.48Mpa). Therefore, the basic principle of the spark plug cooling channel design is that driven by the differential pressure, the low-temperature gas flows through the spark plug and flows out from the firing end face, in order to reduce the temperature of the ignition end of the spark plug exposed to the high temperature environment.

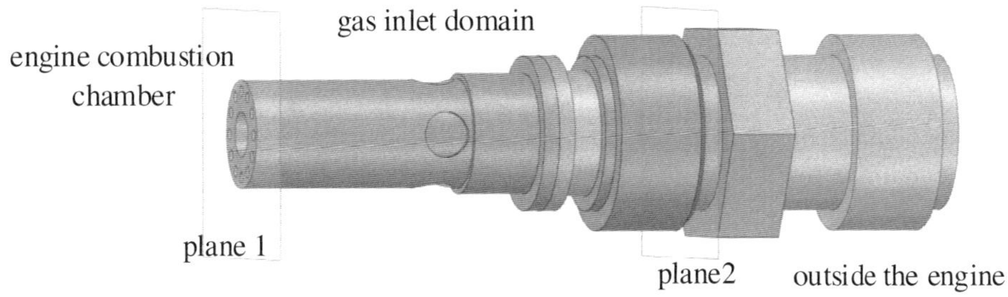

Figure 1. Complete geometric model of the spark plug

Since the axial heat conduction from the outside of the engine to the firing end face of the spark plug has little effect on the temperature of the firing end, this paper only intercepts the spark plug head containing the complete cooling flow channel for calculation. Its geometric model is shown in figure.2. The truncated end surface formed by simplification is set as the adiabatic surface, and the outer surface of the spark plug shell is the low temperature wall surface. Driven by differential pressure, the gas flows in from the four small holes opened in the spark plug shell, flows through the spark plug shell, the insulating tube, and the cooling channel formed by the electrode, The gas flows in from the four small holes opened in the spark plug shell under the pressure difference, flows through the spark plug shell, the insulating tube, and the cooling channel formed by the electrode, and then flows out from the small holes around the ignition surface, thereby cooling the ignition end surface. The green part in the figure is the fluid domain.

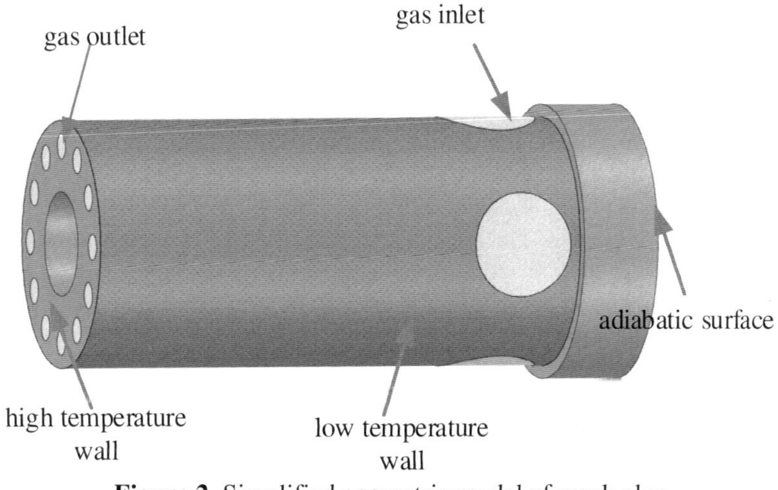

Figure 2. Simplified geometric model of spark plug

2.2. Governing equations and boundary conditions

This paper uses the commercial CFD software FLUENT 2020 for simulation. In order to establish the mathematical model, the following assumptions are made: (1) steady state. (2) single-phase, turbulent flow. (3) ideal gas compressible flow. (4) constant solid material properties. (5) considering the gas thermodynamic parameters temperature variation. (6) gravitational, viscous heat dissipation and radiation heat transfer are ignored. Based on these assumptions, the governing equations can be given as follows.

Continuity equation:

$$\frac{\partial \rho}{\partial t}+\nabla\cdot\left(\rho \boldsymbol{V}\right)=0 \tag{1}$$

Momentum equation:

$$\frac{\partial(\rho u)}{\partial t}+\nabla\cdot(\rho V u)=\nabla\cdot(\mu\nabla u)+S_{\mathrm{u}} \tag{2}$$

Energy equation:

$$\frac{\partial(\rho T)}{\partial t}+\nabla\cdot(\rho V T)=\nabla\cdot\left(\frac{\lambda}{c_p}\nabla T\right)+S_{\mathrm{t}} \tag{3}$$

Ideal gas equation of state:

$$\rho=\frac{p_{\mathrm{op}}+p}{RT} \tag{4}$$

where V is the velocity vector, u is the velocity component in the x direction, ρ is the density, μ is the dynamic viscosity, cp is the specific heat capacity, and λ is the thermal conductivity, T is the temperature, p is the pressure, pop is the system operating pressure, and R is the ideal gas constant. Su and St denote the source term and the viscous dissipation term.

According to the "Handbook of Thermophysical Properties of Materials Commonly Used in Engineering" [19], in the temperature range of 200-1000 °C and pressure of 0.48-0.50MPa, the isobaric heat capacity and molar mass of air remain almost unchanged, while thermal conductivity and dynamic viscosity vary with The temperature exhibits a linear change. Therefore, a linear interpolation function is used to describe the mathematical relationship of thermal conductivity and dynamic viscosity to temperature.

$$\lambda(T)=10^{-5}(5.581T+1146) \tag{5}$$

$$\mu(T)=10^{-5}(2775T+1.358) \tag{6}$$

The basic conditions of this simulation are that, driven by differential pressure, the low-temperature gas flows through the spark plug gap to cool the high-temperature ignition end face, and the pressure and temperature of the fluid inlet and outlet are all known values. Therefore, the fluid inlet is set as the pressure inlet, the given pressure value is 0.5MPa, and the initial temperature is 300°C. the fluid outlet is set as the pressure outlet, the given pressure value is 0.48MPa, and the initial temperature is 1100°C. The low temperature wall is set as the convective boundary condition, and the given convective coefficients are 15 , respectively. Due to the symmetry plane and adiabatic plane of the model formed by cutting, the boundary conditions are set to be symmetrical and adiabatic.

The heat exchange in the combustion chamber of an aero-engine depends on various factors such as combustion and convection, so there is a large change in different strokes of the engine [20], which needs to be determined based on experimental values. In the absence of experimental data, according to the author's calculation, for the same spark plug model, the average temperature of the firing end face of the engine is proportional to the given convection heat transfer coefficient at the firing end. This coefficient does not affect the objective law that the temperature of the ignition end face changes with the spark plug structure, so 300W/m^2·K is used for calculation in the following text.

2.3. Numerical solution and grid independence test

Considering that the thickness of the flow field area enclosed by the shell and the central electrode is about 0.8 mm, and the strong convection characteristics of the fluid in the tube driven by high pressure difference, the turbulence model chooses the k-ω SST shear transport model (SST k-ω turbulence model for short) proposed by Menter [21]. The model combines the k-ω and k-ε models using the mixing function F1, and can handle free flow and wall-confined flow well. At the same time, the turbulent viscosity in the turbulence model considers the propagation of turbulent shear stress, which has good accuracy for the flow in such a narrow area.

The solver is set as a steady-state solver based on pressure, and the Coupled algorithm is used to solve the coupled equation of pressure and velocity. The energy equation convergence criteria are set to 10-6, and the other equations convergence criteria are set to 10^{-3}.

To ensure the solution is independent of the grid size, the grid number is varied from 0.5 million to 1.5 million . It is possible to get the temperature value of a point as a function of the number of grids (see in figure 3). According to the simulation analysis, the simulated results of the micro heat sink with 1.1 million grids are more accurate. Consequently, a grid system with 1.1 million cells has been adopted as the optimum grid system for further simulations.

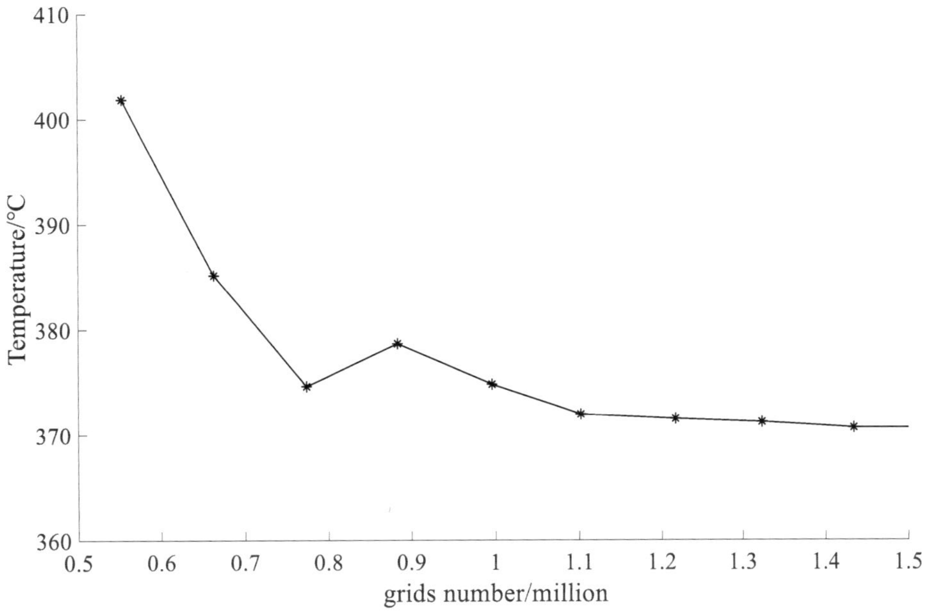

Figure 3. Plot of temperature at a point as a function of the number of grids

3. Results and discussion

Based on the above theories, methods and simulation settings, the obtained simulation results are shown in figure 4.

From the analysis of the streamline diagram, it can be seen that, driven by the pressure difference, the fluid velocity increases significantly from the inlet to the outlet, and the maximum velocity appears in the narrow flow channel close to the outlet, which is about 87.6m/s. In addition, the volume flow of this cooling channel gas is about 0.113L/s.

From the results of the temperature cloud map, it can be seen that the surface temperature of the firing end presents a ring-shaped distribution, and the temperature gradually increases from the inner ring to the outer ring. When the cooling runner is not added, the temperature of the firing end is 920.1-923.3 °C, after adding the cooling channel, the temperature of the firing end is 387.4-404.7 °C, and the temperature is lowered by about 550 °C, which shows that the design of the cooling channel can significantly reduce the temperature of the firing end. In addition, after adding the cooling runner, the temperature difference of the ignition end surface changed from 3.3 °C to 16.9 °C, and the temperature difference increased.

4. Optimized design

As shown in Figure.2, the flow area of the gas in the tube can be divided into three sections: the inlet section, the middle section and the outlet section. In the middle section, the gas flows approximately in the annular tube, and the flow rate follows the formula:

$$q_0 = \frac{\pi d_1 h^3}{12ul} \Delta p \tag{7}$$

In this study, since the differential pressure is a fixed value, the flow rate in the middle section is a fixed value. Therefore, the gas flow of the spark plug mainly depends on the structural design of the inlet and outlet. The combined design of different inlet diameters, outlet diameters and the number of end holes will change the flow resistance of the inlet and outlet sections, thereby affecting the gas flow of the spark plug.

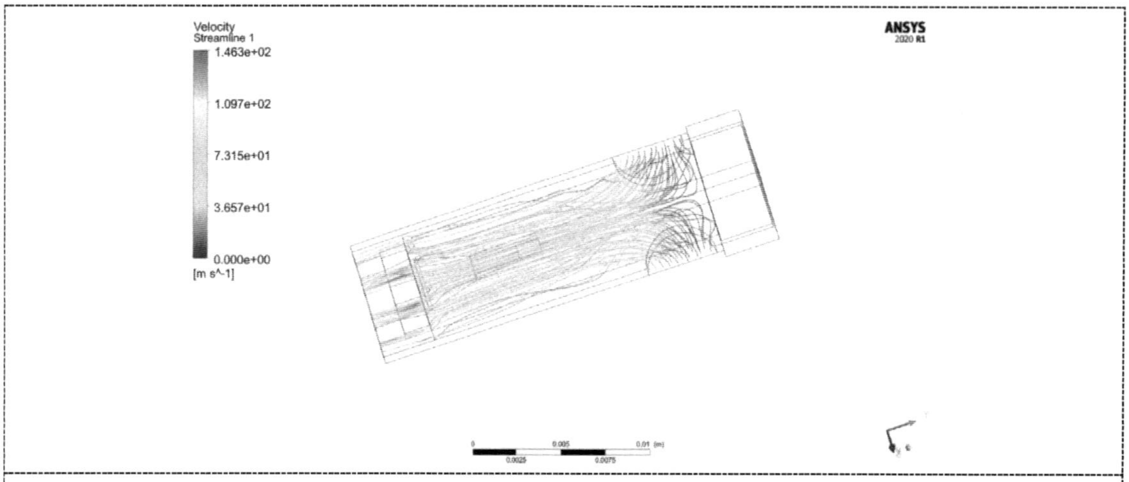

(a) Streamline diagram of spark plug fluid domain with cooling channel

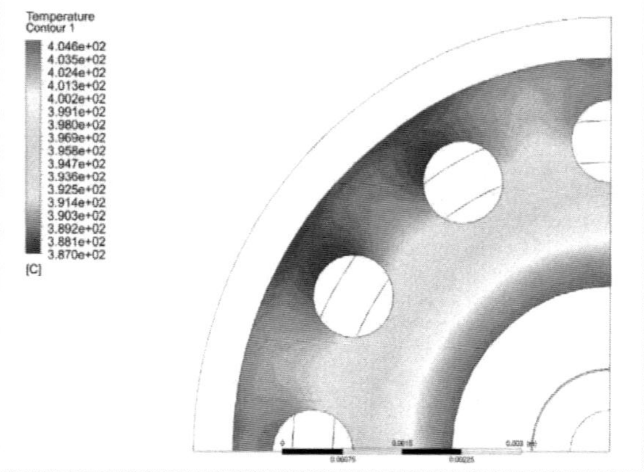

(b) Temperature distribution of the firing end of the spark plug with cooling runner.

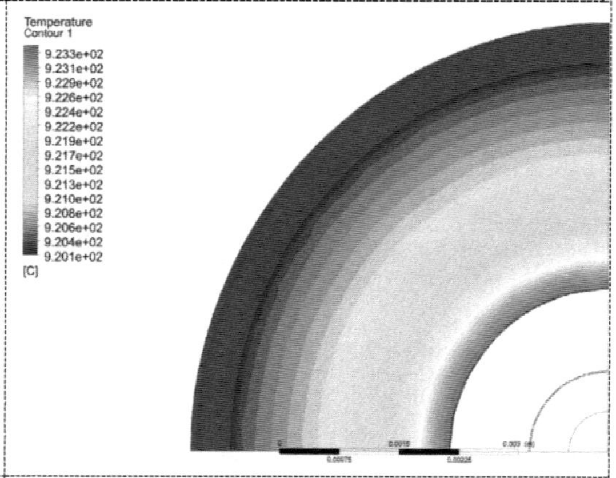

(c) Temperature distribution of the firing end of the spark plug without cooling runner.

Figure 4. Simulation results

In this paper, the method of orthogonal experimental design and variance analysis is used to optimize the design of the spark plug cooling channel, so as to reduce the maximum temperature of the spark plug firing end and improve the cooling channel flow. The optimization goal is to reduce the maximum temperature of the ignition end of the spark plug by 3%, and increase the volume flow of the cooling runner by 30%.

4.1. Orthogonal experiment

Orthogonal experimental design is another design method to study multi-factor and multi-level. It selects some representative points from the comprehensive test according to the orthogonality. It is an efficient, fast and economical experimental design method [22]. This paper adopts this method to optimize the design of the spark plug cooling channel.

In order to consider the inlet diameter (D, unit mm), outlet diameter (d, unit mm) and the number of end holes (N, unit "piece") of the ignition spark plug, the maximum temperature of the ignition end face (Tmax, unit °C) and the cooling flow channel For the influence of volume flow (L, unit L/s), a four-level and three-factor orthogonal experiment table is designed. The input three factors are the inlet diameter(3mm-6mm), the outlet diameter(0.8mm-1.4mm) and the number of end faces(4 pieces-16 pieces). The level value of each factor is within its possible range. The orthogonal experiment and simulation data are shown in table 1.

Table 1. Orthogonal experimental data

Number	D(mm)	d(mm)	N	T_{max}(°C)	L(L/s)
1	3	0.8	4	468.60	0.022
2	3	1	8	368.40	0.076
3	3	1.2	12	378.70	0.143
4	3	1.4	16	405.20	0.207
5	4	0.8	8	417.40	0.051
6	4	1	4	430.90	0.037
7	4	1.2	16	371.90	0.188
8	4	1.4	12	421.60	0.170
9	5	0.8	12	395.10	0.077
10	5	1	16	372.60	0.150
11	5	1.2	4	438.40	0.048
12	5	1.4	8	430.90	0.114
13	6	0.8	16	381.90	0.102
14	6	1	12	381.90	0.115
15	6	1.2	8	399.20	0.098
16	6	1.4	4	455.60	0.056

4.2. Analysis of variance

Multivariate analysis of variance is to decompose the total variance into components of each variance, and then use the significance test for analysis and judgment to study whether two or more control variables have a significant impact on the observed variables.

Multivariate analysis of variance is performed on the obtained orthogonal experimental data using the commercial software SPSS, and the obtained results are shown in table 2.

Table 2. SPSS analysis of variance results

independent variable	dependent variable	F	P
D	T_{max}	0.215	0.883
D	L	3.393	0.095
d	T_{max}	8.351	0.015
d	L	37.149	0.000
N	T_{max}	21.113	0.001
N	L	101.044	0.000

The default significance level of SPSS is 0.05. From the analysis of variance results, it can be seen that for the outlet diameter and the number of holes at the end face, the associated probability (P value) is less than 0.05, and it can be considered that it has a significant impact on the maximum temperature and volume flow of the end face. The P value corresponding to the number of holes at the end face is close to zero, indicating that it has the most significant effect on the volume flow and the maximum temperature at the end face. For the inlet diameter, the accompanying probability value is greater than 0.05, and it can be considered that it has no significant effect on the maximum temperature and volume flow of the end face.

Using the post-hoc multiple comparison function of SPSS, the linear effect of a single factor on the dependent variable can be determined. The basic idea is to take other factors as the mean value when judging the degree of influence of a single factor. The single factor effect table obtained from the analysis is shown in table 3.

It can be seen from table 3 that the maximum temperature of the end face is negatively correlated with the number of end faces. The more the end faces are, the lower the maximum temperature is. The maximum temperature of the end face is not linearly related to the diameter of the inlet and the outlet. Under the current level, taking D=6mm, d=1mm, the maximum temperature of the end face can be minimized.

For volume flow, it has a negative correlation with the inlet diameter and a positive correlation with the outlet diameter and the number of holes at the end face.

Table 3. Single factor effect table

factor	level	Tmax	volume flow
D	3	405.22	0.112
D	4	410.45	0.111
D	5	409.25	0.098
D	6	404.65	0.093
d	0.8	415.75	0.063
d	1	388.45	0.096
d	1.2	397.05	0.119
d	1.4	428.33	0.137
N	4	382.90	0.041
N	8	394.33	0.084
N	12	403.98	0.126
N	16	448.38	0.163

In the priority of parameter design, the maximum temperature of the end face should be guaranteed to be the lowest first, and then the volume flow should be made as large as possible. Since the influence of the inlet diameter on the maximum temperature of the end face is not significant, the single factor effect value of the optimal value of 6 mm and the sub-optimal value of 3 mm is only 0.57°C apart, and the volume flow is negatively correlated with the inlet diameter. Therefore, the value of the inlet diameter is 3 mm. A lower maximum end face temperature and maximum volume flow can be obtained. The outlet diameter has a significant effect on the maximum temperature of the end face, and the optimal value is 1mm. Because the maximum temperature of the end face is negatively correlated with the number of end faces, and the volume flow is positively correlated with the number of end faces, the more end faces, the better. In the case of a fixed outlet diameter, the maximum number of end holes is 20 (minimum 0.15mm processing spacing between the two holes). From this, the optimal structural parameters of a set of electric spark plugs are obtained: D=3mm, d=1mm, N=20.

Figure 5. The temperature cloud map of the ignition end face of the spark plug after optimization.

4.3. Simulation verification

The model was established according to the optimized design parameters obtained from the orthogonal experiment, and Fluent was used for verification. As shown in figure 5.

The maximum temperature of the ignition end face of the spark plug is only 368.2℃, and the calculated value of the volume flow is 0.172L/s. In the 16 sets of orthogonal experiments, the temperature of the ignition end face of this design is the lowest value, and the volume flow rate ranks third. Compared with the initial cooling flow channel, the maximum temperature is reduced by 3.84%, and the volume flow rate is increased by 52.21%, reaching the set value. The optimization goal proves that the results of the orthogonal test and variance analysis are reasonable and reliable, and there is a set of optimal structural designs for the spark plug cooling channel.

5. Conclusion

In this paper, for a specific semiconductor spark plug model, under the condition of high temperature and high pressure, the coupled simulation model of flow and heat transfer in the cooling channel of the ignition spark plug is established by Fluent, and the optimization design is carried out on the basis of the simulation. In the research process, the orthogonal experiment method was adopted, and a three-factor and four-level orthogonal experimental design was carried out to study the influence of the side cooling aperture, the end face cooling aperture, and the number of end face cooling holes on the gas flow rate and the temperature of the ignition end face. The main conclusions are as follows:

(1) The cooling runner has a good heat dissipation effect. After the cooling runner is added, the temperature of the ignition end face of the spark plug is reduced by about 550℃.

(2) Multi-factor analysis of variance was used to process the orthogonal experimental data. It can be found that the number of end faces has the most significant effect on the maximum temperature and volume flow of the end face, followed by the outlet diameter, while the inlet diameter is not significant.

(3) After the optimized design of the spark plug cooling channel, the maximum temperature of the ignition end face can be reduced by 3.84%, and the volume flow rate can be increased by 52.21%.

Through the work of this paper, the influence law of the spark plug cooling channel structure parameters on the spark plug cooling performance and air flow has been preliminarily mastered. It is worth mentioning that the analysis method used in this paper and the final optimization conclusion obtained are not only applicable to aero-engine spark plugs, but also have certain reference value for other types of air-domain cooling channels.

6. References

[1] YUAN Chun, Liang Haitao. Discussion on Maintenance of Aviation Piston Engine Electric Spark plug[J]. Science and Technology Innovation Herald, 2012,25(02):79-80.

[2] PENG Zeyan, LIU Gang,GUI Xingming,ea al. Principles of Aviation Gas Turbine. Beijing[M]: National Defence Industry Press，2008:108-122.

[3] Yan Jun. Analysis of the malfunction of the electric spark plug of the aviation piston engine[J]. Forum of the Association for Science and Technology: Second half of the month, 2010,11(2):30-31.

[4] YANG Yufeng. Operation and maintenance of the ignition spark plug of the aviation piston engine[J]. Science & Technology Information, 2012, 36 (1):85-86.

[5] LI Luoyang. Selection and Maintenance of Electric Spark plugs for Aviation Piston Engines[J]. Science and Technology Innovation Herald, 2011,14(1):54-55.

[6] Dong Ying. The application of laser welding technology in the repair of electric spark plug wires[J]. Aviation Maintenance and Engineering, 2012,4(2): 70-71.

[7] Ma Hongwei. Development of the ignition failure detection system for aviation piston engines [D]. University of Electronic Science and Technology of China, 2011:16-20.

[8] Xiao Zunwen. High-energy ignition system semiconductor spark plug (1)[J]. Spark plug and special ceramics, 1995,4(3):7-10.

[9] Xiao Zunwen. High-energy ignition system semiconductor spark plug (1)[J]. Spark plug and special ceramics, 1995,3(4):9-13.

[10] Xiao Zunwen. High-energy ignition system semiconductor spark plug (1)[J]. Spark plug and special ceramics, 1996,2(1):10-12.
[11] Xiao Zunwen. High-energy ignition system semiconductor spark plug (1)[J]. Spark plug and special ceramics, 1996,3(3):1-6.
[12] Xiao Zunwen. High-energy ignition system semiconductor spark plug (1)[J]. Spark plug and special ceramics, 1997,2(3):22-26.
[13] Osamura H , Abe N . Development of New Iridium Alloy for Spark Plug Electrodes[C]// International Congress & Exposition. 1999,2688-2696.
[14] ZHAO Jun. The Research of the Semiconductor's Formula and Sintering Technology[J]. New Technology & New Process.,2008,12(2) :116-117.
[15] SHAO Yue, WU Enhui, ZHANG Tao,el al.Research on Semiconductor Igniter Tip's Electrode Gap of the Aircraft Engine Ignition System[J]. Mechanical Science and Technology for Aerospace Engineering. 2017,36(01):161-164.
[16] Abdel-Rehim, Ahmed A . Impact of spark plug number of ground electrodes on engine stability[J]. Ain Shams Engineering Journal, 2013, 4(2):307-316.
[17] Badawy T , Bao X C , Xu H . Impact of spark plug gap on flame kernel propagation and engine performance[J]. Applied Energy, 2017,191(1):311-327.
[18] Li Yixin, TAN Hang, YANG Shuiyin. An initial study on the current development and derivatives of the electric ignition system of aero-engine[J]. Gas Turbine Experiment and Research, 2015,28(06):49-54.
[19] Zhang Jiarong,Zhao Tingyuan. Handbook of Thermophysical Properties of Commonly Used Engineering Substances[M]. New Age Publishing House,1987,79-105.
[20] M. Diviš and R. Tichánek and M. Španiel. Heat Transfer Analysis of a Diesel Engine Head[J]. Acta Polytechnica, 2003, 43(5):27-40.
[21] Menter F R . Two-equation eddy-viscosity turbulence models for engineering applications[J]. AIAA Journal, 1994,32(8):1598-1605.
[22] Zhou Jianmin, Shen Renfang. Dictionary of Soil Science[M]. Science Press, 2013:16-28.

Acknowledgments
This research was supported by the National Fundamental Research Program (No. G1772JC21B009B).

Hollow Pin Optimization of Compression Ignition Aviation Engine Crankshaft

S Galata[1,2,*], E Ünlü[1], Ö Işın[2] and E Tuzlu[1]
[1] TEI - TUSAŞ Engine Industries Inc., R&D Center, Piston Engines Department, Eskisehir, Turkey
[2] Yildiz Technical University, Mechanical Engineering Department, Internal Combustion Engines Laboratory, Istanbul, Turkey

* serkan.galata@gmail.com

Abstract. In this article, a crankshaft design has been made for a compression ignition internal combustion engine that is planned to be used in an aerial vehicle. Alternative models were designed as hollow pins with different diameters. Finite element analysis was performed on 1 reference model and 3 alternative models using AISI 1045 and AISI 4140 materials. As the hollow diameter increases for both materials; maximum of equivalent stress has increased and minimum of safety factor has decreased. As a result of the study, it has been evaluated that the alternative 3 with AISI 4140 material is suitable for use on the crankshaft.

1. Introduction

Internal combustion engines are classified based on their usage areas such as road, sea, air, and off-road. The ultimate goal of such engines is the movement of the wheel for road vehicles, the propeller for sea and aerial vehicles, and the pallet for off-road uses such as snowmobiles. The common point of all is the desire for circular motion. Therefore, the crankshaft, which is a circular motion generator, is one of the most basic components of powertrain systems using internal combustion engines.

The combustion phenomenon in internal combustion engines occurs as a result of the reaction of the fuel sent to the combustion chamber with oxygen. Fuel, whether liquid or gas, contains molecules such as C, H, S, and O in varying amounts and bond types [1]. In Otto engines, the fuel-air mixture is compressed and ignited using a spark plug, and in Diesel engines, combustion occurs by injecting the fuel into the compressed air with increased pressure and temperature; as a result of this combustion, the chemical bonds of the fuel are broken and energy is released; finally, linearity is achieved as a result of the movement of the piston in the expansion process of the engine cycle [2].

This linear work output is equal to the crankshaft revolutions for two-stroke engines, and half the crankshaft revolutions for four-stroke engines. The force obtained as a result of combustion is transmitted to the crank pin by transferring from the piston's upper surface to the connecting rod. The crankshaft, which is mounted on the engine block, transforms the linearity formed in the piston as a result of combustion into a circular motion with the load on the crank pin.

As a result of the tremendous loads to which it is subjected, the crankshaft should have a strong and lasting structure while being light due to inertia and centrifugal forces. Weight is one of the most essential design outputs, particularly in engine parts utilized in the aviation industry, because it directly affects critical parameters such as take-off power, take-off weight, and airtime. Considering the contradicting requirements stated, the significance of the optimization process in crankshaft design is obvious.

In this paper, weight reduction was aimed in a crankshaft that designed for an inline 4 cylinder, compression ignition engine planned to be used in aircraft. The optimization of the drilling process applied to the web region for weight reduction in a crankshaft was carried out in this study with the assistance of a computer generated three dimensional model of the part and finite element software.

2. Crankshaft Weight Reduction

The main goal of internal combustion engine design is to attain low weight and cost values while maintaining good performance and durability. It is seen that fuel consumption and emission values in automobiles are reduced as engine weight is reduced. In several types of aviation engines, the weight of ballast added to the aircraft's nose or tail decreases as engine weight decreases. Instead of this gain, additional fuel is added to extend flight time or as an additional payload. For these reasons, along with strength, lightness is regarded as the most critical design criterion in aviation parts.

A very high level of weight reduction in the head and block system has been achieved as a result of the use of low density aluminum raw material for the cylinder block and the cylinder head. Despite this improvement, there has not been much innovation in the powertrain system. After the transition to the use of titanium raw materials in the connecting rods and intake valves of some mass produced engines, the process of innovation entered a new era [3].

Crankshafts draw attention for weight reduction projects since it accounts for approximately 10% of total engine weight. A cast iron crankshaft is 10% lighter than the forged steel portion due to material and density differences. The fact that the casting process permits components of the item to be manufactured in a hollow structure without machining also presents the possibility of weight gain. The use of forged steel for parts is disadvantageous in terms of weight, and its production and processing costs are high. Regardless, the forging method is commonly utilized in crankshafts. The high elasticity module of the steel improves the NVH values and provides better comfort to the vehicle user [4]. Because the probability of a porous structure in the casting method is great, and there is a risk of creating a notch effect, the forging method comes to the fore [2].

As the number of cylinders, engine volume, and engine power increase, the weight of the crankshaft also increases as a result of the increase in the strength requirement in connection with the forces generated. Figure 1 shows the increase in crankshaft weight with the increase in engine volume. Particularly for 1600 cm^3 and 2000 cm^3 engine volumes, crankshaft weight can vary by 2-3 kilograms for different engines [5].

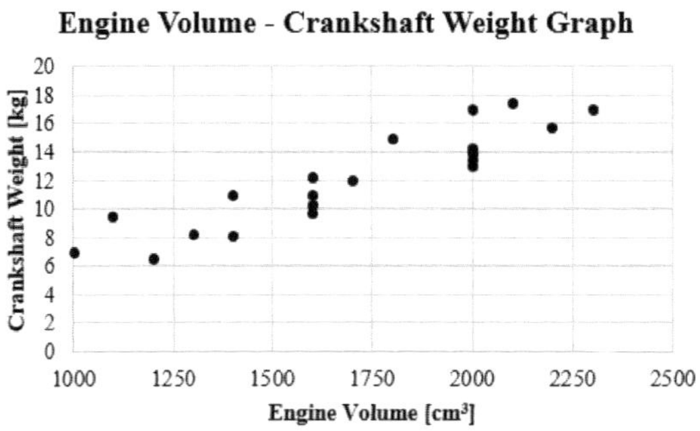

Figure 1. Engine volume – crankshaft weight graph [5]

Crankshaft weight reduction strategies are basically classified as follows: Density shift [4]; hollow main bearings [6], [7], [8], [9]; hollow crank pins [6], [7], [8], [9]; geometry optimization of counter weights [4]; geometry optimization of crank web shape.

3. Crankshaft Hollow Pin Design

A crankshaft was designed within the scope of this study for an aircraft compression ignition engine with a 4 cylinder feature and combustion chamber volume of 2499 cm^3. The technical characteristics of the engine and crankshaft are presented in Table 1. Not all of the parameters presented in the table were used for the calculations.

Table 1. Technical characteristics of engine and crankshaft

Property	Unit	Value
Bore x Stroke	mm x mm	92 x 94
Stroke Volume	cm^3	622
Number of Cylinder	Quantity	4
Combustion Chamber Volume	cm^3	2.499
Engine Power	HP kW	250 @3,500 rpm 186 @3,500 rpm
Number of Counter Weight	Quantity	8
Diameter of Main Bearing Journal	mm	56.00
Diameter of Crank Pin	mm	50.60

3.1. 3D Modeling

Siemens NX software covers all modeling requirements from the conceptual design stage to the final product stage. Designers are able to perform operations such as 2D and 3D wire net, solid modeling, assembly modeling, and drawings via this software. With Siemens NX, solid models can be designed and modified using different part modeling techniques. In addition, physical values such as surface, volume, inertia, and weight can be easily determined.

Within the scope of the corresponding author's thesis, it was evaluated that a crankshaft with 56.00 mm diameter main bearings and 50.60 mm diameter crank pins would be suitable for the powertrain system to transmit the 250 HP power value planned to be generated by the engine.

A 3D model of the part was created in the computer environment using Siemens NX 12.0.0.27 software, as shown in Figure 1. In this context, a value of 23.02 kg was determined by CAD weight measurement.

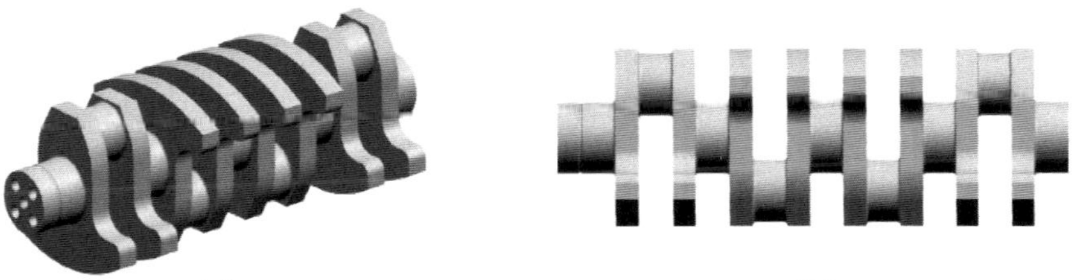

(a) (b)

Figure 2. (a) Isometric **(b)** side view of crankshaft

The crankshaft shown in Figure 2 is the reference part. Weight reduction details were added to the reference model. Therefore, the ideal diameter size of the weight reduction holes to be drilled in the crank pin area was examined in this study.

Due to the symmetrical structure of the designed crankshaft, the quarter slice formed as a result of dividing the part into four was chosen to simulate the whole. The position of the quarter model shown in Figure 3(b) on the crankshaft is shown in Figure 3(a), colored green.

(a) **(b)**

Figure 3. **(a)** Whole **(b)** quarter slice of crankshaft

4. Finite Element Analysis

ANSYS is a general purpose finite element software used in the fields of strength, vibration, fluid mechanics, and heat transfer. The software allows for the simulation of the tests to be performed and the working conditions of the products to be analyzed virtually before producing a prototype. Weak points are identified as the output of the analysis, thus potential risk points can be detected, and improvement actions can be taken [10]. Some advantages of using the finite element method (FEM) are listed below [11]:

- The number of samples is reduced,
- Non-compliant manufacturing for prototype testing is prevented,
- Cost savings are provided, and
- Time savings are provided.

The crankshaft designed in "part" (.prt) file format in the Siemens NX software was converted to "parasolid" (.x_t) format. The geometry shown in Figure 4 was transferred to the ANSYS Workbench 15.0.7 software for stress analysis.

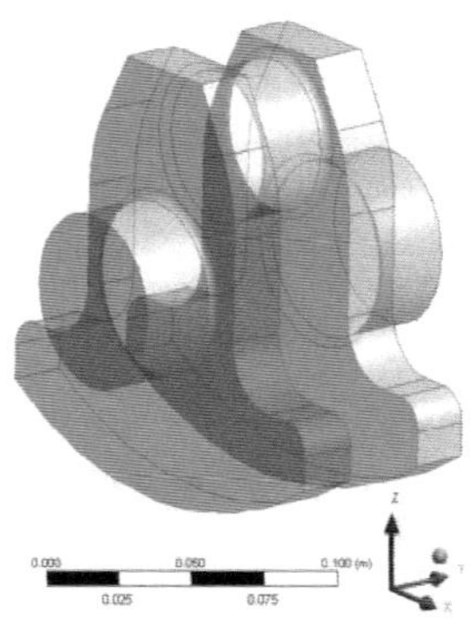

Figure 4. Geometry imported into ANSYS

4.1. Material

The raw material of the designed crankshaft was AISI 1045 or AISI 4140 [12]. 1045 steel stands out with its high strength and impact resistance properties and has good machinability properties. This type of steel has an important place in the industry due to its high wear resistance and is used in engine parts such as gears, shafts, crankshafts, and connecting rods [13]. 4140 steel is a low alloy steel containing chromium, molybdenum, and manganese which is used to improve strength and hardenability parameters [14]. This easy to supply material, which is also commonly used in the industry, is highly suitable for use in a crankshaft due to its toughness, fatigue strength, and resistance to abrasion and impact. The chemical properties of AISI 1045 and AISI 4140 steels are presented in Table 2 and the basic properties of AISI 1045 and AISI 4140 are presented in Table 3 [15], [16].

Table 2. Properties of AISI 1045 and AISI 4140 materials [13], [14], [15], [16]

Element	AISI 1045	AISI 4140	Property	Unit	AISI 1045	AISI 4140
Carbon	0.43-0.50	0.38-0.43	Density	kg/m^3	7870	7850
Manganese	0.50-0.80	0.75-1.00	Tensile Strength	MPa	565	655
Silisium	0.40	0.15-0.35	Yield Strength	MPa	310	415
Phosphorous	0.04	0.02	Bulk Modulus	MPa	140,000	140,000
Sulfur	0.04	0.02	Shear Modulus	MPa	80,000	80,000
Chromium	0.40	0.80-1.10	Elastic Modulus	MPa	200,000	210,000
Molybdenum	0.10	0.15-0.25	Poisson's Ratio	[ ]	0.29	0.30

The mechanical properties presented in Table 2 were identified manually in the engineering data section of the ANSYS software and the analyses were performed based on the properties of AISI 1045 and AISI 4140 materials.

4.2. Meshing

The geometry to be analyzed with the finite element method was divided into infinite-small sub-parts (element) with the mesh structure. The behavior pattern of the nodes of each sub-part was examined, from the behavior of the points to the behavior of the whole.

An element size of 0.20 mm in critical areas and 1.00 mm in non-critical areas was used in the study. The mesh structure was created with 2,727,930 nodes and 1,684,933 elements on the reference model shown in Figure 4.

Skewness and orthogonal quality values are important to examine the quality of the created mesh structure in metric measures. High skewness and low orthogonal quality are not recommended for the analysis model [17]. In the models created for the alternative designs made throughout the study, the skewness was between 0.20-0.25, and the orthogonal quality was between 0.80-0.96. The quality spectrum of the mesh structure and the quality values used in the study are shown with the black box on Figure 5.

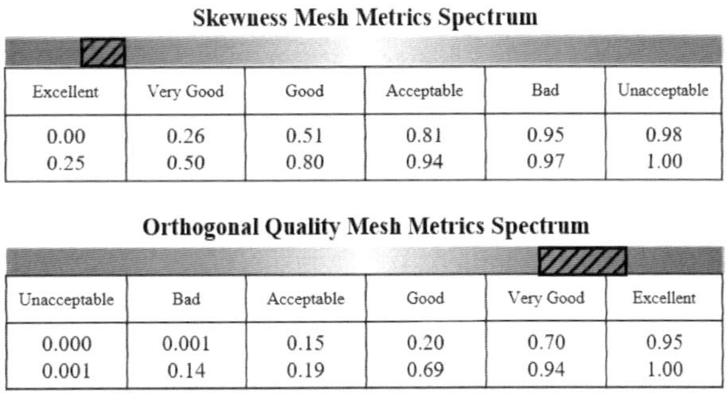

Figure 5. Mesh metrics [17]

4.3. Boundary Conditions

The boundary conditions were defined following the creation of the model. In the literature, there are studies in which the main bearings are limited to fixed support [11], [18], [19], [20]. However, the bearing of the crankshaft was established with a fixed support from the left side and a cylindrical support from two main bearings in order to increase the accuracy of the analysis in the present study [21]. The cylindrical support allows for radial movement and is fixed in the axial and tangential directions.

Studies in which the load is applied as pressure over the crank pin [18], [20] can be seen in the literature. However, it has been interpreted that there is a direct force effect on the crank pin. The scope of the thesis study of responsible author, a force of 90,000 N, calculated considering the combustion values, was presented as a bearing load on the crank pin as shown in Figure 6 [11], [21].

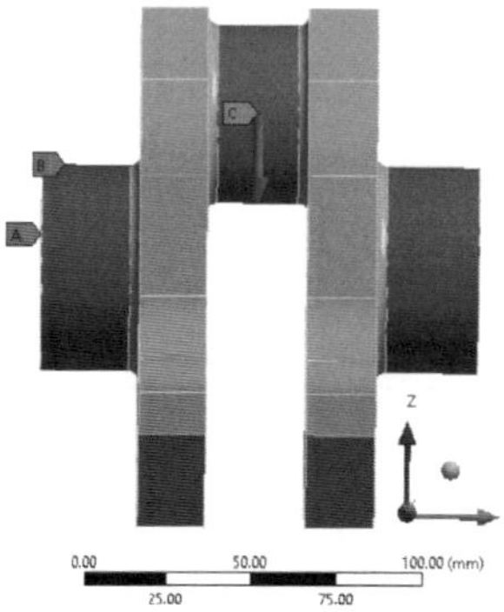

Figure 6. Boundary conditions

5. Alternative Designs

Within the scope of weight reduction, 3 alternative models were created with holes of 10 mm, 20 mm, and 25 mm in diameter in the crank pin. The visuals of the reference model and alternative models are shown in Figure 7.

(a) **(b)** **(c)**

Figure 7. Models of crank pin hollow diameter **(a)** 10 mm **(b)** 20 mm **(c)** 25 mm

Weight measurements of alternative models were performed in the CAD environment and the relative gains obtained compared to the reference model are presented in Table 3. Accordingly, for alternative 3 with a 25 mm crank pin hole, it was observed that there is a potential weight gain of 269.07 grams for the quarter model and 1,076.28 grams for the whole crankshaft.

Table 3. Weight reduction values of alternative designs

Model	Alternative 1 10 mm	Alternative 2 20 mm	Alternative 3 25 mm
Reduction of Quarter Slice [gram]	43.05	172.21	269.07
Reduction of Whole Part [gram]	172.20	688.84	1,076.28

6. Results

A reference model was developed as a part of this study. In order to reduce weight, drilling processes for different diameters were applied to the crank pin areas and a total of 4 models were analyzed using two different materials. Total deformation, equivalent elastic strain, equivalent stress, and safety coefficient values were calculated through static analyses. The obtained results are shown in Table 4.

Table 4. Analysis results of reference and alternative models

Model	Material	Total Deformation [mm]	Equivalent Elastic Strain [ ], [mm/mm]	Equivalent Stress [MPa]	Safety Factor [ ]
Reference	AISI 1045	0.0132	0.0009	181.21	1.7107
	AISI 4140	0.0127	0.0008	181.25	2.2897
Alternative 1 10 mm	AISI 1045	0.0136	0.0009	190.57	1.6267
	AISI 4140	0.0130	0.0009	190.69	2.1763
Alternative 2 20 mm	AISI 1045	0.0147	0.0010	207.65	1.4929
	AISI 4140	0.0141	0.0009	207.65	1.9986
Alternative 3 25 mm	AISI 1045	0.0160	0.0012	226.63	1.3679
	AISI 4140	0.0153	0.0011	226.72	1.8304

In the analyses, the statuses of 1 reference model and 3 alternative models were examined by using AISI 1045 and AISI 4140 materials, and the following results were obtained:

The maximum value of total deformation for AISI 1045 was 0.0132 mm in the reference model, 0.0136 mm in the alternative 1 model with a 10 mm crank pin hole, 0.0147 mm in the alternative 2 model with a 20 mm crank pin hole, and 0.0160 mm in the alternative 3 model with a 25 mm crank pin hole. The maximum value of total deformation for AISI 4140 was 0.0127 mm in the reference model, 0.0130 mm in alternative 1 with a 10 mm crank pin hole, 0.0141 mm in alternative 2 with a 20 mm crank pin hole, and 0.0153 mm in alternative 3 with a 25 mm crank pin hole. Figure 8 shows the maximum total deformation of alternative 3. For both materials, it was observed that the total maximum deformation increased as the diameter of the crank pin hole increased.

The maximum value of equivalent elastic strain for AISI 1045 was 0.0009 mm in the reference model, 0.0009 mm in alternative 1 with a 10 mm crank pin hole, 0.0010 mm in alternative 2 with a 20 mm crank pin hole, and 0.0012 mm in alternative 3 with a 25 mm crank pin hole. The maximum value of equivalent elastic strain for AISI 4140 was 0.0008 mm in the reference model, 0.0009 mm in alternative 1 with a 10 mm crank pin hole, 0.0009 mm in alternative 2 with a 20 mm crank pin hole, and 0.0011 mm in alternative 3 with a 25 mm crank pin hole. For both materials, it was observed that the equivalent elastic strain increased as the crank pin hole diameter increased.

The maximum value of equivalent stress for AISI 1045 and AISI 1040 was 181 MPa in the reference model, 190 MPa in the alternative 1 model with a 10 mm crank pin hole, 207 MPa in the alternative 2 model with a 20 mm crank pin hole, and 226 MPa in the alternative 3 model with a 25 mm crank pin hole. For both materials, it was observed that the equivalent stress increased as the diameter of the crank pin hole increased.

The maximum value of safety factor for AISI 1045 was 1.7107 in the reference model, 1.6267 in the alternative 1 model with a 10 mm crank pin hole, 1.4929 in alternative 2 with a 20 mm crank pin hole, and 1.3679 in alternative 3 with a 25 mm crank pin hole. The minimum value of safety factor for AISI 4140 was 2.2897 in the reference model, 2.1763 mm in alternative 1 with a 10 mm crank pin hole, 1.9986 mm in alternative 2 with a 20 mm crank pin hole, and 1.8304 mm in alternative 3 with a 25 mm crank pin hole. It was observed that the minimum safety factor decreased as the crank pin hole diameter increased for both materials, but this value was higher for AISI 4140 compared to AISI 1045. The results obtained here are consistent with the results of crank pin hole studies in the literature [21].

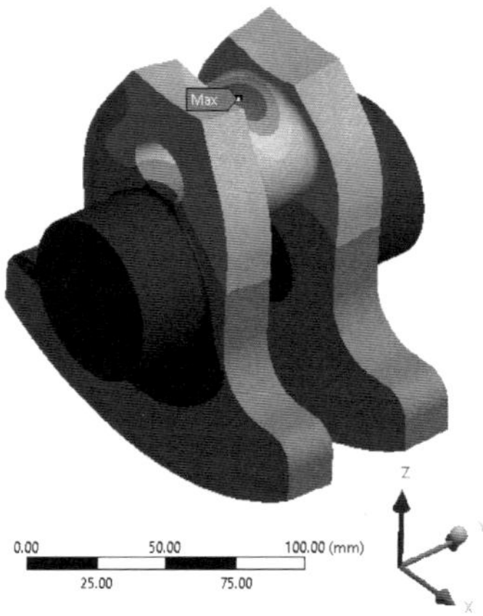

Figure 8. Maximum total deformation point of alternative 3

7. Conclusions

The operations and results of the study are as follows:

- Reference crankshaft modeling was performed with Siemens NX software via computer.
- Three alternative designs were created by drilling the inside of the crank pin diameter on the crankshaft in different sizes.
- It was determined that there is a weight reduction potential of up to 1,076.28 grams for the alternative models prepared.
- AISI 1045 and AISI 4140 were selected as candidate materials for the crankshaft.
- Total deformation, equivalent elastic strain, equivalent stress, and safety factor were calculated for the reference model and alternative models by using ANSYS software via computer.
- It was determined that the maximum values of total deformation and equivalent elastic strain were lower for AISI 4140.
- For both materials, it was observed that the equivalent stress increased as the diameter of the crank pin hole increased.
- It was determined that the safety factor was higher for the use of AISI 4140 compared to the use of AISI 1045.
- In the study, the minimum safety factor value for moving parts was determined as 1.40.
- In the model made of AISI 1045 with a 25 mm crank pin hole, since the safety factor dropped below 1.40 (1.3679), it will be necessary to switch to a smaller step and make a 20 mm crank pin hole. On the other hand, since the safety factor was 1.8304 for the model made of AISI 4140 with a 25 mm crank pin hole, it has been interpreted that this model will be fit to use.
- Since the safety factor falls below the critical limit for the use of AISI 1045 in the alternative 3 model, it would be possible to use the alternative 3 model with AISI 4140 or to use alternative 2 with AISI 1045 and AISI 4140.
- Considering that the 25 mm hole has a higher weight reduction benefit, it has been evaluated that alternative 3 with AISI 4140 is suitable for use for the crankshaft.

7.1. Future Studies

Within the scope of the study, an approach for yield stress was applied in the selection of the crank pin hole diameter value. After future studies, it is planned to perform fatigue analysis on the final model, which will be revealed after other parts of the crankshaft such as counterweight, web, and main bearing are optimized.

For the scope of the study, the connecting rod lubrication hole was excluded for the selection of the crank pin hole diameter value. As a future study, it is planned to clarify the position, angle, and diameter values of the connecting rod lubrication hole.

8. Acknowledgement
This technical paper was supported by TEI – TUSAŞ Engine Industries Inc. with the project number 2020-PMM-3885-01 and was created within the scope of the corresponding author's PhD thesis.

9. References

[1] Galata S 2017 *Experimental Investigation of Water Injection on the Performance and Exhaust Emissions of an Internal Combustion Engine* Yildiz Technical University, Graduate School of Science and Engineering, Mechanical Engineering Department MSc Thesis, Istanbul, Turkey.
[2] Galata S 2020 *Parçalı Krank Milinde Ölçüm Adımları* TEI – TUSAŞ Engine Industries, Inc. Technical Report Number: TBP-0209-01, Eskisehir, Turkey.
[3] Warrick R J, Ellis G G, Grupke C C, Khamseh A R, McLachlan, T H and Gerkits C 1999 *Development and Application of Enhanced Compacted Graphite Iron for the Bedplate of the New Chrysler 4.7 Liter V-8 Engine* SAE – American Society of Automotive Engineers Technical Paper Number: 1999-01-1225.
[4] Druschitz A P, Fitzgerald D C and Hoegfeldt I 2006 *Lightweight Crankshaft* SAE – American Society of Automotive Engineers World Congress, 03-06 April 2006, Detroit, Michigan, USA.
[5] IFP – French Institute of Petroleum 2018 *Moving Parts* IFP Base Engine Design Training Notes, Lillebonne, France.
[6] Pathak U, Shingade V, Senthilkumar V, Patil S and Gokhale K 2011 *Hollow Crankshaft for IC Engines* World Intellectual Property Organization International Patent Publication Number: WO 2011/121611 A1, Mumbai, Maharashtra, India.
[7] URL 1: FEV Europe GmbH *Cranktrain Design Optimized for Reducing Weight and Friction Supported by Coupled CAE Tools* https://www.fev.com/fileadmin/user_upload/Media/TechnicalPublications/Design/OptimizedCranktrainDesignSupportedbyCAE.pdf, Date: 29 June 2020.
[8] URL 2: Honda Racing Formula One, Mizoue K, Kawahito Y and Mizogawa K *Development of Hollow Crankshaft* http://www.f1-forecast.com/pdf/F1-Files/Honda/F1-SP2_31e.pdf, Date: 01 April 2021.
[9] Papadimitriou I and Track K 2018 *Lightweight Potential of Crankshafts with Hollow Design* Georg Fischer Automotive AG, Schaffhausen, Switzerland.
[10] Kovaci H and Albayrak O 2008 *Ankastre Kiriş Tasarımı için MatLab ve ANSYS Optimizasyonu* Atatürk University Mechanical Engineering Department, Erzurum, Turkey.
[11] Shelke S S, Dhamejani C L and Gadhave A S 2016 *Modelling and Analysis of the Crankshaft Using ANSYS Software* IJESC – International Journal of Engineering Science and Computing Journal, **6**:6, p.6796-6800.
[12] Stojanovic B and Glisovic J 2016 *Automotive Engine Materials* Reference Module in Materials Science and Materials Engineering, University of Kragujevac, Kragujevac, Serbia.
[13] URL 3: SAE (AISI) 1045 Steel Data Page, kanatcelik.com/1045, Date: 24 April 2021.
[14] AMS6382 Revision R 2018 *Aerospace Material Certification AMS6382 Steel, Bars, Forging, and Rings – SAE 4140* SAE – American Society of Automotive Engineers Aerospace Material Certification.
[15] URL 4: AZO Materials, *AISI 1045 Medium Carbon Steel Material* azom.com/article.aspx?ArticleID=6130, Date: 23 April 2021.
[16] URL 5: AZO Materials, *AISI 4140 Alloy Steel (UNS G41400) Material* azom.com/article.aspx?ArticleID=6769, Date: 23 April 2021.
[17] Özen M 2014 *Meshing Workshop* ANSYS Meshing Workshop Training Notes, Istanbul, Turkey.
[18] Thriveni K and Jaya-Chandraiah B 2013 *Modelling and Analysis of the Crankshaft Using Ansys Software* International Journal of Computational Engineering Research, (2013):**03**:5, p.83-89.
[19] Deshbhratar R J and Suple Y R 2012 *Analysis & Optimization of Crankshaft Using FEM* IJMER – International Journal of Modern Engineering Research, (2012):**02**:5, p.3086-3088.
[20] Sandya K, Keerthi M and Srinivas K 2016 *Modelling and Stress Analysis of Crankshaft Using FEM Package ANSYS* IRJET – International Research Journal of Engineering and Technology, (2016):**03**:01, p.687-693.

[21] Shahane V C and Pawar R S 2017 *Optimization of the Crankshaft Using Finite Element Analysis Approach* Automotive and Engine Technology, (2017):**2**:1-23.

Optimal Design of Shock Absorption for Hexapod Robot based on Magnetorheological Fluid

Sihao Huang[1,a*], Boyang Chen[2,b], Maoxin Yin[1,c], Yiming Ren[3,d], Zikang Xie[4,e]

[1]Department of Materials Science and Engineering, Materials Science and Engineering, Shenzhen MSU-BIT University, Shenzhen, 518172, Guangdong, China
[2]School of Intelligent Manufacturing and Equipment, Mechanical Design Manufacture, Shenzhen Institute Of Information Technology, Shenzhen, Guangdong,518000, China
[3]Department of Computational Mathematics and Cybernetics, Mathematics and Applied Mathematics, Shenzhen MSU-BIT University, Shenzhen, 518172, Guangdong, China
[4]Department of Engineering, Electrical and Computer Engineering, Shenzhen MSU-BIT University, Shenzhen, 518172, Guangdong, China

[a] huangsihao2021@163.com; [b] OliverChen.work@outlook.com;
[c]1315379512@qq.com; [d] ren4102212002@163.com; [e] xie13318782539@163.com

Abstract. To achieve the best shock absorption effect of the hexapod robot, the influence of three leg structures on the stability of the hexapod robot under different path environments was studied, and a set of highly implementable solutions was designed. Use SolidWorks to build a hexapod robot model, import it into ADAMS to build a dynamic model and build different path environments, and conduct forward and inverse kinematics analysis at the same time. Using the Simulink plug-in in MATLAB and ADAMS joint kinematics simulation, the spring damping and magnetorheological fluid components are established in Simulink, which are connected in parallel with the model in ADMAS, and the stability of the hexapod robot under three-leg structures in different environments is obtained. According to the results, the optimal shock-absorbing leg structure is obtained.

1. Introduction

In the face of extreme conditions, such as space exploration, military reconnaissance, rescue, and disaster relief, there are difficulties and dangers in human behavior. With the advancement of technology, mobile robots provide an efficient solution to the above pain points. But in the extreme conditions described above, terrain irregularities and ruggedness are common features of these environments. Therefore, to meet the requirements of the robot to replace the human operation in special environmental conditions, the performance requirements of the mobile robot are in the following order: Stability over obstacles > maneuverability > battery life > speed.[1][2]

To meet the high stability of mobile robots, hexapod robots came into being. The hexapod robot's unique multi-redundant limb structure supports the ground separately to achieve the crossing and avoidance of obstacles in complex terrain, This enables the hexapod robot to complete operations in unstructured environments that cannot be done by wheeled and tracked robots, and has high adaptability and passability to complex and special terrains. The mechanical structure of only hexapods cannot fully meet the high requirements for robot stability in extreme conditions. To improve the stability of the hexapod robot, Pan Xixiang et al. designed a hexapod wheel-legged robot with a spring buffer mechanism at the foot end, Liu Yiqun et al. proposed a foot-end trajectory planning method to reduce the impact of foot-ground contact, and Yao Mingchao et al. optimized the comprehensive performance by optimizing the length of the leg. However, for the optimization process, only the longitudinal comparison before and after the optimization of its body performance was considered, and the horizontal comparison of various optimization effects was not carried out. Therefore, based on the proposed suspension optimization based on magnetorheological dampers, this paper uses SolidWorks to model, Import ADAMS to establish a virtual prototype, and finally uses MATLAB to analyze the co-simulation scheme of the obtained data, By comparing the vertical speed indicators of the common air spring damper type suspension optimization and magnetorheological damper type suspension optimization before and after optimization, and horizontally, the shock absorption optimization results are compared, Thereby a set of practical shock absorption scheme is obtained. To provide a reference for the follow-up research on the hexapod robot.

2. Overall organization

Due to the structural design of the robot itself, such as the length of the joints of the legs and the weight of the body, the vibration of the body during travel will affect the Sperling stability index, so a hexapod robot is used in all kinds of simulations. The conceptual diagram of the specific structural relationship is shown in the following Figure 1.

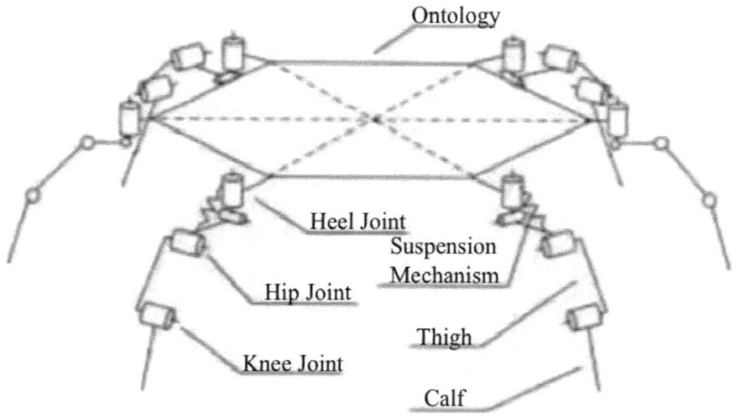

Figure 1. Schematic diagram of the driving structure and chassis of the hexapod robot

It is mainly composed of a body and six mechanical legs, and the legs are distributed around the body in an axis-symmetric manner, providing all-around flexibility. The mechanical leg is mainly composed of the basal segment, the femoral segment, and the tibia. The basal segment is connected to the bottom base of the robot and swings back and forth around the axis perpendicular to the bottom plane of the robot; the femur is connected to the basal segment and rotates around the axis perpendicular to the rotation axis of the basal segment; The shaft of the femur rotates on an axis parallel to the axis of rotation. To make the robot turn on the spot during the process of traveling, the chassis of the robot is designed as a regular hexagon, and the six legs of the robot's kinematic structure have the same structure and are evenly distributed on the six corners of the regular hexagon of the body.[3]

3. Structural design improvements

In this paper, through the improvement scheme with the suspension damping system, the independent suspension system of the robot is composed of six suspension mechanisms, each of which mainly includes a heel joint connection, a link hip joint connection, and a shock absorber. The suspension mechanism is shown in the Figure 2.

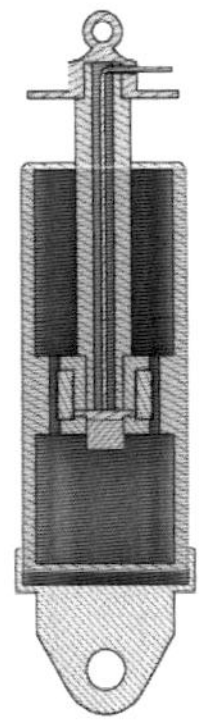

Figure 2. Structural sectional view of magnetorheological damper

Each suspension mechanism is installed between the heel joint of each leg of the robot and the main body, as shown in the Figure 3.

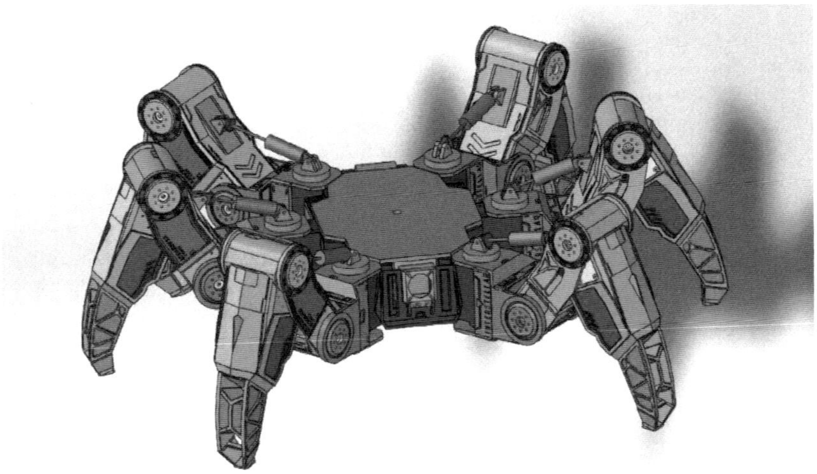

Figure 3. The final SolidWorks model rendering of the hexapod robot structure

In the natural straight state, there is a mechanical limit inside the shock absorber. At this time, its length is fixed and will not be elongated due to tension. Its internal structure is a damping system. This paper proposes to replace the type of damping system inside the shock absorber and compares the shock absorption effect of the spring damper suspension and the magnetorheological damper suspension on the traveling time of the hexapod robot.[4]

3.1. Dynamic model of spring damper suspension
In the process of establishing the dynamic model of the spring damper, its damping force characteristic is generally linear, so to ensure its effectiveness and stability in the control process, an accurate dynamic model of the damper model used will be established. The following Figure 4 is a simplified linear spring physical model established

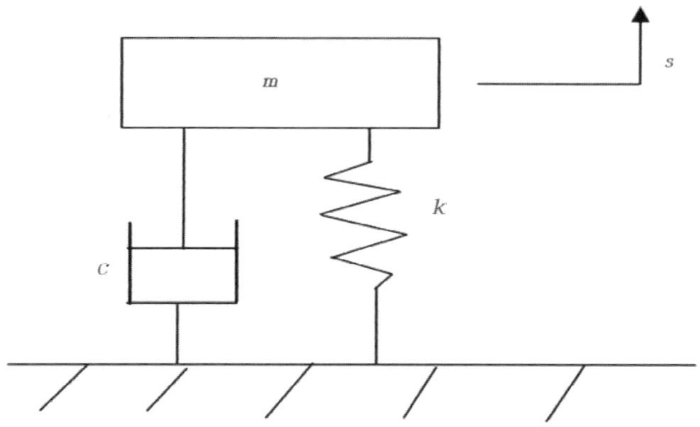

Figure 4. Schematic diagram of a single-degree-of-freedom spring system

During the movement of the spring, the damping force of the damper is always opposite to the movement direction of the spring block, as shown in the following formula:

$$F_c = -c\dot{s} \quad (1)$$

where c is the damping coefficient of the damper and is the derivative of the movement displacement of the spring block. When the spring is subjected to external action, in the linear range, the magnitude of the spring force is proportional to the deformation of the spring, and the direction of the spring force is always opposite to the direction of the deformation. The heat provided by the spring is as follows:

$$F_k = -ks \quad (2)$$

Where k is the spring coefficient of the damper, and s is the movement displacement of the spring block. The spring block itself has mass, and because of the acceleration, there is also a force, as shown below:

$$F_m = -ms \quad (3)$$

where m is the mass of the spring block, s is the second derivative of the motion displacement, According to Newton's law of motion, the force analysis of the spring block is carried out to establish a mechanical balance, and the equation of motion of the system can be obtained, As shown in the following formula:

$$m\dot{s} + c\dot{s} + ks = 0 \quad (4)$$

make

$$2d = \frac{c}{m} \quad (5)$$

$$\omega_0^2 = \frac{k}{m} \quad (6)$$

After some transformation of the above formula, the equation of motion is shown below. This is a second-order homogeneous linear ordinary differential equation. The characteristic root of the differential equation can be obtained first, and the solution of the equation can be obtained according to the properties of the differential equation.

$$\dot{s} + 2d\dot{s} + \omega_0^2 s = 0 \quad (7)$$

in,ω_0 is called the natural circular frequency of the system; The effect of damping on motion depends on the ratio of m and ω_0. This is called the damping ratio, as follows the formula

$$\Omega = \frac{d}{\omega_0} = \frac{c}{2\sqrt{mk}} \quad (8)$$

Build a simulation model for formula (7) in MATLAB/SIMULINK, formula (7) involves the differentiation of the function, Considering the complex usage conditions of the differential module in SIMULINK, Therefore, the simulation model is built in the form of integral, and the Simulink block diagram is shown in the following Figure 5.

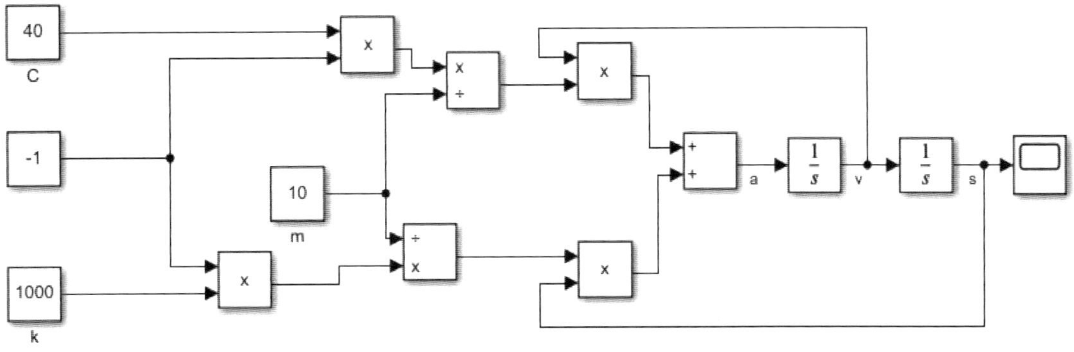

Figure 5. Simulink block diagram of the single-degree-of-freedom spring system module

This spring damping system will be used as a subsystem in the following co-simulation for experimental purposes

3.2. Magnetorheological damper suspension

3.2.1 Kinetic model. In the process of establishing the dynamic model of the magnetorheological damper, not only its good damping force characteristics, but also strong nonlinear hysteresis characteristics should be shown, Therefore, to ensure its effectiveness and stability in the control process, it is necessary to establish an accurate dynamic model of the damper.
Among them, the Bingham model is the most widely used and has a dynamic relationship.

$$F = F_i + F_t = \frac{12iLA_p^2}{\pi Dh^3} v + \frac{3Lt_yA_p}{h} sgn(v) \quad (9)$$

$$t_y = A_1 e^{-1} + A_2 \ln(I + e) + A_3 I \quad (10)$$

where F_i is the viscous damping force, v is the velocity between the piston and the cylinder, $F_t$is the Coulomb damping force, t_y is the shear yield strength, A_1 , A_2 , and A_3 are the coefficient of performance of magnetorheological fluid,I is the applied current, L is the length of the piston rod, D is the wall thickness of the outer cylinder, and h is the width of the liquid flow channel, $A_p$is the effective area of the piston.
In a magnetorheological damper, its damping force can be represented by the Bingham model

$$F_d = F_y sgn(\dot{x}_d) + C_{po}\dot{x}_d \quad (11)$$

In the formula, F_y is the yield force of the damperC_{p0}, is the damping coefficient of the damper after yielding, and x_d is the speed of the damper piston. However, the Bingham model can well simulate the response relationship between yield force and displacement, however, the fitting results of the relationship between yield force and velocity are not ideal. So Spencer introduced the Bouc-Wen link to better describe the mechanical properties of the magnetorheological damper, as shown in Figure 4. (Modeling and simulation analysis of magnetorheological damper based on semi-active control)
The mathematical expression of the Spencer model is:

$$F_d(t) = c_1\dot{y}_d + k_1(x_d - x_0) \quad (12)$$

In

$$\dot{y}_d = \frac{1}{c_0+c_1}[a_d\dot{z} + k_0(x_d - y_d) + c_0\dot{x}_d] \quad (13)$$

$$\dot{z} = -\gamma|\dot{x}_d - \dot{y}_d||z|^{n_d-1}z - \beta(\dot{x}_d - \dot{y}_d)|z|^{n_d} + A(\dot{x}_d - \dot{y}_d) \quad (14)$$

$$a_d = a_a + a_a u_d; c_1 = c_{1a} + c_{1b}u_d \quad (15)$$

$$c_0 = c_{0a} + c_{0b}u_d; \dot{u}_d = -\Omega(u_d - V_0) \quad (16)$$

where a_d, c_1, c_0 are functions of the applied voltage u to the damper, x_d is the instantaneous displacement of the magnetorheological damper piston, x_d is the instantaneous speed of the magnetorheological damper piston, z is the hysteresis characteristic value, and γ, β, and A are the adjustment parameters of the hysteresis curve.

3.2.2. Magnetorheological fuzzy pid control strategy for suspension system

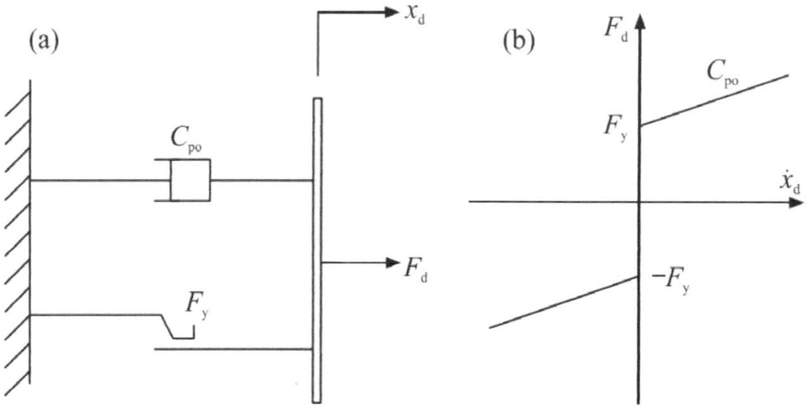

Figure 6. (a) Physical model of magnetorheological damper
(b) Relationship between damping force and shock absorber displacement

Fuzzy PID control is a compound control method that combines the principle of fuzzy adaptive control with classical PID. This control method takes the relative acceleration deviation e and the rate of change of the deviation ec as input, The magnetorheological current i is the output. The two sets of variables establish a mapping relationship through the processes of fuzzification, fuzzy reasoning, and defuzzification.

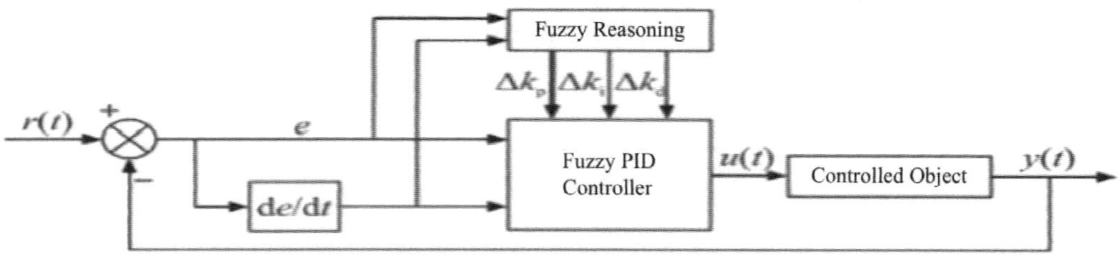

Figure 7. Schematic diagram of fuzzy PID control

Among them, the adaptive optimization formula of fuzzy PID controller gain parameters K_p , K_i , and K_d

$$\begin{cases} K_p = K_{p0} + \Delta K_p \\ K_i = K_{i0} + \Delta K_i \\ K_d = K_{d0} + \Delta K_d \end{cases} \quad (17)$$

In the formula, K_{p0}, K_{i0} , and K_{d0} are the initial parameters respectively; ΔK_p , ΔK_i , and ΔK_d are the changes of K_p , K_i , and K_d values obtained by fuzzy inference, respectively. When the input variables of the controller change irregularly, the fuzzy controller uses the function of fuzzy inference to adjust the gain parameters and establishes a suitable fuzzy rule table based on the summed-up expert experience. In addition, in the application of the fuzzy PID control algorithm, the Mandani algorithm of multi-rule, multi-input, and single-output fuzzy reasoning is used, and the precise calculation method of the center of gravity method is used. (i.e. de-fuzzification),Accurately calculate the parameters obtained by fuzzy inference to obtain the corresponding precise values, and then obtain the output control quantity through the relevant calculation.[5]

4. Build a simulation model

4.1 Solid works modeling[6]

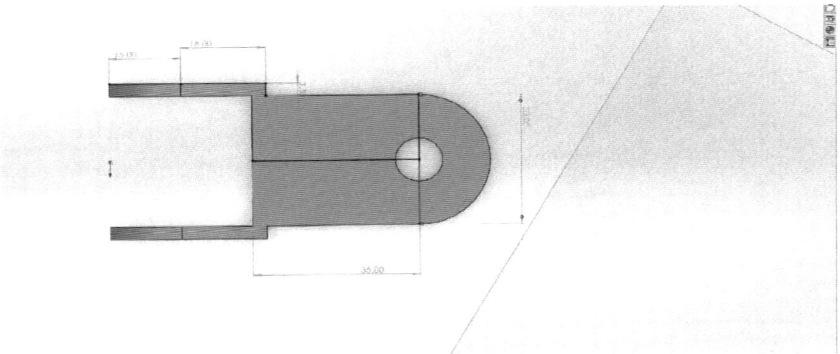

Figure 8. Model diagram of base section module

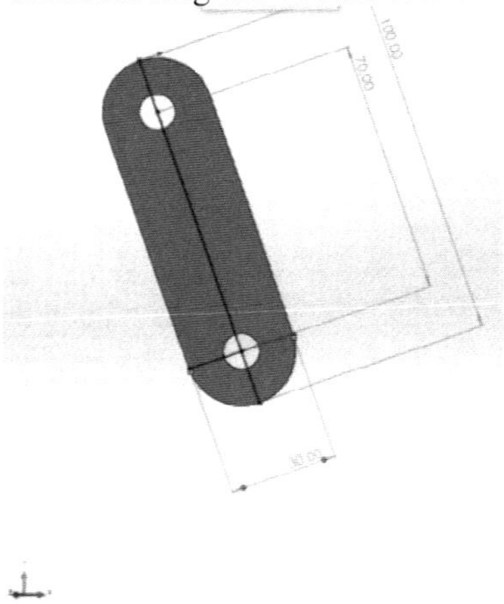

Figure 9. Model diagram of thigh module

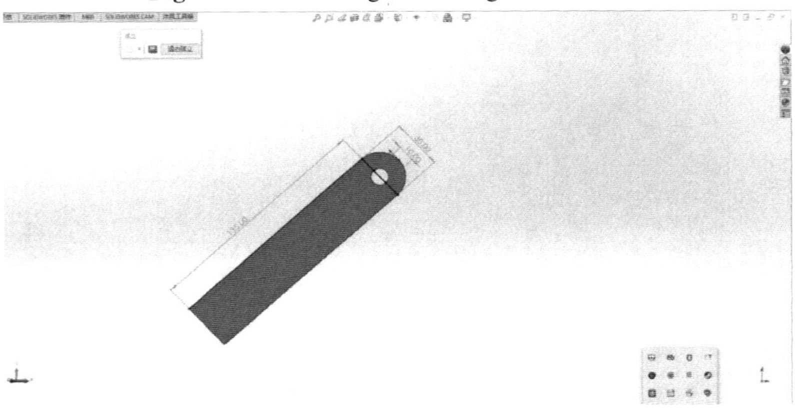

Figure 10. Model of the calf module

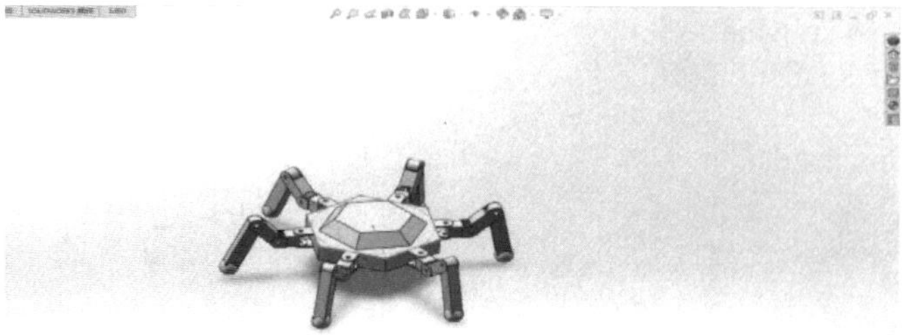

Figure 11. Simplified model after each module is combined

Table 1. Diagram of related parameters and some physical properties of the hexapod robot model

parameter name	Numerica l value	parameter name	Numerical value
Empty car body mass/kg	5.7	Chassis wheelbase/mm	490.0
Empty car body volume/cubic centimeter	422.3	Thigh length/mm	318.0
Base length/mm	80.0	Calf length/mm	444.5
Base section width/mm	33.0	Range of motion of hip joint/°	100.0
Femoral length/mm	100.0	Chassis height/mm	85.0
Section width/mm	30.0	Main material thermal conductivity W/(m*K)	0.2
Tibia length/mm	150.0	Impact strength of main material/KJ*m^-2	108.0
Tibia width/mm	30.0	Main material yield strength/MPa	70.0

4.2 Adams imports the model to create a virtual prototype

Import the mechanical structure in SolidWorks into ADAMS to get the simulation model:

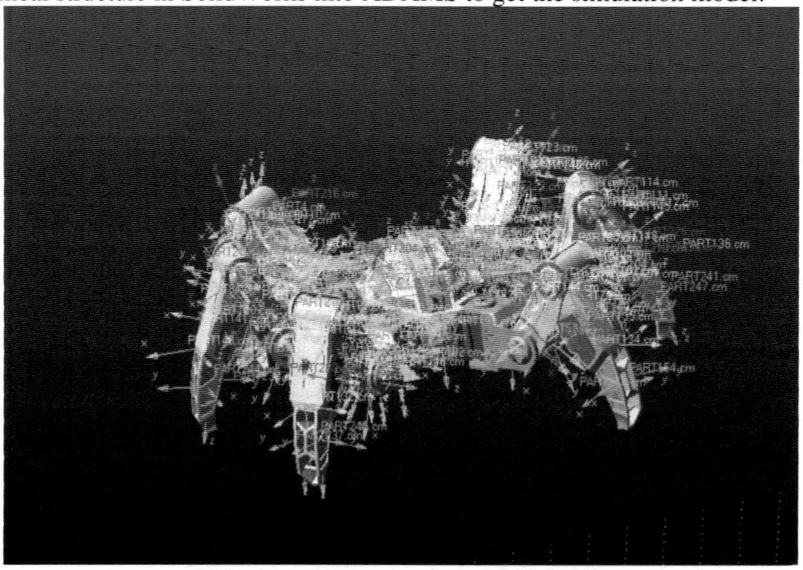

Figure 12. Import model diagram in ADAMS

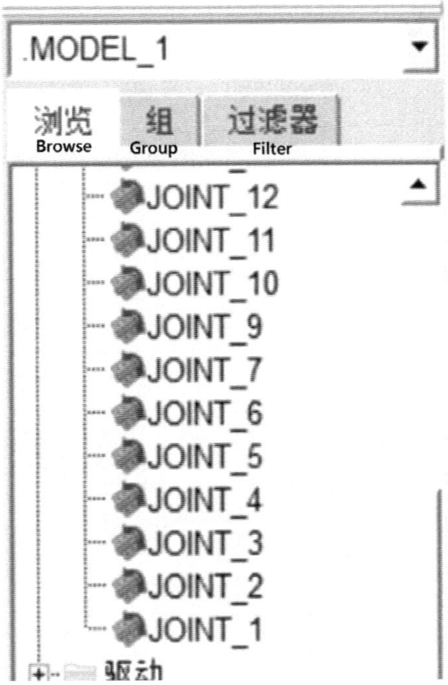

Figure 13. Each rotating pair and contact force setting of the model in ADAMS

4.3 Adams/Simulink co-simulation

In the simulation of the traveling process, the following three situations are analyzed: 1. The robot does not use a suspension system; 2. The robot uses a magnetorheological damping suspension; 3. The robot uses a spring damping suspension. At the same time, the simulation site element: the ground will be set, and the above motion simulation analysis will be carried out in the case of flat and rough ground, respectively. Then compare the effect of different forms of suspension on different ground conditions.

4.3.1 Build a shock-absorbing suspension simulation model in MATLAB. Establish the Simulink simulation module according to the mechanical model equation of the magnetorheological damper.

$$F = K_0 x + C_0 \dot{x} + F_d Z - f_0 \quad (18)$$

$$\dot{Z} = A\dot{x}[1 - Z sgn(\dot{x})] \quad (19)$$

The above figure is the mechanical equation of the magnetorheological damper. According to this equation, the simulation model of the corresponding shock-absorber characteristics in Simulink is established as shown in the figure 14.

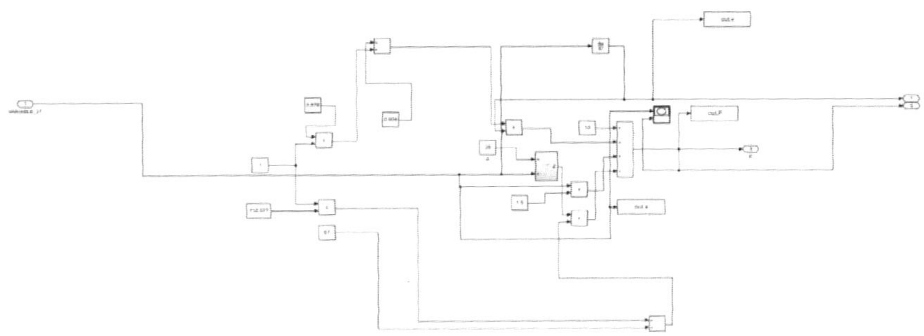

Figure 14. Simulink block diagram of the effect of magnetorheological dampers

Finally, connect the established model to the data input source of ADAMS, and use the speed of the simulation object as the output.[7]

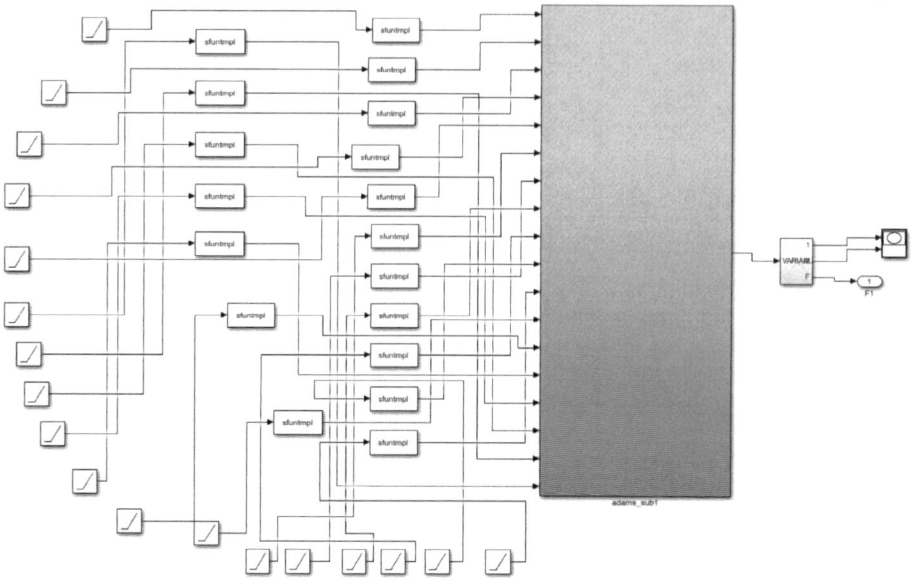

Figure 15. Block diagram of ADAMS/Simulink co-simulation implementation

The step signal (far left) is used as the initial input signal source of the hexapod robot's 18 joint rotation pairs, and then the S-Function class module is set to process the step signal. After processing, the signal enters the corresponding input interface of the adams_sys module and performs data exchange on ADAMS. The 16 input ports of the module correspond to the 16 input system units of the simulation model in ADAMS, and these signals will drive 16 joints. The internal function setting of the S-Function is obtained by analyzing the different joints of the hexapod robot through forward and inverse kinematics.

4.3.2 Forward and inverse kinematics analysis. Firstly, the single-leg coordinate system of the hexapod robot is established, and the base joint, thigh, and calf are respectively simplified into connecting rods of equivalent length, and the rotation center of mass of the hip joint is set as the coordinate origin. As shown in Figure 16.[8]

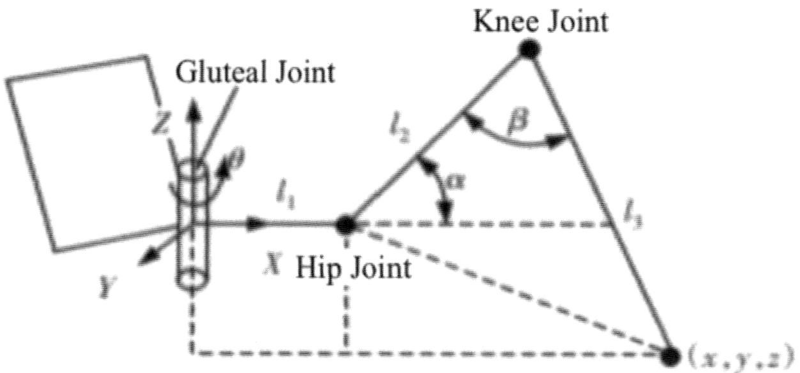

Figure 16. Schematic diagram of a single leg structure.

By giving the coordinates of the foot endpoint (x, y, z) and the lengths l_1, l_2, and l_3 of the basal segment, thigh, and calf, According to the inverse kinematics of the robot, the specific relationship expressions of the hip joint θ, hip joint α, knee joint β and each joint can be solved, As shown in formula (20) to formula (22). Assign the required joint rotation angle to the corresponding rotary joint drive respectively, Then, through the coordinates of the endpoints of each foot, debug and set the initial position of each leg of the hexapod robot, As the initial state of the three-legged gait simulation, Among them, α=30°, β=75°, each basal segment is symmetrically distributed in the center, The included angle between the basal segments is 60°.[9]

$$\theta = \arctan\left(\frac{y}{x}\right) \quad (20)$$

$$\alpha = \arctan\left(\frac{z}{\sqrt{x^2+y^2-1}}\right) + arccos\frac{l_2^2+(\sqrt{x^2+y^2}-l_1)^2+z^2-l_3^2}{2*l_2*\sqrt{(\sqrt{x^2+y^2}-l_1)^2+z^2}} \quad (21)$$

$$\beta = \arccos\left(\frac{l_2^2+l_3^2-[(\sqrt{x^2+y^2}-l_1)^2+z^2]}{2*l_2*l_3}\right) \quad (22)$$

Where: θ—hip joint angle; α—hip joint angle; β—knee joint angle; l_1—base segment length; l_2—thigh length; l_3—calf length. According to this, the driving function of each joint of the hexapod robot in the s-function is established.

4.3.3 Joint motion gait simulation. The advancement of the hexapod robot body is coordinated through the movement of each leg, and the robot gait refers to the process in which each leg of the robot moves in a certain order and trajectory. Common gaits are "3+3" gait, "4+2" gait, and "5+1" gait. Among them, "3+3" is the most common gait, so it is used for simulation analysis.[10]

By establishing an s-function function in MATLAB to drive the trajectory of each joint, the trajectory of the robot foot end satisfies the motion equation, and the overall forward motion trajectory of the robot is maintained. The simulation time is set to 10s and the number of simulation steps is 50. Through ADAMS simulation, the speed v of the robot center of mass in the y-axis direction is obtained in the flat environment and the rugged environment.

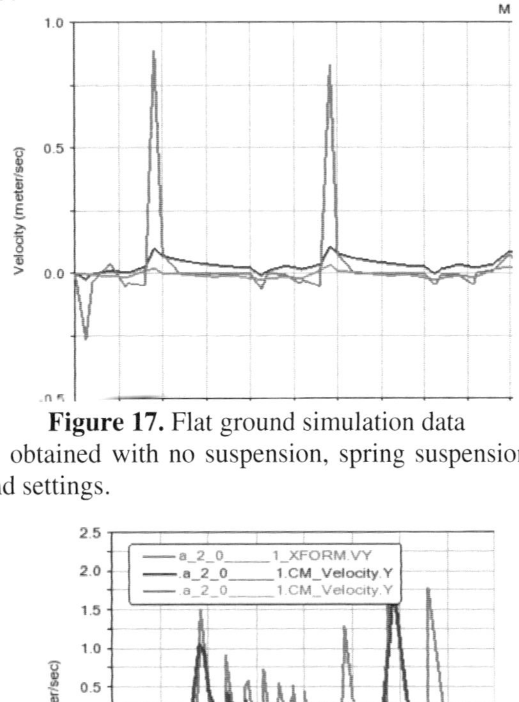

Figure 17. Flat ground simulation data

It shows the data curves obtained with no suspension, spring suspension, and magnetorheological suspension under flat ground settings.

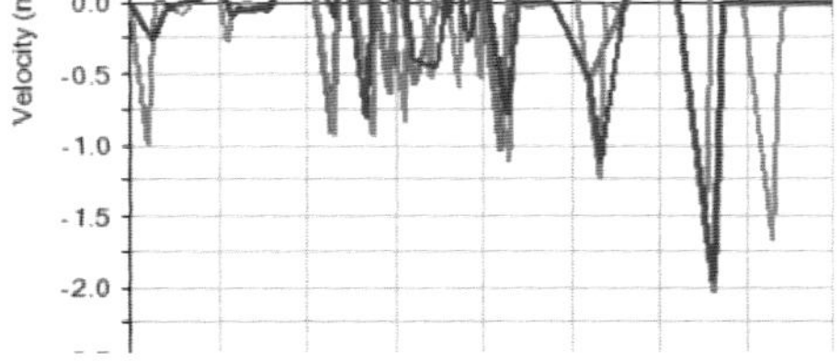

Figure 18. Rough ground simulation data

It shows the data curves of no suspension, spring suspension, and magnetorheological suspension under rough ground settings
By comparing the data of the two images, the red line is without suspension, the blue line is with spring suspension, and the green line is with magnetorheological suspension. When the hexapod robot moves under the parameter setting of rough ground, the body will be disturbed by greater resistance and impact, This is reflected in the larger and more frequent changes in speed. In the case of shock absorption by the suspension system, the vibration situation is greatly reduced compared with the state without suspension, which is reduced to 20% to 10% of the original. In the comparison of the shock absorption effect of the magnetorheological suspension and the spring suspension, it is obvious that the shock absorption effect is that the magnetorheological suspension is greater than the spring suspension, and the vibration amplitude is reduced by about half.

5. Conclusion

Combining the above-mentioned oscillation frequency and amplitude of the velocity of the center of mass of the hexapod robot with the three leg structures in the vertical direction in different environments, It can be concluded that among the three leg structures of blank, spring, and MR fluid, the hexapod robot with MR fluid leg structure is on flat ground or rough paths, The oscillation frequency of the velocity of the center of mass in the vertical direction is the lowest, and the amplitude is the smallest. It can be seen that the design of magnetorheological fluid shock absorption is very necessary, and it can play an important role in the stability of the hexapod robot. important role

6. References

[1] Raibert M,Blankespoor K,Nelson G et al.*BigDog,the rough-terrain quadruped robot*[J].IFAC ProceedingsVolumes,2008,41(2):10822-10825.

[2] Coelho Joana ,Ribeiro Fernando ,Dias Bruno et al.*Trends in the Control df Hexapod Robots:A Survey*[J].Robotics,2021,10(3):100

[3] Zhuang H C, Gao H B, Ding L, et al.*Method for analyzing articulated torques of the heavy-duty six-legged robot*[J].Chinese Journal of Mechanical Engineering,2013,26(4):801-812.

[4] Muller J, Schneider M, Hiller M.*Modeling, simulation, and model-based control of the walking machine ALDURO*[J].IEEE/ASME Transactions on Mechatronics,2000,5(2):142-152

[5] GÖKREM Levent, CAN Mehmet Serhat, AYDIN SefJa , *Hexapod Robot Design and Performance Comparison of Fuzzy and PID Control Methods*[J]. Balkan Journal of Electrical and Computer Engineering,2020,8(1):88-97

[6] Yizhao Hou, Guangyi Liu, Zhiyuan Wang, *Design and Application of Wheel-legged Hexapod Robot*[J].International Core Journal of Engineering,2021,7(12):733-736

[7] Luiz Fernando Pinto Oliveira, Flavio Luiz Rossini, *Modeling, Simulation and Analysis of Locomotion Patterns for Hexapod robots*[J].IEEE Latin America Transactions,2018,16(2):376-378

[8] Tomas Luneckas, *Analysis of Hexapod Robot Locomotion*[J].Science;Future of Lithuania,2011,2(1):37-39

[9] Dai Owaki, Masashi Goda, Sakiko Miyazawa, et al.*A Minimal Model Describing Hexapedal Interlimb Coordination: The Tegotae-Based Approach*[J].Journal of Intelligent & Robotic Systems,2017,11:39-46

[10] Zhou Zequan, Xu Fenghui, Jia Zhaidong, et al.*Research on gait design of Hexapod Bionic Robor under damaged condition*[J].Journal of Physics: Conference Series,2020,1550(4):3-5

Experimental Study Based on Fuel Nozzles Capable of Achieving Wide Fuel Supply

Jiaxin Wang*

School of Mechatronics Engineering, Shenyang Aerospace University, Shenyang, China

Corresponding author's e-mail: 1031526092@qq.com

Abstract. In order to overcome the limitation of fuel supply to the nozzle of an aero-engine, a nozzle with two main and secondary fuel paths and accompanied by air-assisted atomization can be used to solve this kind of problem. The two fuel paths make the fuel centrifugal and direct injection respectively, which can achieve the purpose of combining the advantages of centrifugal and direct injection nozzles, thus realizing a wide working range of fuel flow. The results show that when the fuel pressure reach a certain range, the main and secondary fuel lines start to work simultaneously, the atomization cone angle and fuel flow rate increase with the increase of pressure, and when the pressure reaches a certain value, continuing to increase the fuel pressure has little effect on the atomization angle. For the case of air supply and simultaneous operation of the main and secondary fuel lines, with the increase of fuel pressure, the atomization particle size changes gradually and evenly and has a tendency to increase, which is in line with the expected goal.

1. Introduction

The fuel nozzle is the core component of the combustion chamber of the aero-engine, and has a great impact on the performance of the combustion chamber [1]. The atomization effect of the fuel nozzle mainly depends on the working condition of the nozzle, and the performance of the fuel nozzle can determine the sprayed atomization particle size, the smaller the atomization particle size and the larger the atomization cone angle means the atomization effect is better, and also means the fuel combustion effect is better. If the injection of fuel oil beam is too concentrated, it will produce black smoke, and in serious cases will also occur ablation flame barrel and other bad conditions [2-4]. The performance of fuel nozzles is a key issue in the development of aero engines. Dual fuel nozzles can provide a wide operating range of fuel flow and small atomization particle size, and are widely used in the combustion chamber of aero engines [5]. Liu Qi et al. [6] conducted spray experiments on aero-engine nozzles and also quantified the spray characteristics such as atomization cone angle and liquid film breaking length; Wang Honglian et al. [7] conducted a simulation study on the atomization effect of the nozzle and concluded that the sprayed atomized droplets were cone-shaped distribution without air assistance; Durdina et al. [8] applied the phase Doppler analyzer to obtain the The particle size values of the sprayed fuel spray were obtained by applying a phase Doppler analyzer, and the droplet size distribution was measured, and it was found that the atomized particle size near the axial center of the conical spray was smaller than the droplet size at the edges; Ashgriz et al. [9] studied the process of liquid film fragmentation by using high-speed camera technology, and concluded that the liquid film was first fragmented into a ribbon-like structure and then into fine droplets. Sivakuma et al. [10] used high-speed camera observed the process of polymerization and separation of the inner and outer liquid film, and concluded that the polymerization point of the liquid film shifts outward and vice versa when the momentum ratio of the liquid film in the main and secondary oil paths is increased.

The nozzle under study has two air-assisted channels and two oil paths, which make the fuel centrifugal and direct injection respectively. The nozzle can well combine the advantages of centrifugal and direct injection nozzles and can achieve a wide operating range of fuel flow. There is little literature on the atomization characteristics of this structure type of dual oil path nozzle, here based on this dual oil path nozzle, the atomization cone angle and atomization particle size change law when the oil pressure is large and small, respectively, to provide a reference basis for this type of nozzle research.

2. Research object and test system

2.1. Research subjects

The main structure of the dual oil circuit nozzle is nozzle housing, fuel nozzle, cyclone, oil filter, spring and spring support, and the structure design diagram, as shown in Fig. 1. When the oil pressure is small, the main fuel line works alone and the fuel is centrifugally sprayed at the nozzle of the main fuel line. When the fuel enters the centrifugal nozzle, a hollow oil film is formed at the outlet of the nozzle by the centrifugal force, and the oil film is destabilized into fine droplets by the combined action of the oil film surface wave and aerodynamic force [11]. When the oil pressure is large, depending on the pressure of the oil, the spring inside the nozzle is pressed down by the oil pressure, making the secondary oil channel is opened, the main and secondary oil channels can work at the same time, the fuel is sprayed straight out at the nozzle of the secondary oil channel, the multi-hole nozzle can also make the spray process faster [12]. The nozzle works with air-assisted atomization, regardless of whether the oil pressure is high or low. The fuel is aerodynamically influenced in the cyclonic air ring cavity of the direct injection air atomizing nozzle and sprayed into the combustion chamber under the action of high-speed airflow, which makes the droplets have good penetration and good atomization, in addition to reducing the complexity of the nozzle system to a large extent [13]. In order to determine the atomization performance of the nozzle can meet the requirements of the actual work, spray test to test its atomization effect is essential, including the measurement of atomization particle size and atomization cone angle, but also to analyze the factors affecting its atomization characteristics.

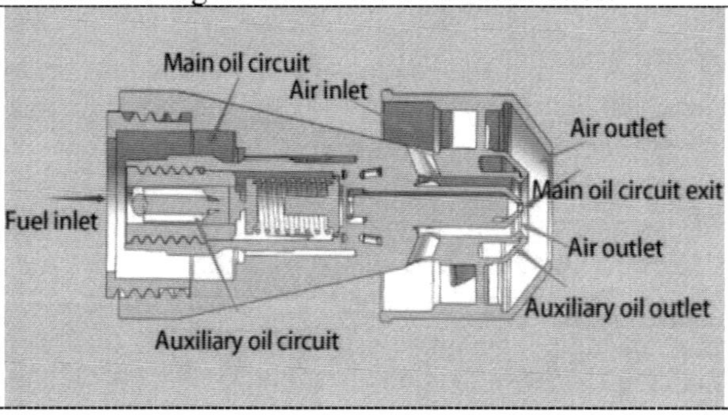

Figure 1. Nozzle structure diagram.

2.2. Test system

The atomization effect of the nozzle at different oil pressures can be investigated based on the laser fuel test system shown in Fig. 2, where different oil pressures are set to yield atomization particle size and spray cone angle. The atomization particle size was measured using a phase Doppler particle analyzer from TSI (USA) (the equipment consists of: laser emitter, laser receiver, signal processor, and grating rotation motor controller). The emitter consists of a standard 2 W nitrogen-neon laser source, a focusing lens, and a spectroscopic grating [14]. The test is carried out at room temperature and pressure, and the test system consists of an atomization test stand, a fuel supply system, a measurement system and a fuel recovery system.

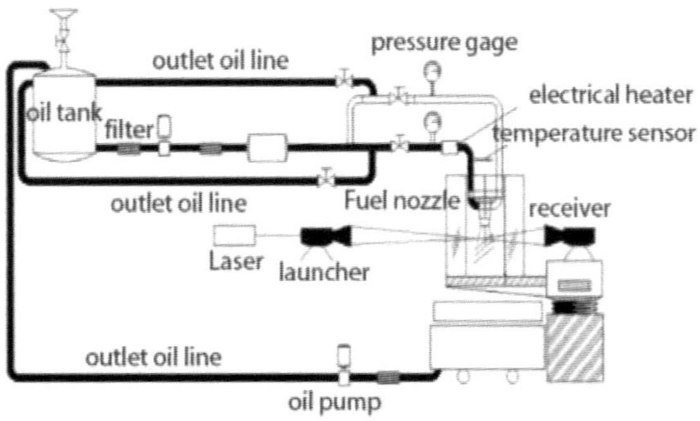

Figure 2. Test content.

3. Test Methodology

The general parameters used to characterize the atomization effect of aviation fuel are spray cone angle, liquid film breakage length, average diameter of soltaire, and droplet velocity distribution. The larger the spray cone angle, the smaller the liquid film break-up length, the smaller the average diameter of the soltaire, and the larger the droplet velocity, the better the atomization effect [15]. A high-speed camera system can be used to photograph the entire atomization process of the fuel being sprayed from the nozzle to the flow field, forming liquid film, liquid filament and finally droplets. The test is divided into two parts, one is to measure the fuel flow rate and atomization particle size at different pressures. The purpose of the fuel flow measurement is to know when the fuel pressure reaches a value where the flow starts to increase significantly and the main and secondary fuel lines start to work at the same time. The [16] size of the droplet atomization particle size will be related to the hot spot of the outlet temperature field and the smoke level. First, we adjust the appropriate supply pressure, measure the supply volume with a float flow meter, and then adjust the different supply pressures. The measurement range of the atomization particle size is 0.5-600 μ m, and the experimental effect is shown in Fig. 3. The other part is the measurement of the atomization cone angle at different pressures. The measurement was carried out by taking pictures of the atomization of the jet and using the software to identify the angle. Considering the rolled suction of the surrounding air when the liquid mist is jetted, the spray cone angle can be measured by the mechanical probe method, and the position coordinates at the intersection of the spray wrap line through the mechanical probe at a distance of 50 mm from the lower end surface of the nozzle are converted as the calculation method of the spray cone angle, as shown in Fig. 4 below.

Figure 3. Experimental diagram of SMD measurement of atomized particle size.

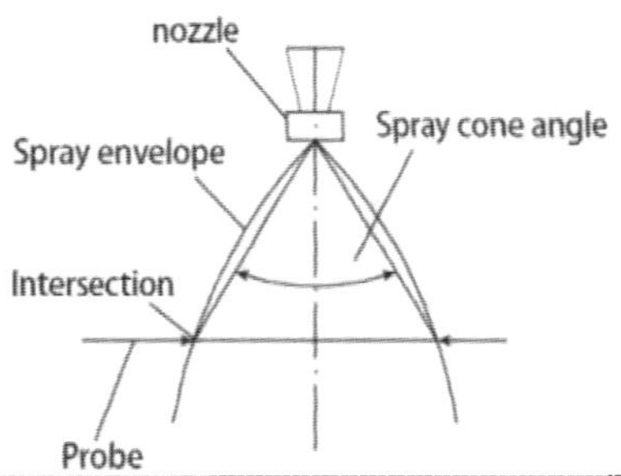

Figure 4. Schematic diagram of atomization cone angle measurement method.

4. Experimental results and analysis

In order to obtain the fuel supply characteristics and atomization of the nozzle, several fuel supply pressures were selected for the test, and the fuel pressure was set to 0.3MPa, 0.6MPa, 0.9MPa, 1.2MPa, 1.5MPa, 1.8MPa, 2.1MPa, 2.4MPa, 2.7MPa, and 3.0MPa, respectively. the variation curve of fuel flow with pressure is shown in Figure 5 below as shown in Fig. 5, when the fuel supply pressure is at the position of A (about 1.2MPa), the flow rate increases sharply, and the valve of the secondary fuel line is opened under the action of fuel pressure, and the main and secondary fuel lines start to work simultaneously, and the fuel flow rate supplied by the secondary fuel line is obviously higher than that of the main fuel line. This shows that when the fuel pressure is greater than 1.2MPa, the main and secondary fuel lines are supplied simultaneously, and conversely, when the fuel pressure is less than this value, only the main fuel line is supplied separately.

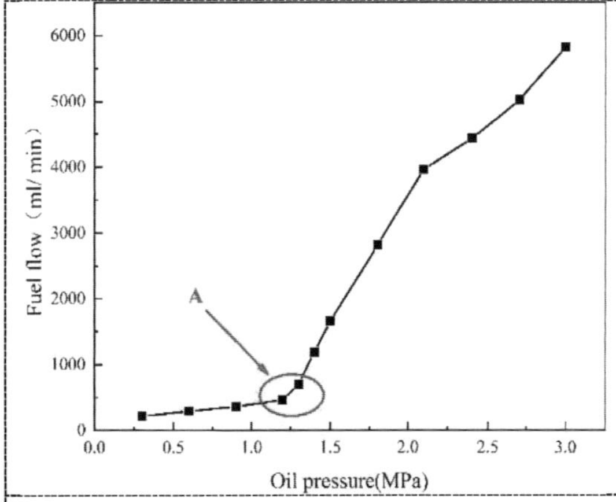

Figure 5. Variation curve of flow rate with pressure.

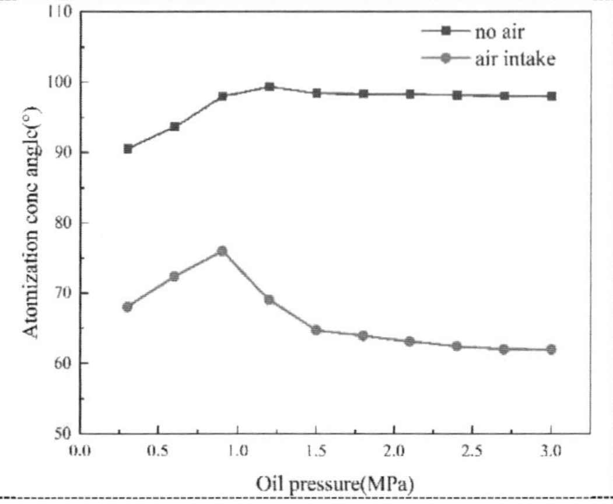

Figure 6. Variation curve of atomization cone angle with pressure.

In actual operation of the aero-engine, if the atomization cone angle of the nozzle spray is too large, the oil droplets will penetrate the most turbulent airflow area thus leading to poor oil and gas mixing[17] ; if the atomization angle cone angle is too small, the droplets cannot spread to the whole combustion chamber, which will make the air and fuel not fully mixed, and the fuel will be concentrated to the oxygen-deficient reflux area, thus forming a large amount of thermal decomposition. Study of spray cone angle at different fuel pressures, considering two cases when no air is added and when air is added, the size of the atomization cone angle mainly depends on the liquid viscosity and surface tension, as well as the degree of turbulence of the jet [18], fuel nozzle with or without the action of auxiliary air atomization droplet distribution are cone-shaped distribution. Increased pressure will increase the axial velocity of the droplets and lead to an increase in the droplet spray distance, fuel in the spray will affect the movement of the gas, which affects the droplet spray trajectory, resulting in a smaller atomization cone angle. The change curve of atomization cone angle with pressure is shown in Fig. 6 below, through the trend of change can be seen, when no air is added, the atomization cone angle will increase at first, the reason for the change of cone angle at this time is that when the fuel supply pressure is low, the fuel flow is small, at this time the atomization cone angle is positively correlated with the fuel supply pressure, when the fuel pressure continues to increase to a certain value, the atomization cone angle tends to be stable, this is due to the main fuel circuit centrifugal spraying atomization cone angle dominates dominant position, the pressure continues to increase, the rate of increase in fuel flow in the main fuel circuit becomes slower. When air is introduced, the air circuit provides a sufficient amount of atomized air, and the sprayed droplets come together in the middle, making the overall atomization cone angle lower than the atomization cone angle when no air is added. Initially, the atomization cone angle gradually becomes larger, and then the spray cone angle gradually decreases, and the cone angle changes more smoothly throughout the process [19]. This is due to the increase in oil supply pressure will lead to an increase in the speed of the droplets ejected from the nozzle, which will increase the negative pressure of the air outside the nozzle, and then the spray cone angle decreases.

The variation of atomization particle size at different pressure is studied, and the variation curve of atomization particle size with pressure is shown in Fig. 7 below. In the initial stage, the atomization particle size is decreasing with the increase of fuel pressure when the main fuel line is supplied alone. This is because the increase of pressure will increase the speed of fuel leaving the nozzle, while the air in the environment is at rest, the velocity difference between gas and liquid becomes larger, and then the effect of shear force becomes obvious, and the droplets are fully broken under the disturbance of air, and the atomization particle size decreases. As the pressure continues to increase, the secondary oil circuit also began to work, the main and secondary oil circuit work at the same time, the size of the atomization particle size increased significantly. Not pass into the air, when the atomization particle size increased to a certain value, atomization particle size began to decrease, and nozzle particle size reduction is also slowing down, the reason for this phenomenon is that when the pressure is small, the primary atomization dominant role, increasing pressure on the role of primary atomization is more obvious; with the increase in pressure, the role of primary atomization is limited, the core flow area is gradually stabilized, while the role of secondary atomization The effect of secondary atomization begins to show, because the external air flow rate is low, the effect of secondary atomization is still lower than that of primary atomization[20] . When air is introduced to assist, the particle size gradually becomes uniform and larger as the fuel pressure rises under the simultaneous operation of the primary and secondary fuel lines, due to the collision, fragmentation and aggregation of small droplets under the action of air pressure, so the atomization particle size gradually increases.

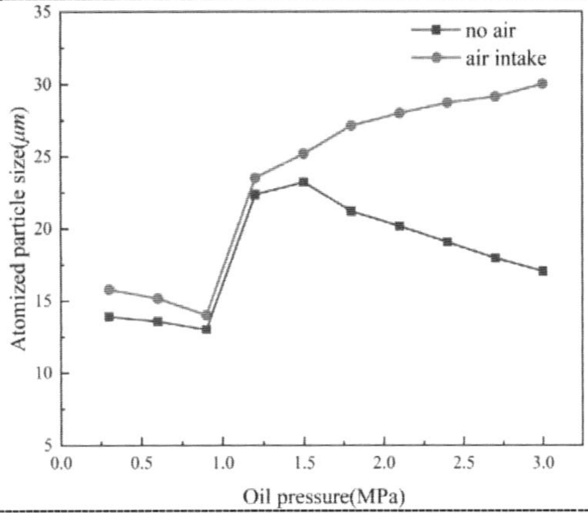

Figure 7. Variation curve of atomization particle size with pressure.

5. Conclusion

(1) When the fuel pressure reaches 1.2MPa, the main and secondary fuel lines are supplied at the same time, and when the fuel pressure is less than this value, only the main fuel line is supplied separately.
(2) Without air, with the increase in pressure atomization cone angle also increases, when the fuel pressure continues to increase to a certain value, the atomization cone angle tends to stabilize; when the air is introduced, the air circuit provides a sufficient amount of atomized air, the sprayed droplets to the middle, the overall atomization cone angle is lower than the atomization cone angle when air is not added.
(3) In the initial stage, when only the main fuel line is supplied, the atomization particle size decreases with the increase of fuel pressure. When the main and secondary fuel lines work at the same time and air is introduced, the size of atomization particle size tends to stabilize and become larger as the fuel pressure increases.

6. References

[1] Zhang T, Dong B, Chen X. Spray characteristics of pressure-swirl nozzles at different nozzle diameters. *Applied Thermal Engineering*,2017, pp. 984-991.
[2] Liu Q, Xia Z, Huang Z, Qian Y, Gou De-Hao. Macroscopic spray characteristics of centrifugal nozzles for aero engines. *Propulsion Technology*, 2021, pp. 362-371.
[3] Ma H, Zhang B, Liu K. Comparative experimental study of fuel nozzles for a type of aero-engine. *Aeroengine*, 2009, pp. 54-57.
[4] Han Z, Parrish S, Farrell P V. Modeling atomization processes of pressure-swirl-hollow-cone fuel spray. *Atomization and Sprays*, 1997, pp. 663- 684.
[5] Gan X. Aviation Gas Turbine Fuel Nozzle Technology. *Beijing: National Defense Industry Press*, 2006.

[6] Liu Q. Experimental and numerical simulation of n-decane spray from centrifugal nozzle of aero-engine. *Shanghai Jiao Tong University*, 2020.
[7] Wang H, Zhao D, Shao F, Zheng G. Simulation of centrifugal fuel atomization nozzle for gas turbine. *Gas Turbine Technology*, 2020, pp. 17-23.
[8] Durdina L, Jedelsky J, Jicha M. Investigation and Comparison of Spray Characteristics of Pressure-Swirl Atomizers for a Small-Sized Aircraft Turbine Engine. *International Journal of Heat and Mass Transfer*, 2014, pp. 892-900.
[9] Ashgriz, Nasser. Handbook of Atomization and Sprays. US: Springer, 2011.
[10] Sivakumar D, Raghunandan B N. Formation and separation of merged liquid sheets developed from the mixing of coaxial swirling liquid sheets. *physics of Fluids*, 2003, pp. 3443-3451.
[11] Huang Y, Lin YZ, Fan WJ. Combustion and combustion chambers. *Beijing: Beijing University of Aeronautics and Astronautics Press*, 2014.
[12] Sanghoon L, Sungwook P. Spray atomization char-acteristics of a GDI injector equipped with a group-hole nozzle, *Fuel*, 2004, pp. 50- 59
[13] Zhang Z, Zhang RW, Xu GQ, Lin YZ. Atomization effect of direct-jet dual-cyclonic air atomization nozzle. *Journal of Aerodynamics*, 2006, pp. 805-809.
[14] Yu H, Liu A, Chen S, Liu K, Zeng W. Study on the effect of structural parameters on atomization characteristics of pneumatic atomization nozzle. *Journal of Shenyang University of Aeronautics and Astronautics*, 2020, pp. 1-9.
[15] Yan H, Chen F-Z. Research progress on fuel atomization characteristics of aero-engine. *Propulsion Technology*, 2020, pp. 2038-2058.
[16] Zhang Z, Wang C, Lin YZ. Combustion characteristics of self-stabilized double-rotor cups in unrestricted space. *Journal of Aerodynamics*, 2008, pp. 1778-1782.
[17] Zhang S. Experimental study of fluid-dynamic ultrasonic nozzle. *Beijing University of Technology*, 2007.
[18] Wang W Y, Zhu S Z. An experimental study on the atomization of a cyclonic nozzle. *Fire Science and Technology*, 2021, pp. 1222-1224+1262.
[19] Wang Y, Wang X, Li Z. Study on atomization characteristics of aero-engine nozzles based on image recognition technology. *Machine Tools and Hydraulics*, 2010, pp. 52-55.
[20] Wu J, Chang G, Qian X. Mechanistic study on the effect of pressure on atomization of direct injection nozzle. *Modern Machinery*, 2019, pp. 46-49.

Parametric Research and Optimization of Ducted Propeller Aerodynamic Configuration

Zeyu Li, Jianghao Wu, Haiying Lin

Beihang University, Beijing 100191, P.R. China

haiying@buaa.edu.cn

Abstract. In this paper, the ducted propeller configuration is optimized using the geometry parameterization method and proposed Newton method with the total system thrust as the optimization objective and the maximum torque as the constraint. Based on the initial model, the propeller geometry is optimized by decoupling the ducted propeller, and iterations are carried out until the optimization converges. The propeller optimization primarily consists of an increase in the thickness of the propeller root wing and the torsion angle of the middle wing, and a reduction in the curvature of the propeller tip. The ducted propeller system has a 6.29% improvement in thrust after aerodynamic optimization.

1. Introduction

Ducted unmanned aerial vehicles have the capability of vertical take-off and landing and high hovering efficiency [1]. Studying the influence of ducted propeller geometric parameters on its flight performance, mastering the mutual coupling relationship between the duct and the propeller [2], and improving the aerodynamic efficiency of the duct and propeller geometric shapes by optimizing the design [3] are the key points to improving the aerodynamic performance of the ducted propeller system [4].

The modification of the leading-edge lip shape allows the ducted propeller thrust to be enhanced [5]. And a preferred interval of duct expansion angle exists to enhance the duct thrust [6][7]. The application of the lobe element theory [8] and the lamellar strip theory could somewhat simplify the optimization iterations of the ducted propeller [9] but lack adaptation improvements for the duct flow field. The response surface [10] and neural network models have practical optimization design enhancements for specific duct geometry profiles [11].

This paper proposes an optimized design process with decoupled duct and propeller based on the gradient method with a ducted propeller as the primary research object. Based on the standard MATLAB-Ansys simulation optimization method, the optimized design of ducted propeller is accomplished with the improvement of thrust performance.

2. Numerical Method and Verification

2.1. Numerical validation

This paper implements Computational Fluid Dynamics (CFD) using ANSYS Fluent 19R1 Test experiments are carried out based on a standalone propeller to verify the computational model. A six-component box-type strain gauge balance was used for measuring the thrust and torque, and a Hall coil signal was applied to obtain the rotational speed with a 1% speed error. A DC power supplier was used to control and measure the input voltage and current to calculate the power, considering a 95% efficiency of the motor system. Numerical results of power and thrust are displayed in Table 1, compared with the obtained experimental data. The calculated thrust is underestimated by about 5% and the power by 4% but within the measurement uncertainty.

Table 1. Comparisons of aerodynamic characteristics between experimental and numerical results.

Speed/rpm	CT(Exp)	CT(Num)	CP (Exp)	CP (Num)
3500	0.13345881	0.12831064	0.057968079	0.056116903
4000	0.135845447	0.132590942	0.056401885	0.054686729
4500	0.128334936	0.129049231	0.055522924	0.053075317
5000	0.128521605	0.1292159	0.052784767	0.052482951
5500	0.129021446	0.127510447	0.054062892	0.05235113

2.2. Baseline design and numerical model

The ducted propeller is used as the power system of the UAV, as shown in Figure 1. The propeller design adopts the radially occupied variable airfoil method with the airfoil of ARA-D6%, ARA-D10%, and ARA-D13%, with the pitch angle ranging from 52° to 20° from 25% radial position to the tip.

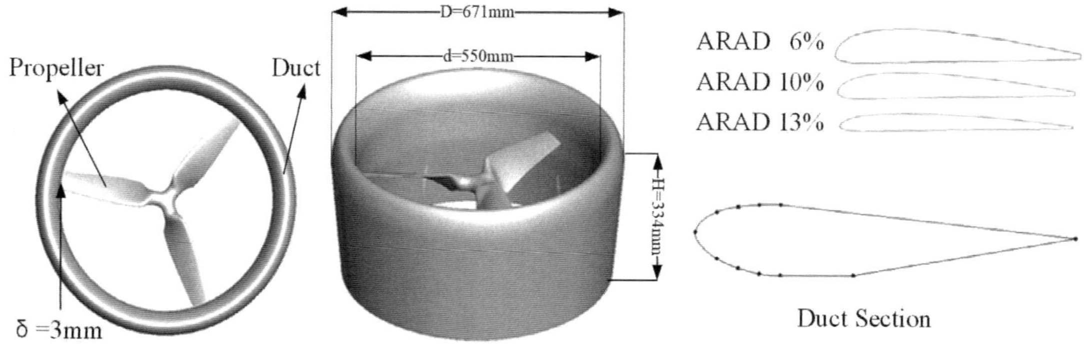

Figure 1. Numerical model in of baseline design.

In this paper, an MRF model is used to calculate the aerodynamic performance of the ducted propeller. The entire calculation basin is shown in Figure 2. The propeller diameter is denoted as D. The basin consists of two cylinders. The outer flow field extends radially to 4D. The rotating region is axially 6D away from the inlet and 10D away from the outlet.

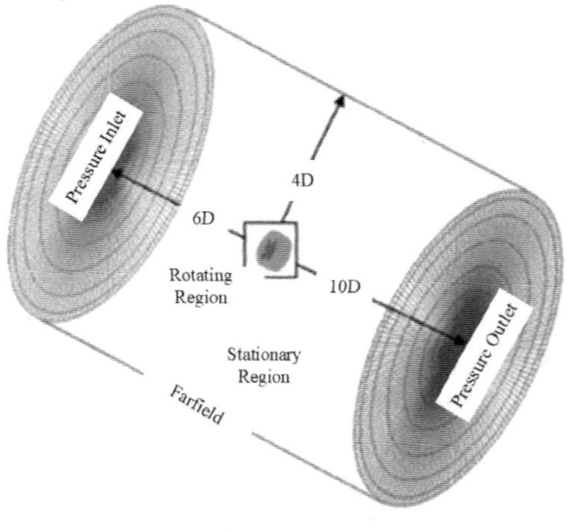

Figure 2. Computational region of the ducted propeller.

The meshes of stationary and rotating regions are generated by ANSYS ICEM CFD separately. They are densified at the leading-edge and trailing-edges of the duct, propeller blade tip and propeller blade, and inner duct wall to improve the calculation accuracy. The meshing of the propeller and the outflow field is shown in Figure 3.

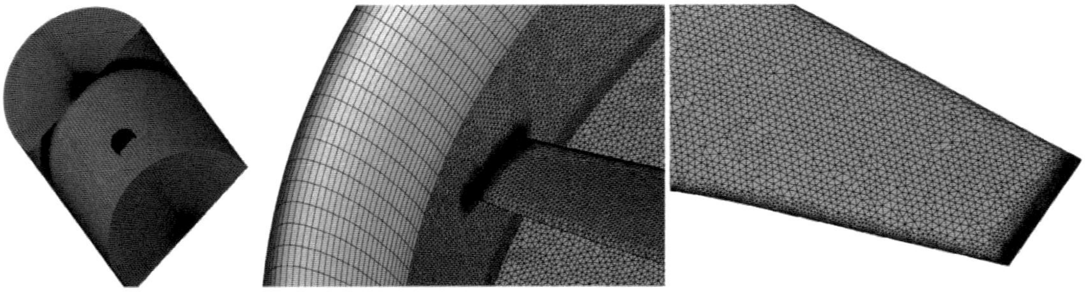

Figure 3. Schematic of mesh system.

2.3. Grid-independent validation

The grid volumes of 4.7 million, 6.2 million, 9.5 million, and 11.7 million grids were used for the calculation, and the results of the aerodynamic characteristics of the four scale grids were calculated as shown in Table 2. Considering the accuracy of calculation results and the limitation of calculation time and resources, the grid density chosen in this paper is 9.5 million.

Table 2. Comparison of aerodynamic results with different mesh sizes.

Mesh number	CT	CM	CP	Efficiency
4.7 million	0.45352	0.03876	0.24298	1.86648
6.2 million	0.45684	0.03832	0.24079	1.89726
9.5 million	0.46218	0.03785	0.23784	1.94326
11.7 million	0.46271	0.03773	0.23708	1.95165

2.4. Aerodynamic performance parameters

The ducted propeller thrust direction is aligned with the incoming flow direction. Concerning the propeller aerodynamic evaluation, its aerodynamic performance is judged by considering its dimensionless thrust, torque, power, and efficiency parameters.

Thrust coefficient:

$$C_T = \frac{T}{\rho n_s^2 D^4} \tag{1}$$

Torque coefficient:

$$C_M = \frac{M}{\rho n_s^2 D^5} \tag{2}$$

Propeller power:

$$P = 2\pi M \times n_s \tag{3}$$

Power coefficient:

$$C_P = \frac{P}{\rho n_s^3 D^5} \tag{4}$$

The figure of merit:

$$FM = \frac{P_i}{P} = \frac{P_i}{\kappa P_i + P_0} = \frac{C_T^{\frac{3}{2}}}{\sqrt{2} C_P} \tag{5}$$

Disk load:

$$DL = \frac{T}{A} \tag{6}$$

Power load:

$$PL = \frac{T}{P} = \sqrt{\frac{2\rho}{DL}} \times FM \tag{7}$$

FM was generally considered to reflect the propeller efficiency but only applicable to different propellers with the duplicate DL. In the case of κ and P0 being the same, the greater the propeller with the more significant disc load, the greater the *FM*, at which point the two propellers produce the same pull, while the more efficient propeller consumes more of the total power. For a propeller with the same thrust coefficient and power coefficient, the greater the speed, the higher the thrust (proportional to the speed squared), and the higher the propeller disc load, but its power load is lower because the power is proportional to the speed cube.

Therefore, the efficiency can be expressed by the thrust ratio, i.e., the thrust, to the power that a ducted propeller can generate at a given power. The balance of the thrust coefficient to the power coefficient, *CT*/*CP*, a dimensionless quantity, is considered for judging the propeller efficiency.

3. Geometric Parameterization and Optimization Methods

3.1. Propeller geometry parameterization

The two-dimensional shape of the control section airfoil is determined by a CST-type function. Generating the propeller blade model is implemented through MATLAB and ICEM script files. The airfoil cross-section is represented by 1-7 in order from the propeller centre outward, and the leading edge points of the control airfoil cross-section are represented by A*i*(*i*=1,2,3...7), while the trailing edge points are represented by B*i*, C*i*(*i*=1,2,3...7). The equations of the relationship between the coordinates of the leading-edge and trailing-edge points of the guideline and the geometric shape parameters such as the airfoil chord length c*i*(*i*=1,2,3...7), the span b*i*(*i*=1,2,3...7) and the torsion angle θ*i*(*i*=1,2,3...7) of the propeller are as follows.

$$\begin{aligned} b_i &= y_i \\ c_i &= \sqrt{\left(x_{Ai} - x_{Bi}\right)^2 + \left(z_{Ai} - z_{Bi}\right)^2} \\ \theta_i &= \arctan\frac{x_{Ai} - x_{Bi}}{z_{Ai} - z_{Bi}} \end{aligned} \tag{8}$$

Based on the type function weights of the base shape function, the consequences are perturbed to change the airfoil shape of the control section.

$$\begin{aligned} y_{\text{new}} &= y_{\text{old}} + \Delta y(x) \\ \Delta y(x) &= \sum_{i=1}^{n} \delta_i f_i(x) \end{aligned} \tag{9}$$

Where δ_i is the amount of change in the weights of each type function.

The CST method represents the wing types by superimposing Bernstein polynomials, and any nth-order Bernstein polynomial consists of the following n terms. The images of the propeller section and Bernstein polynomial are shown in Figure 4.

$$S_{r,n}(x) = \frac{n!}{r!(n-r)!} x^r (1-x)^{n-r} \tag{10}$$

Where r=0,1,...,n.

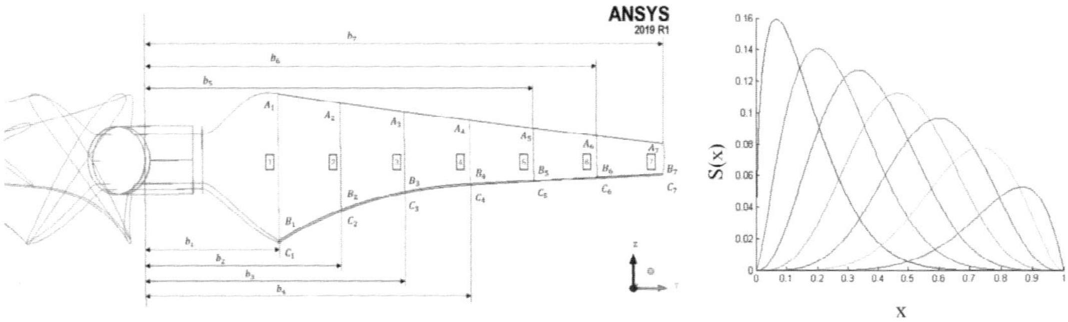

Figure 4. Propeller section position and Bernstein polynomial.

The airfoils of the 1st, 4th and 7th cross-sections are selected as the optimization objects. A total of 42 optimization variables are used, and the fitting effect of the initial airfoil shape is shown in Figure 5.

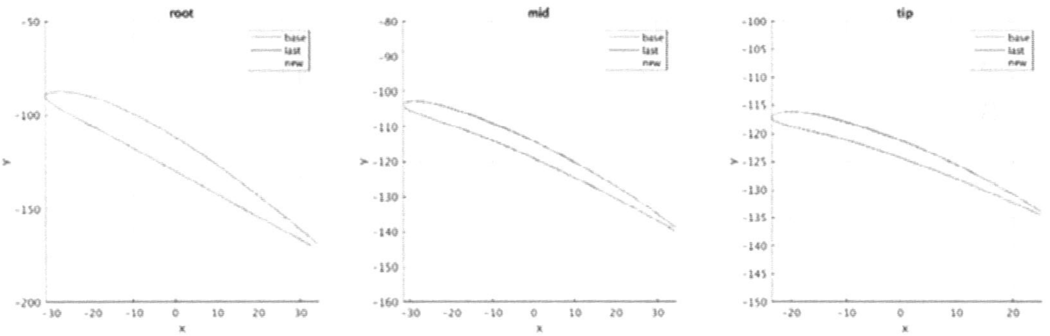

Figure 5. Propeller section airfoil fitting.

3.2. Optimization algorithm

The proposed Newton method is described with the following iteration formula.

$$x^{(k+1)} = x^{(k)} - \alpha^{(k)} A^{(k)} \nabla f\left(x^{(k)}\right) \tag{11}$$

The Fmincon function from the Matlab optimization toolkit is used in the optimization process, and its general form is as follows.

$$x = f\,\mathrm{mincon}\left(\mathrm{fun}, x_0, A, b, \mathrm{A_{eq}}, \mathrm{b_{eq}}, \mathrm{l_b}, \mathrm{u_b}, \mathrm{options}\right) \tag{12}$$

Where x_0 is the initial optimization value, the corresponding optimization problem of this function is represented as follows.

$$\begin{aligned} &\min_x f(x) \\ &c(x) \le 0 \\ &A \cdot x \le b \\ &c_{eq}(x) = 0 \\ &l_b \le x \le u_b \\ &\mathrm{A}_{eq}\ x \le b_{eq} \end{aligned} \tag{13}$$

Where x is the minimal value of the objective function, $c(x) \leqslant 0$ is a nonlinear inequality constraint, and $c_{eq}(x) = 0$ is a nonlinear equation constraint. $a \cdot x \leqslant b$ is a linear inequality constraint, and $A_{eq} \cdot x \leqslant b_{eq}$ is a nonlinear equation constraint. u_b, l_b is the upper and lower bounds of the solution variables, respectively. The gradient can be provided in the fun function, and GradObj of the options parameter is set to "on". The optimization algorithm is selected as an Activist, and the Hessian matrix is updated using the BFGS method.

3.3. Optimization process

The design flow-optimized in this paper is shown in Figure 6 and can be described as follows.

- For a given initial geometry shape, or shape y_n obtained from the previous optimization iteration, the mesh required for the flow field solution is generated by the ICEM parameterization method, and the Fluent flow field solution is performed to obtain the aerodynamic characteristics parameters such as CL, CM, C*P*.
- For a given shape perturbation, the sensitivity matrix of the objective function and constraint function is obtained by the flow field solution method.
- The nonlinear function with constraints is solved, and thus the step size δ_k is obtained, resulting in a new shape y_{n+1}.
- Whether the objective function change in two iterations is a slight error constant is judged, i.e., F_{n+1}-$F_n \leqslant$Error. If satisfied, the optimization is completed, and if not, the iteration is repeated by returning to step 1.

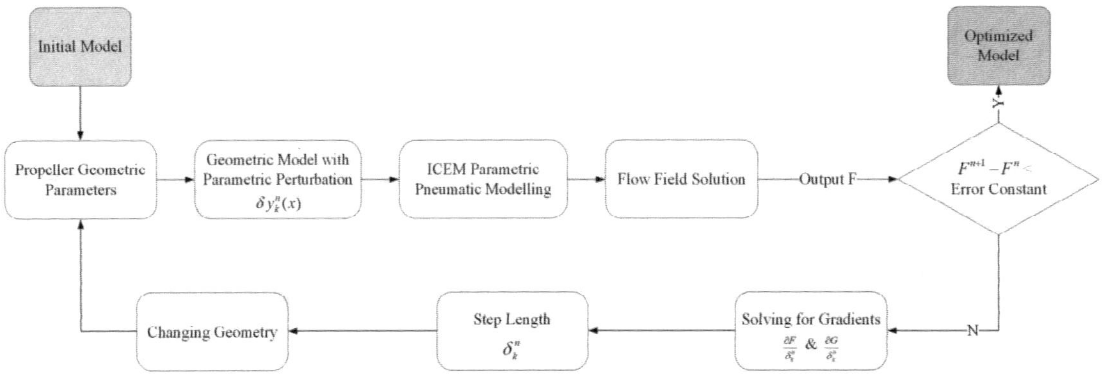

Figure 6. Optimization flowchart.

4. Aerodynamic Shape Optimization Design and Aerodynamic Characteristics

The propeller is optimized with the torque coefficient constraint CM≤0.03716 given to allow the maximum total thrust of the duct propeller system as the optimization target. Table 3 compares the thrust resistance characteristics of the ducted propeller system before and after the propeller optimization at 6000 rpm. After optimization, the ducted propeller system thrust is reduced at the same rotation speed. In contrast, the propeller torque and power coefficient decrease significantly, improving overall efficiency.

Table 4 indicates that the propeller thrust decreases after optimization while the additional duct thrust stays almost constant, increasing the duct propeller thrust ratio and enhancing inefficiency.

Considering no improvement in the overall system thrust but an increase in efficiency, the overall thrust could be improved with increased propeller speed within the torque constraint. Hence it is considered that the optimization obtains a better propeller geometry profile.

Table 3. Comparison of ducted propeller aerodynamic characteristics.

Model	CT	CM	CP	Efficiency
Original	0.45221	0.036994	0.23244	1.9455
Optimized	0.44148	0.033571	0.21654	2.0388

Table 4. Comparison of ducted propeller aerodynamic forces.

Model	Total Thrust (N)	Duct Thrust (N)	Propeller Thrust (N)	Duct-propeller Thrust Ratio
Original	662.20	337.41	324.78	1.0389
Optimized	646.48	335.88	310.61	1.0814

Figure 7 demonstrates that the flattening of the trailing edge on the upper surface of the optimized propeller root section airfoil has increased. In contrast, the lower surface has an overall convexity on the belly. The optimized mid-profile airfoil has a concave upper surface leading edge and middle region. In contrast, the lower surface leading edge is significantly convex with an open trailing edge, and the trailing edge point is shifted upward due to the optimized guideline. The airfoil shape of the optimized propeller tip section has changed significantly compared with the original airfoil shape, in which the upper surface is concave. In contrast, the lower surface leading edge and belly are convex. After optimization, the blade's trailing edge is shifted upward, and the displacement is gradually reduced from the blade root to the tip.

Figure 8 demonstrates that, compared to the initial pressure distribution of the model, the area of the negative pressure region on the upper surface of the optimized propeller is reduced and moves toward the propeller tip—the negative pressure at the upper surface of the central propeller decreases and the pressure gradient.

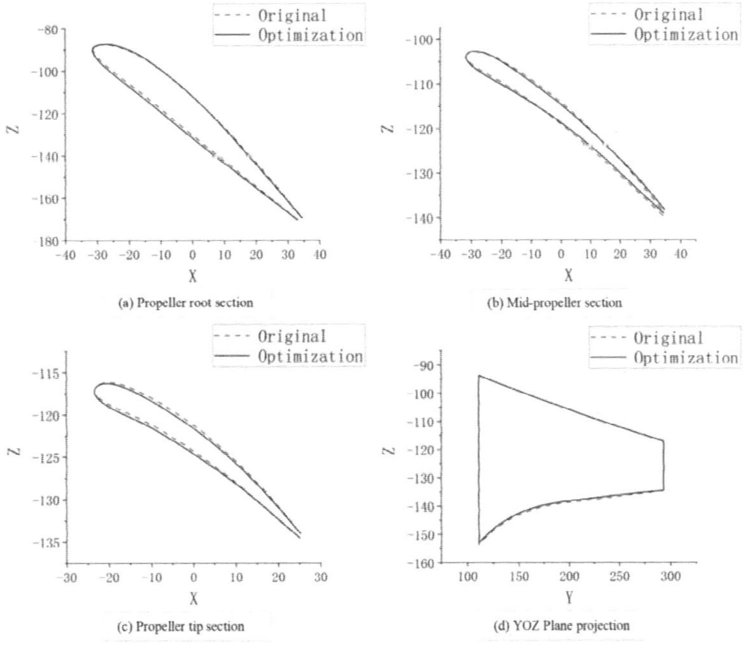

Figure 7. Comparison of propeller section airfoil and YOZ plane projection.

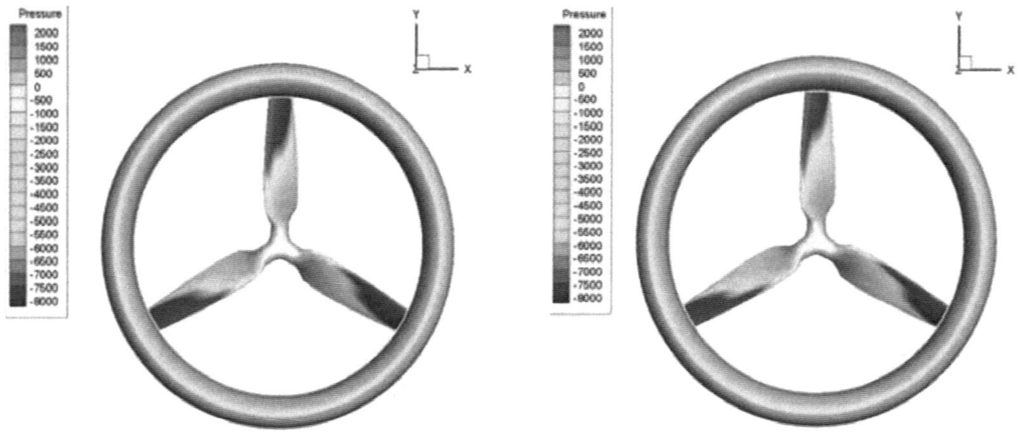

Figure 8. Comparison of ducted propeller surface pressure distribution in top view.

Figure 9 indicates that the pressure coefficient distribution before and after optimization follows a similar trend, i.e., the pressure difference between the upper and lower surfaces of the airfoil leading edge area reaches the maximum. The pressure distribution on the upper surface becomes flattered after optimization, with a higher pressure at the trailing edge, which helps to reduce the differential pressure drag on the propeller blade. The maximum negative pressure point on the lower surface moves forward, and the negative pressure on the rear side of the belly and trailing edge area decreases. The pressure distribution on the upper surface of the propeller tip section is almost unchanged, and the negative pressure on the lower surface is significantly reduced in the front section and belly area. In contrast, the pressure on the trailing edge area keeps the same. As can be seen, the differential pressure resistance of the propeller blade decreases after optimization, resulting in a reduction in the overall propeller power consumption.

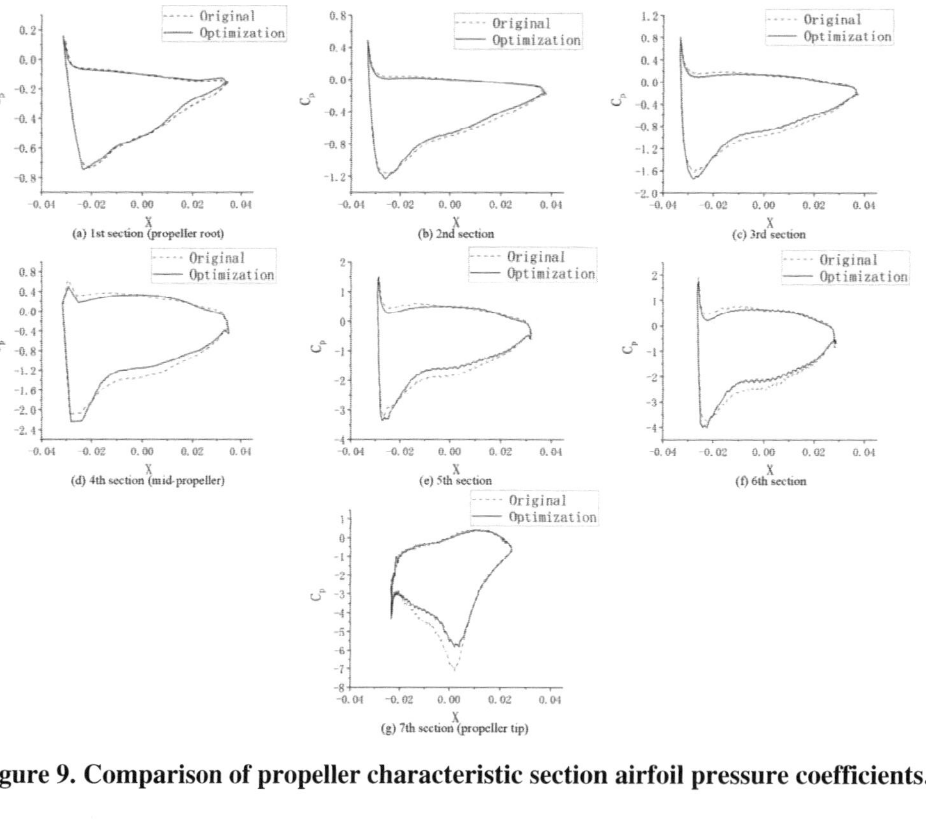

Figure 9. Comparison of propeller characteristic section airfoil pressure coefficients.

Journal of Physics: Conference Series **2361** (2022) 012027 doi:10.1088/1742-6596/2361/1/012027

5. Conclusions

In this paper, the geometric shape optimization design is carried out for a ducted propeller model. The optimization is defined as a single-point optimization problem with the overall system thrust as the optimization objective with a maximum torque as the constraint. The conclusions of the study are as follows.

Under constant negative twist value, the optimization mainly consists of changing the airfoil shape of the characteristic section and adjusting the radial distribution of the negative twist angle. The thickness of the root airfoil increases, and the middle airfoil twists downward after the optimization. The curvature of the propeller tip is decreased, which effectively reduces the differential pressure drag. Therefore, the propeller thrust drops while the additional duct thrust remains constant. The power decreases significantly with the duct propeller thrust ratio. Thus, the aerodynamic efficiency increases. The inner duct wall near the propeller tip is configured as a straight section, allowing the interaction between the duct and the propeller to be decoupled and separately analysed. It guarantees the accuracy of the iterative convergence optimization results and the aerodynamic stability of the ducted propeller system. Eventually, a better geometric shape of the duct propeller is derived by parametric modelling and gradient method-based optimization algorithm. Under the maximum torque limit, the overall thrust of the optimized system increases from 644.61N to 685.14N, which is about 6.29%.

6. References

[1] Yilmaz S, Erdem D and Kavsaoglu M 2013 Effects of duct shape on a ducted propeller performance. In: *51st AIAA aerospace sciences meeting including the new horizons forum and aerospace exposition,* p 803

[2] Du S, Lu Z and Tang Z 2016 Numerical Simulation of Lift Enhancement Method of Ducted Propeller Under Axial Flow *Journal of Projectiles, Rockets, Missiles and Guidance* **48** 173-9

[3] Xu J, Fan N and Zhao P 2009 The Study on Aerodynamic Characteristics of Duct Body of Ducted Fan Aircraft *Journal of Projectiles, Rockets, Missiles and Guidance* **29** 174-8

[4] Toleos Jr L R, Luna N, Manuel M C E, Chua J M R, Sangalang E M A and So P C 2020 Feasibility study for Fused Deposition Modeling (FDM) 3D-printed propellers for unmanned aerial vehicles *International Journal of Mechanical Engineering and Robotics Research* **9** 548-58

[5] Fan C, Adjei R A, Wu Y and Wang A 2020 Parametric study on the aerodynamic performance of a ducted-fan rotor using free-form method *AEROSP SCI TECHNOL* **101** 105842

[6] Syal M 2008 *Contributions to the aerodynamic optimization of a coaxial rotor system*: University of Maryland, College Park)

[7] Lee T E 2010 *Design and performance of a ducted coaxial rotor in hover and forward flight*: University of Maryland, College Park)

[8] Misiorowski M P, Gandhi F S and Oberai A A 2019 Computational Study of Diffuser Length on Ducted Rotor Performance in Edgewise Flight *AIAA J* **57** 796-808

[9] Fu J and Zhou Z 2012 Research on the Characteristics of Computational Methods of Ducted Fan System of VTOL UAV *Science Technology and Engineering* **20** 1294-300

[10] Xu H and Ye Z 2011 Numerical simulation and comparison of aerodynamic characteristics between ducted and isolated propellers *Journal of Aerospace Power* **26** 2820-5

[11] Bich V N and Hoa N T N 2022 Numerical Evaluation of Roughness Influences on Open Water Propeller Characteristics Using RANSE Method *International Journal of Mechanical Engineering and Robotics Research* **11** 15-21

Sensitivity analysis regarding the structural parameters of the inner cylinder embedded fluidlastic base

Zhixiong Dai*, Xi Yuan[a], Feng Qian[b], Zhizhuang Feng[c], Zhenkun Li[d]

Key Laboratory of Rotor Dynamics, China Helicopter Design Institute, Jingdezhen, Jiangxi, 333001

[a]yuanx033@avic.com
[b]qianf002@avic.com
[c]fengzz@avic.com
[d]lizk023@avic.com
*Corresponding author's email: 848232546@qq.com

Abstract. The inner cylinder embedded fluidlastic base is a device installed on the main reduction transmission channel of the helicopter to isolate the vibration load of the fuselage from the rotor system, whose vibration isolation efficiency is quite high. Compared with the traditional combined fluidlatic isolator, the fluidlastic base has higher reliability and safety in engineering application prospects. In this paper, taking a UAV as the research object, the structure of the fluidlastic base is preliminarily designed in light of structure space of its main reduction system, and its dynamic equation is established according to the working principle of the fluidlastic base to deduce the force transmission rate expression. Finally, taking the minimum force transmission rate as the research target, sensitivity analysis of the structural parameters is carried out for the fluidlastic base, which provides an optimization method for the optimal design of the vibration isolator.

1. Introduction

The vibration of helicopter is a quite prominent problem in its development and application, which has been hindering the development and use of helicopter models. Excessive vibration will not only lead to the decline of helicopter flight performance, shortened life of structural components, reduced accuracy of electronic equipment, and increased maintenance costs, but also result in declined central function of the human brain, making pilot prone to dizziness, fatigue and other physical discomfort, thus affecting the normal work of pilots. Aircrews who have been in a high-vibration environment for a long time generally suffer from spinal, waist and gastrointestinal diseases. Therefore, research on helicopter vibration control is particularly important. With the increasingly strict requirements on the vibration level of helicopters in relevant specifications, vibration level will become the biggest technical bottleneck in future model improvement and new aircraft development, which will in turn affect the design finalization and development progress [1].

The traditional combined fluidlastic vibration isolator is a vibration isolator installed in the main transmission channel and mainly used to isolate the KNΩ vibration load from the rotor [2]. The simplest form of such vibration isolation device is mainly composed of four parts: metal outer cylinder, metal inner cylinder, rubber structural parts, and filling liquid. The specific structure is shown in the following Figure 1:

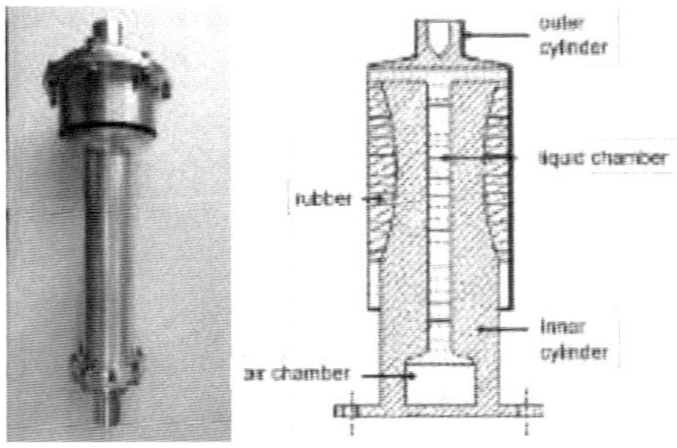

Figure 1. Model of combined fluidlatic isolator

The fluidlatic isolator follows the dynamic anti-resonance principle [3]. When the vibration load from the rotor is transmitted to the main reduction transmission channel, the filling liquid in the vibration isolator will be excited by the load and reciprocate, and the inertial force generated during the movement will counteract external excitation loads. Such vibration isolation device is not only simple in structure, small in mass and compact in structure, but also has a significant vibration isolation effect at a specific frequency point.

Despite its high vibration isolation efficiency and compact space, the traditional combined fluidlatic isolator has two shortcomings in structural design. One is that the connection between the inner cylinder and the outer cylinder completely depends on rubber. Once the rubber is broken, it means the rotor system is separated from the fuselage, so the safety and reliability are relatively poor. Secondly, at present, there is very compact connection between the main reduction system and the fuselage of most helicopters. Such rod-type shock absorber has limited vertical installation space and relatively poor engineering applicability.

In order to solve problem of safety and reliability, foreign researchers have proposed an improved form of fluidlatic isolator, as shown in Figure 2. Such base-type fluidlatic isolator can effectively solve the traditional problems of combined type hydro-elastic isolator.

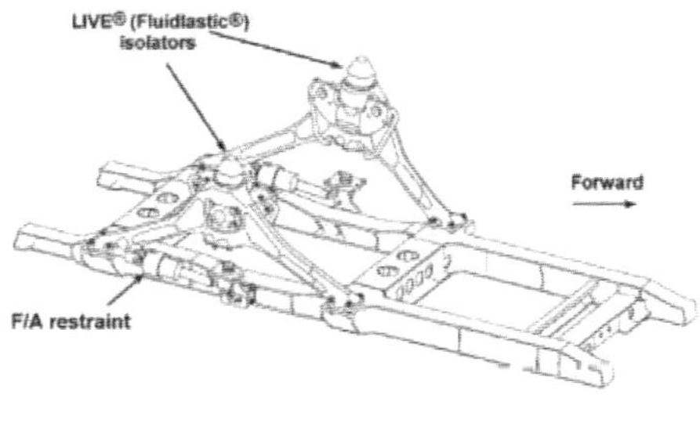

Figure 2. The Model 427 Pylon Isolation System

As early as the 1970s and 1980s, researches have been carried out abroad on this type of base-type fluidlatic isolator [4], for example installation application of fluidlastic base in Figure 3. After explorations into appearance optimization [5], material optimization [6] and control technology [7-8] improvement, hydro-elastic isolator has been applied to many types of helicopters [9-11].

Figure 3. Installation application of fluidlastic base

In China, only traditional combined liquid-elastic vibration isolator has a good research foundation. From 2010 to 2016, scholars from Nanjing University of Aeronautics and Astronautics [12-16] successively began to explore the principle and design method of the liquid-elastic vibration isolator, but sample development and related experimental work did not start. In 2016, researchers such as Deng Jinghui and Cheng Qiyou from China Helicopter Design and Research Institute [17-22] carried out systematic research on fluidlatic isolators, filling the research gap of liquid-elastic vibration isolators in China from principle exploration, design method, test method and installation of fluidlatic isolator. However, there are few domestic researches on the base-type fluidlatic isolator.

In this paper, structure design is carried out on the basis of the existing research foundation of the traditional fluidlatic isolator, and a set of fluidlastic base with reasonable structure, safety and reliability is designed. The dynamics modeling is established according to the physical model of the fluidlastic base and the vibration isolation principle to deduce the key parameter expressions. Based on structural parameters of a certain machine type, analysis is made on the parameter sensitivity of key structural parameters to the vibration isolation performance of the fluidlastic base, and the direction for structural optimization design is proposed.

2. Model of the inner cylinder embedded fluidlastic base

As shown in Figure 1, the traditional combined fluidlatic isolator is mainly composed of four parts, and the inner and outer cylinders are connected together by rubber. Without the guarantee of other safety measures, breakage and quality and safety problems are prominent. Based on this consideration, this paper changes the combined structure to the inner cylinder embedded structure, with the specific structure diagram shown in Figure 4:

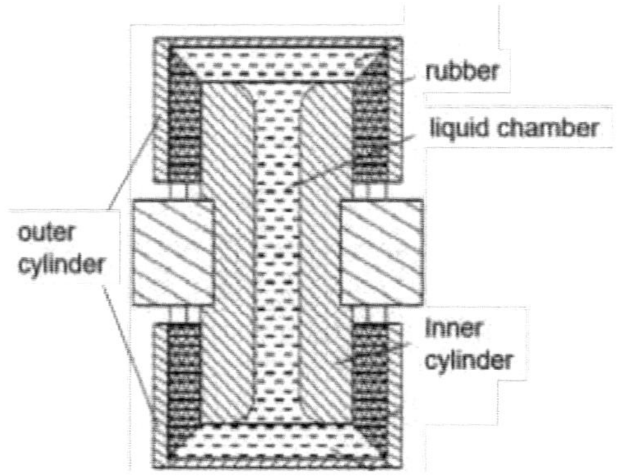

Figure 4. Physical model of the fluidlastic base

In this structure, the outer cylinder is divided into upper and lower parts rigidly connected together. The inner cylinder is embedded between the upper and lower outer cylinders, and the space is sealed by rubber, so that even if the inner and outer cylinders are separated due to rubber aging or overloading, there will only be loss of vibration, but no safety issues will occur.

In the installation, the outer cylinder is connected to the fuselage, and the inner cylinder is connected to the main reduction system through the spherical hinges. The working principle is the dynamic anti-resonance principle. When the fluidlastic base is subjected to external excitation force, the excitation force will first act on the inner cylinder, so that the inner cylinder will drive the rubber movement, causing rubber expansion and contraction and then generating a force. When the liquid inside the inertial channel is squeezed, it will flow in the inertial channel, and the flowing liquid will generate inertial force that counteracts with the rubber force, thus creating a vibration isolation effect.

3. Dynamic equation of the inner cylinder embedded fluidlastic base

According to the motion process and working principle of the fluidlastic base, the simplified mechanical model of the fluidlastic base is established, as shown in Figure 5:

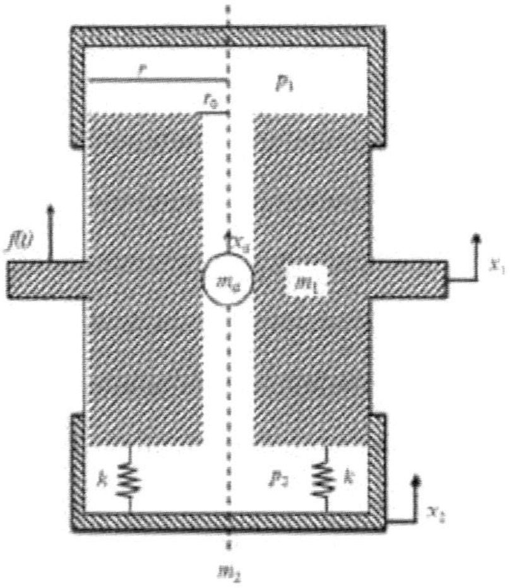

Figure 5. Dynamics model of the fluidlastic base

m_1 represents the mass of the vibration-isolated part composed of the helicopter rotor and the main reduction system, m_2 represents the fuselage mass, m_a represents the mass of the liquid in the inertial channel of the fluidlastic base, *r* represents the cross-sectional radius of the upper and lower liquid chambers, r_0 represents cross-sectional radius of the inertial channel, *k* represents the rubber stiffness, f_1 represents the periodic excitation force acting on the fluidlastic base, x_1 , x_2 represent the displacement of the inner and outer cylinder under the action of external excitation force respectively, x_a represents the displacement of the inertial liquid in the fluidlastic base. p_1,p_2 respectively represent the pressure of the upper and lower liquid chambers. Let the damping of the fluidlastic base be *c*, A is the cross-sectional area of the upper and lower liquid chambers, and A_0 is the cross-sectional area of the inertial channel. Considering the achievability in the experimental design process, it is assumed that the outer cylinder is fixed, so x_2 is 0, and the support reaction force at the bottom of the outer cylinder is f_2.

A mechanical equilibrium equation for the inner cylinder is established as follows:

$$m_1\ddot{x}_1 = f_1 - kx_1 - c(\dot{x}_1 - \dot{x}_a) + \Delta p(A - A_0) \tag{1}$$

Where $\Delta p = p_2 - p_1$

Equilibrium equation is established for the outer cylinder:

$$f_2 + kx_1 - \Delta p \cdot A = 0 \tag{2}$$

Equilibrium equation is established for the inertial liquid:

$$m_a\ddot{x}_a = c(\dot{x}_1 - \dot{x}_a) + \Delta p \cdot A_0 \tag{3}$$

From the liquid continuity equation, there is:

$$-(A - A_0)x_1 = A_0 x_a \tag{4}$$

The magnification ratio R is defined for the fluidlastic base, and $R = \dfrac{A}{A_0}$, then equation (4) can be transformed into:

$$x_a = (1 - R)x_1 \tag{5}$$

From equation (3), there is

$$\Delta p = \frac{1}{A_0}[m_a\ddot{x}_a - c(\dot{x}_1 - \dot{x}_a)] \tag{6}$$

By substituting equation (6) into equation (1), there is:

$$m_1\ddot{x}_1 = f_1 - kx_1 - c(\dot{x}_1 - \dot{x}_a) + (R-1)\left[m_a\ddot{x}_a - c(\dot{x}_1 - \dot{x}_a)\right] \tag{7}$$

After simplification, there is:

$$m_1\ddot{x}_1 - (R-1)m_a\ddot{x}_a + Rc(\dot{x}_1 - \dot{x}_a) + kx_1 = f_1 \tag{8}$$

At the same time, from equation (5), there is:

$$\begin{aligned} \dot{x}_a &= (1-R)\dot{x}_1 \\ \ddot{x}_a &= (1-R)\ddot{x}_1 \end{aligned} \tag{9}$$

By substituting equation (9) into equation (8), there is:

$$(m_1 + m_a(R-1)^2)\ddot{x}_1 + cR^2\dot{x}_1 + kx_1 = f_1 \tag{10}$$

Let the periodic excitation force on the fluidlastic base be $f_1 = F_1e^{j\omega t}$, there is:

$$\begin{cases} x_1 = X_1e^{j\omega t} \\ x_a = X_ae^{j\omega t} \\ f_2 = F_2e^{j\omega t} \end{cases} \tag{11}$$

From equation (2) and equation (9), there is:

$$f_2 = m_aR(1-R)\ddot{x}_1 - cR^2\dot{x}_1 - kx_1 \tag{12}$$

The force transfer rate is defined as $T = \left|\frac{F_2}{F_1}\right|$:

$$T = \left|\frac{F_2}{F_1}\right| = \left|\frac{m_aR(R-1)\omega^2 - jc\omega R^2 - k}{-(m_a(R-1)^2 + m_1)\omega^2 + jc\omega R^2 + k}\right| \tag{13}$$

It is worth pointing out that the mass m_a of the inertial liquid can be expressed as: $m_a = \pi r_0^2 l\rho$

4. Parameter sensitivity calculation and analysis

Transmission rate is an important index to measure the vibration isolation effect of the isolator. The study regarding the effect of structural parameters of the fluidlastic base on transmission rate is one key research content of this paper. By derivation of the dynamic formula of the fluidlastic base, it can be seen that the vibration isolation effect of the fluidlastic base mainly concerns the liquid chamber radius r, the inertial channel radius r_0, and the inertial channel length l. It is worth noting that rubber stiffness, liquid density and damping also impact force transmission rate, but these factors are attributed to material properties. This paper intends to study the effect of vibration isolator structure size on the vibration isolation effect, so these factors will not be analyzed. In this paper, an unmanned helicopter is used as the background machine. After dimensionless processing, the mass of the background machine's rotor system is m_1=0.1 and the rotor rotation frequency is ω=1.95. The structure of the fluidlastic base is designed according to the background machine structure to derive design parameters. After dimensionless processing, the following data Table 1 is obtained:

Table 1. Structural parameters and related material properties of the fluidlastic base

Structure name	Dimensionless value
Upper and lower liquid chamber radius r	0.04
Inertia channel radius r_0	0.007
Inertial channel length l	0.3
Liquid density ρ	41.2
Rubber stiffness k	4.66
Liquid damping c	0.5

The transmission formula was programmed using Matlab, and the parameter sensitivity analysis was made on the liquid chamber radius r, the inertial channel radius r_0, and the inertial channel length l. The following results are obtained:

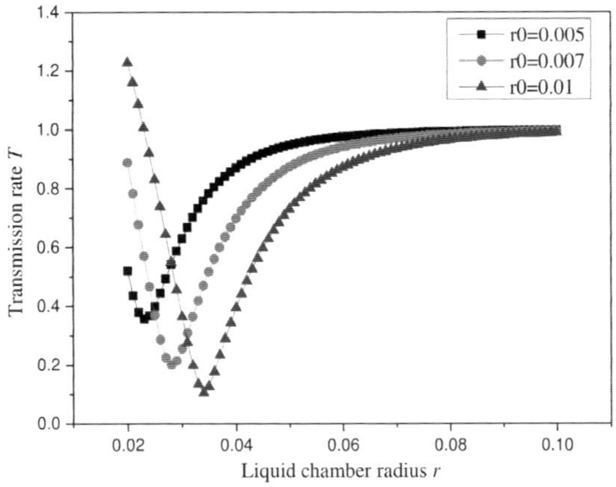

Figure 6. Variation curve of transmission rate with the liquid chamber radius

Figure 6 shows that the transmission rate of the fluidlastic base first decreases with the increase of the liquid chamber radius, resulting in higher vibration isolation efficiency, and then the figure decreases with the increase of the liquid chamber radius, with the vibration isolation efficiency getting lower and finally tending be balanced at T=1. That is, f_1=f_2,, which means that the vibration isolator does not have any vibration isolation effect at this time. In addition, change of the inertia channel radius will also affect the maximum vibration isolation efficiency of the isolator and the optimal vibration isolation design point of the liquid chamber radius, but the change trend of the transmission rate remains the same.

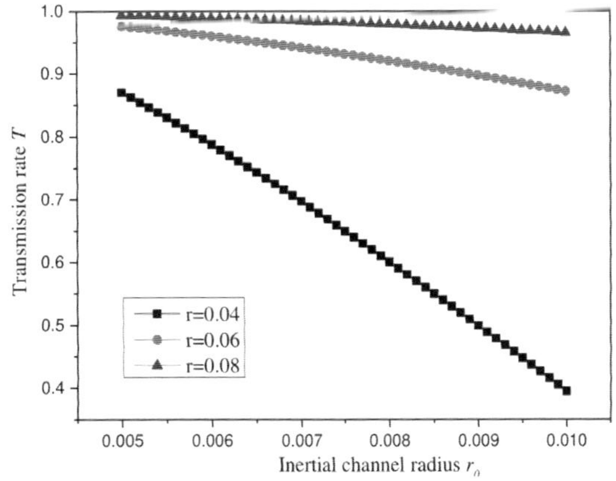

Figure 7. Variation curve of transmission rate with the inertial channel radius

Figure 7 reveals that the transmission rate of the fluidlastic base decreases with the increase of the inertial channel radius, resulting in higher vibration isolation efficiency. In addition, by comparing the three curves, it is found that within the design range of the inertial channel radius, the size of the liquid chamber radius greatly affects the vibration isolation efficiency. For the two curves r=0.06 and r=0.08, the transmission rate is very little affected by the inertial channel radius, but for r = 0.04, the transmission rate varies by more than 50% in the research scope.

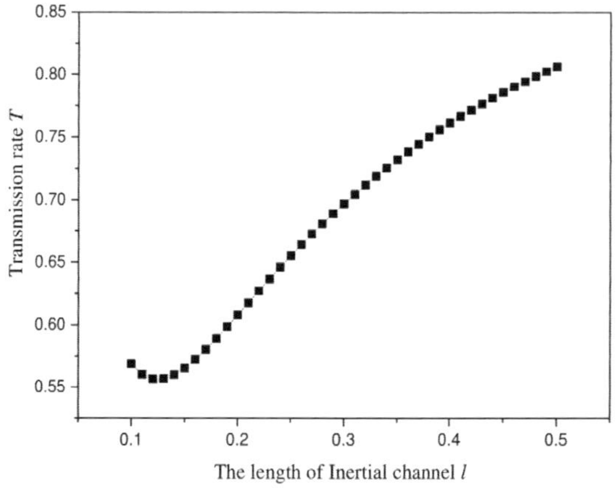

Figure 8. Variation curve of transmission rate with the length of the inertial channel

According to Figure 8, the transmission rate of the fluidlastic base will first increase and then decrease with the increase of the inertial channel length. That is, when the liquid chamber radius and the inertial channel radius are fixed, vibration efficiency of the fluidlastic base will first reach an optimum point and then keep getting worse.

According to the expression regarding the mass of the inertial channel liquid, the mass concerns the radius and length of the inertial channel. As can be seen from Figure 7, the shorter the inertial channel length, the smaller the mass and the better the vibration isolation effect. However, it can be seen from any curve in Figure 6 that as the radius increases, the mass increases and the vibration isolation effect is better. For its reason, the magnification ratio R also changes when the radius increases, which affects the vibration isolation effect, and the magnification ratio exerts greater effect than liquid mass.

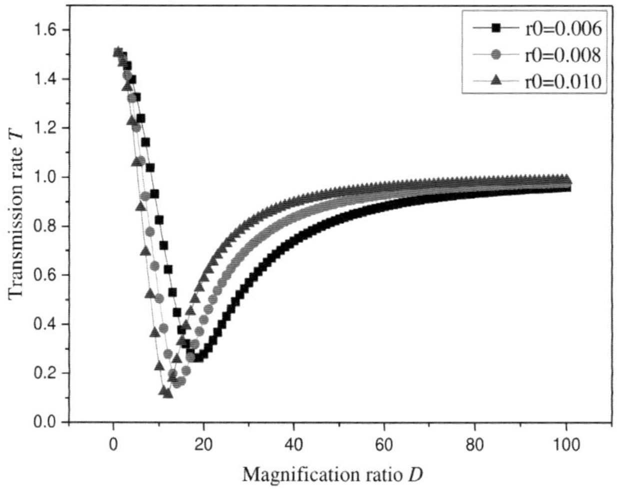

Figure 9. Variation curve of transmission rate with magnification ratio

Figure 9 shows that, under the premise that the inertial liquid mass remains unchanged, the transmission rate of the fluidlastic base will first increase and then decrease with the increase of the magnification ratio. In this changing process, there is an optimal design point for vibration isolation effect, and the optimal design range of this magnification ratio is about 12~18. Namely, within a reasonable range of the inertial channel radius, the optimal liquid chamber radius can be designed based on the optimal magnification ratio.

5. Conclusion

In this paper, taking a UAV as the research object, the structure of the fluidlastic base is preliminarily designed in light of structure space of its main reduction system, and its dynamic equation is established according to the working principle of the fluidlastic base to deduce the force transmission rate expression. Finally, taking the minimum force transmission rate as the research target, sensitivity analysis of the structural parameters is carried out for the fluidlastic base, and the following conclusions are drawn:

1) Compared with the combined fluidlatic isolator, the design of the fluidlastic base solves the safety defect caused by the rubber breakage of the vibration isolator during operation, and the structure of the fluidlastic base can adapt to various main reduction structure forms, demonstrating broader engineering application prospects;
2) For a certain UAV, the vibration isolation efficiency of the fluidlastic base concerns the liquid mass of the inertial channel. By changing the inertial channel length l, it can be seen that as the liquid mass increases, the vibration isolation efficiency of the vibration isolator will first increase and then decrease;
3) For a certain UAV, the vibration isolation efficiency of the fluidlastic base concerns the radius of the upper and lower liquid chambers. As the radius r of the upper and lower liquid chambers increases, the vibration isolation efficiency of the vibration isolator will first increase and then decrease, eventually tending to be balanced, and the vibration isolator has no vibration isolation effect in a balanced state;
4) For a certain UAV, the vibration isolation efficiency of the fluidlastic base will be affected by the magnification ratio R and the inertial channel mass. By comparison, it can be found that the magnification ratio R has a greater impact on the vibration isolation efficiency. Moreover, by studying the magnification ratio R, it can be seen that the optimal design range of the magnification ratio is about 12~18. That is to say, within the reasonable range of the inertial channel radius, the optimal liquid chamber radius can be designed based on the optimal magnification ratio.

6. References

[1] Zhang C L, Zhang X G, Guo S L. Helicopter component design. Nanjing; Nanjing University of Aeronautics and Astronautics. 1986. (In Chinese)
[2] Zhang X G. Helicopter Dynamic Design. Beijing; Aviation Industry Press. 1995. (In Chinese)
[3] Gu Z Q. Dynamic anti-resonance vibration isolation. Noise and Vibration Control. 1989 (06): 36-40.9. (In Chinese)
[4] Flanneley W G. The dynamic anti-resonant vibration isolator. American Helicopter Society 22th Annual Forum. 1966.
[5] Sutton T J, Elliott S J, Brennan M J, et al. Active Isolation Of Multiple Structural Waves On A Helicopter Gearbox Support Strut. Journal of Sound & Vibration, 1997, 205(1):81-101.
[6] Smith M R. Results From The Dynamically Tailored Airframe Structures Program. The American Helicopter Society 58th Annual Forum. Montréal, Canada. June 11–13, 2002.
[7] Braun D. Development of Anti-resonance Force Isolators for Helicopter Vibration Reduction. Journal of the American Helicopter Society, 1982, volume 27 (4):37-44 (8).
[8] Halwes D R. "LIVE – Liquid Inertia Vibration Eliminator". Washington, D.C.; American Helicopter Society 36th Annual Forum. 1980.
[9] MC Guire D P. Fluidlastic Dampers and Isolators for Vibration Control in Helicopters. American Helicopter Society 50th Annual Forum. Washington, DC. May, 1994.
[10] Smith M R, Redinger W S. The Model 427 Pylon Isolation System. American Helicopter Society 55th Annual Forum. Montreal, Quebec, Canada. May 1999.
[11] Balke R W. Development of the Kinematic Focal Isolation System for Helicopter Rotors. 38th Shock and Vibration Symposium. 1968
[12] Deng X D. Analysis and Simulation of Research on Simulation for the Advanced Helicopter Gearbox Isolation System; Nanjing University of Aeronautics and Astronautics, 2011. (In Chinese)
[13] Gong L. Design and Testing of the Fluid Elastomer Isolator; Nanjing University of Aeronautics and Astronautics, 2012. (In Chinese)

[14] Zhang Y. The dynamic modeling and simulation of the helicopter gearbox isolation system with fluide lastomer isolator; Nanjing University of Aeronautics and Astronautics, 2013. (In Chinese)
[15] Lei S B, Zhang Y. Transmission Characteristics of Focal Fluid Elastomer Vibration Isolation System. Journal of Nanjing University of Aeronautics and Astronautics, 2015, 47(2): 275-278. (In Chinese)
[16] Zhang X. Development of a fluid-elastomer vibration isolator for semi-active helicopter pylon isolation; Nanjing University of Aeronautics and Astronautics, 2016. (In Chinese)
[17] Cheng Q Y, Deng J H, Feng Z Z, et al. Analysis of material parameter effects on fluidlastic isolators performance. 2018
[18] Deng, Jing H, and Cheng Q Y. "Analysis of Design Parameters Effects on Vibration Characteristics of Fluidlastic Isolators." IOP Conference Series: Materials Science and Engineering 22(2017) 012012.
[19] Sutton T J, Elliott S J , Brennan M J, et al. Active Isolation Of Multiple Structural Waves On A Helicopter Gearbox Support Strut. Journal of Sound & Vibration, 1997, 205(1):81-101.
[20] Cheng Q Y, Deng J H, Feng Z Z, et al. Analysis of material parameter effects on fluidlastic isolators performance. 2018.
[21] Deng J H. Analysis and experimental study on transmission characteristics of combined hydro elastic vibration isolator. Journal of Northwest University of technology, 2022, 40(1):7
[22] Dai Z X, Qian F, Feng Z Z. Development status and trend of vibration isolation technology of helicopter main hydraulic damper Aviation science and technology, 2020, 31(6):8.

AUTHOR INDEX